# Metaheuristic Optimization in Power Engineering, 2nd Edition

## Volume 2: Distribution systems using MATLAB®-based software

Jordan Radosavljević

The Institution of Engineering and Technology

## About the IET

This book is published by the Institution of Engineering and Technology (The IET).

We inspire, inform, and influence the global engineering community to engineer a better world. As a diverse home across engineering and technology, we share knowledge that helps make better sense of the world, accelerate innovation, and solve the global challenges that matter.

The IET is a not-for-profit organization. The surplus we make from our books is used to support activities and products for the engineering community and promote the positive role of science, engineering, and technology in the world. This includes education resources and outreach, scholarships and awards, events and courses, publications, professional development and mentoring, and advocacy to governments.

To discover more about the IET, please visit https://www.theiet.org/.

## About IET books

The IET publishes books across many engineering and technology disciplines. Our authors and editors offer fresh perspectives from universities and industry. Within our subject areas, we have several book series steered by editorial boards made up of leading subject experts.

We peer review each book at the proposal stage to ensure the quality and relevance of our publications.

## Get involved

If you are interested in becoming an author, editor, series advisor, or peer reviewer, please visit https://www.theiet.org/publishing/publishing-with-iet-books/ or contact author_support@theiet.org.

## Discovering our electronic content

All of our books are available online via the IET's Digital Library. Our Digital Library is the home of technical documents, eBooks, conference publications, real-life case studies, and journal articles. To find out more, please visit https://digital-library.theiet.org.

In collaboration with the United Nations and the International Publishers Association, the IET is a Signatory member of the SDG Publishers Compact. The Compact aims to accelerate progress to achieve the Sustainable Development Goals (SDGs) by 2030. Signatories aspire to develop sustainable practices and act as champions of the SDGs during the Decade of Action (2020–2030), publishing books and journals that will help inform, develop, and inspire action in that direction.

In line with our sustainable goals, our UK printing partner has FSC accreditation, which is reducing our environmental impact on the planet. We use a print-on-demand model to further reduce our carbon footprint.

# Contents

**British Library Cataloguing in Publication Data**
A catalogue record for this product is available from the British Library

**ISBN 978-1-83724-137-8 (Volume 1 hardback)**

**ISBN 978-1-83724-138-5 (Volume 1 PDF)**

**ISBN 978-1-83724-131-6 (Volume 2 hardback)**

**ISBN 978-1-83724-132-3 (Volume 2 PDF)**

Typeset in India by MPS Limited

Cover image credit: halbergman/E+ via Getty Images

Published by The Institution of Engineering and Technology, London, United Kingdom

The Institution of Engineering and Technology (the "**Publisher**") is registered as a Charity in England & Wales (no. 211014) and Scotland (no. SC038698).



First published 2024



The Institution of Engineering and Technology
Futures Place
Kings Way, Stevenage
Herts, SG1 2UA, United Kingdom

www.theiet.org

# Metaheuristic Optimization in Power Engineering, 2nd Edition

## Other related titles:

You may also like

- PBPO131 | Jordon Radosavljević | Metaheuristic Optimization in Power Engineering | 2018
- PBSP008 | Mohamed Bakr | Nonlinear Optimization in Electrical Engineering with Applications in MATLAB® | 2013
- PBPO086 | Federico Milano | Advances in Power System Modelling, Control and Stability Analysis | 2016
- PBPO147 | Juan Manuel Gers | Distribution Systems Analysis and Automation, 2nd Edition | 2020
- PBPO091 | Ismail H. Altaş | Fuzzy Logic Control in Energy Systems with Design Applications in MATLAB®/Simulink® | 2017

We also publish a wide range of books on the following topics:
Computing and Networks
Control, Robotics and Sensors
Electrical Regulations
Electromagnetics and Radar
Energy Engineering
Healthcare Technologies
History and Management of Technology
IET Codes and Guidance
Materials, Circuits and Devices
Model Forms
Nanomaterials and Nanotechnologies
Optics, Photonics and Lasers
Production, Design and Manufacturing
Security
Telecommunications
Transportation

All books are available in print via https://shop.theiet.org or as eBooks via our Digital Library https://digital-library.theiet.org.

IET ENERGY ENGINEERING SERIES 274

# Excerpt from the preface to the 1st edition (2018)

The metaheuristic paradigm is a way of solving complex optimization problems using simple mathematical rules that copy some well-known processes, phenomena, or rules in nature or society. Generally, a metaheuristic procedure is designed to find, generate, or select a sufficiently good solution to an optimization problem, especially with incomplete or imperfect information or limited computation capacity. Metaheuristic optimization methods have become significant tools and are often the only way to solve practical optimization problems. The basic requirement of these methods is to obtain the global solution or a solution close to the global optimum in a reasonable time. Modern and emerging power systems are an important application of such methods.

The book is intended primarily for researchers who are dealing with various aspects of power systems optimization. The aim of the book is to explain the basic principles of metaheuristic optimization methods and their applications in power engineering. The first part of the book gives a brief description of selected metaheuristic optimization methods, whereas the second part covers the specific application of the metaheuristic optimization methods to solve various optimization problems in power systems. Each chapter contains a comprehensive review of recent literature.

# Preface to the 2nd edition (2024)

The rapid transformation of power systems due to a large share of renewable energy sources poses new challenges in their analysis and optimization. Solving the optimal operation and planning problems of such power systems includes adequate modeling of the specificities of integrated renewable energy sources and energy storage systems, as well as the use of robust optimization techniques. As an attempt to respond to some of these challenges, the first edition of the monograph, published in 2018, has been modified and supplemented with the most important results that the author has achieved in dealing with these issues in the last 6 years.

The new edition uses a MATLAB®-based software package to assist with testing, comparing, and applying the methods. The chapters have been revised, and new ones added, to account for technological developments.

With the addition of several new chapters, as well as changes and division of chapters from the first edition, this expanded 2nd edition of the monograph *Metaheuristic Optimization in Power Engineering* has evolved into two volumes, for better visibility and consistency of material.

Volume 1 covers principles of key metaheuristic algorithms and their application for solving various optimal power dispatch problems in transmission power systems using MATLAB-based software.

Volume 2 focuses on power distribution systems, that is, it comprehensively and systematically explains the metaheuristic optimization paradigm on the problems of optimal operation and planning of distribution networks and microgrids. The application of these methods is demonstrated using MATLAB-based software.

**Metaheuristic Optimization in Power Engineering:**
**Volume 2: Distribution systems Using MATLAB®-based software**

Chapter 1 describes the backward/forward sweep algorithm for power flow computation in balanced and unbalanced distribution networks (DNs). Power flow calculation is one of the most important and most used functions in the analysis of DNs. In that context, this chapter presents an introduction to the next chapters dealing with different optimization problems in DNs.

Chapter 2 covers the applications of metaheuristic optimization methods for solving optimal power flow (OPF) problems in distribution systems. A generalized approach for the OPF analysis is presented considering different types of distributed generation (DG) units, uncertainties of the input variables, the multi-objective optimization.

Chapter 3 deals with multi-period OPF related to optimal day-ahead active–reactive power dispatching in unbalanced DNs with integrated single-phase photovoltaic (PV) generation and battery energy storage systems (BESS). To solve this highly constrained non-linear optimization problem, a hybrid particle swarm optimization with sigmoid-based acceleration coefficients (PSOS) and a chaotic gravitational search algorithm (CGSA) called the PSOS–CGSA algorithm is proposed.

Chapter 4 offers two methodologies for optimal voltage control in DNs. The first methodology for determining the optimal transformer tap settings in radial (rural) DNs is based on the space and time decomposition of the voltage control problem. After that, the Volt/Var control is formulated as a non-linear complex optimization problem with constraints and solved using metaheuristic optimization methods.

Chapter 5 discusses the problem of voltage regulation in low-voltage DN with high penetration of rooftop PV sources. A two-stage approach to voltage regulation is proposed. The PSO algorithm was used to solve the optimization problem.

Chapter 6 describes an efficient PSO-based algorithm for energy and operation management (EOM) of microgrids, including various DG units and BESSs. Due to the stochastic nature of renewable energy sources, as well as the uncertainty of loads and market prices, a probabilistic approach is also introduced in EOM.

Chapter 7 presents a two-step procedure for optimal placement and sizing of dispatchable DG units in distribution networks. Two search techniques were used: partial search of variants and genetic algorithm (GA).

Chapter 8 deals with the application of metaheuristic algorithms to solve the problem of optimal placement and sizing of renewable-based DG in DNs, observing the problem from the viewpoints of the DN operator and the DG owners.

Chapter 9 discusses the application of the gravitational search algorithm (GSA) and hybrid GSA–SQP algorithms to optimal coordination of directional overcurrent relays. In this approach, GSA performs global optimization, and SQP is a local optimizer to finetune the solution.

Chapter 10 describes a new three-phase distribution system state estimation (DSSE) paradigm based on metaheuristic optimization. DSSE is formulated as a nonlinear constrained optimization problem where loads and DG outputs are considered as control variables, while real-time measurements are treated as dependent variables.

Chapter 11 shows how a metaheuristic optimization technique such as GA can be applied in solving a non-optimization problem, i.e. in determining the steady-state performances of a self-excited induction generator for general operating conditions. The GA-based optimization procedure was applied in the analysis of SEIG feeding balanced/unbalanced and static/dynamic loads.

The second edition of the book uses MATLAB-based software packages for testing and comparing the proposed methodologies. All chapters of Volume 2 contain descriptions of the MATLAB programs for the problems treated in these chapters. The programs were developed to demonstrate the effectiveness of various metaheuristic optimization methods in solving real optimization problems in distribution systems by testing them on relevant IEEE and other test systems. These programs are available as supplementary files to the corresponding book chapters on the book home page in the IET Digital Library.

*Jordan Radosavljević*
*Kosovska Mitrovica, May 2024*

# About the author

**Jordan Radosavljević** is a professor and dean of the Faculty of Technical Sciences of the University of Priština in Kosovska Mitrovica, Serbia. His research interests include power system analysis and control, power system optimization with renewable energy and distributed generation, and the application of metaheuristic optimization algorithms. He was involved in international projects on energy markets, engineering education, and low carbon economy.

*Chapter 1*

# Power flow calculation in distribution networks

## 1.1 Introduction

It is well known that distribution networks (DNs) have certain specifics in relation to transmission networks. These are the radial structure, the high ratio $r/x$ of the network elements, the large number of buses, as well as the asymmetry of network components and the unbalance of the loads. These properties make most of the classic iterative methods for calculating power flows in transmission networks (Newton–Raphson and others) ineffective in solving the power flow problem in DNs. Starting from the DNs' properties, specialized methods have been developed for the power flow calculation in DNs, including distributed generation (DG) units. These methods, often referred to as *backward/forward sweep*, are branch-oriented methods based on the application of Kirchoff's laws and iterative passage along the branch of the DN. Initial development of such a power flow for radial DNs was addressed in [1].

## 1.2 Power flow calculation in balanced DNs

The first step in power flow calculations is to form an equivalent network model. It is understood here that the DN consists of three-phase symmetrical elements and three-phase and balanced consumers and generators, so that they can be modeled by their single-phase equivalents with positive sequence parameters. It is usual for lines in power flow calculations to be represented by symmetrical $\pi$ equivalents. Lines in DNs are relatively short, so their shunt admittances can be neglected. This means that in power flow calculations, distribution lines are usually represented only with their serial branches. Distribution transformers are modeled with an inverted "Γ" equivalent consisting of series impedance (short-circuit impedance) and shunt admittance (magnetizing branch admittance). In general, the power of an electric load is a function of voltage and frequency. As the frequency is a quantity of a global character whose control is achieved exclusively at the level of the transmission system, it can be considered a constant quantity for DNs. This means that the active and reactive power of a load connected to a bus of the DN are functions only of the voltage of that bus. Accordingly, the complex current of the load is calculated in accordance with the way of modeling the load as a function of voltage. Three basic models are used: the constant power model (constant PQ), the constant current model (constant I), and the constant impedance model (constant Z). In the power flow calculation, three-phase complex power of load is usually specified, which is used for calculating the current of load in accordance with the way they are modeled.

The power flow calculation can be carried out without limitation in the system of absolute units or in the per-unit system. The per-unit system is used here, and therefore, all variables, parameters, and relations are given for the per-unit system (p.u.).

The backward/forward sweep power flow method involves two sweeps of calculations. In the forward sweep, the end voltages are initialized for the first iteration, and currents are calculated starting at the buses at the load end of the radial branch and solved up to the source bus by applying the current summation method. The backward sweep starts at the source bus and calculates voltages using the current calculated from the forward sweep until the load end of the radial branches. The voltages from the backward sweep are used for the next iteration in the forward sweep calculations [2].

The power flow calculation for a radial DN (Figure 1.1) using the backward/forward sweep method is carried out through the following iterative steps:

*Step 1. Read input data*: Define network configuration, parameters ($r$, $x$, and $b$) of network elements, load data, and DG units data;

*Step 2. Initialization*: Set the voltage at the root bus and assuming a flat profile for the initial voltages at all other network buses;

$$V_i^{(0)} = 1; \; \theta_i^{(0)} = 0; \quad i = 1, \; \dots \; , N \tag{1.1}$$

where $V_i$ and $\theta_i$ are the magnitude and angle of voltage at bus $i$ and $N$ is the total number of buses.

*Step 3. Backward sweep*: At iteration $k$, starting from the branch in the last layer (branch corresponding to last bus $N$) and moving toward the branches connected to the root bus, the current in branch $i$, $\underline{J}_i$, according to Figure 1.2, is calculated as follows:

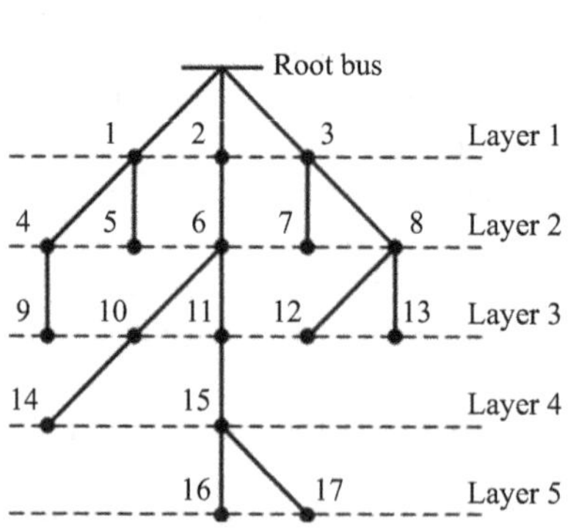


*Figure 1.1 A typical radial configuration of distribution networks*

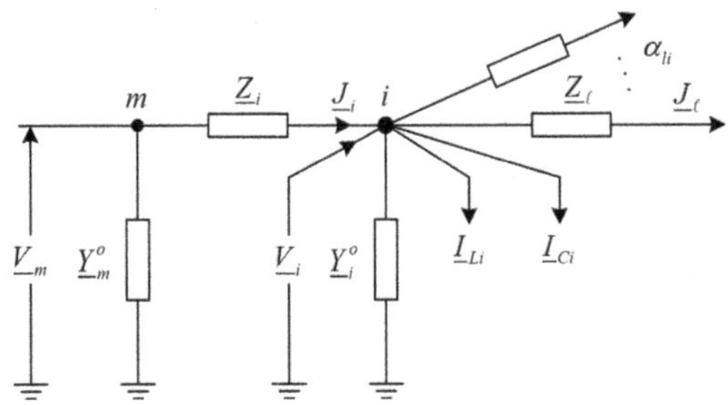


*Figure 1.2 A part of the distribution network*

$$\underline{J}_i^{(k)} = \underline{I}_{Li}^{(k)} + \underline{I}_{Ci}^{(k)} + \underline{Y}_i^o \cdot \underline{V}_i^{(k-1)} + \sum_{\substack{\ell \in \alpha_{\ell i} \\ \ell \neq i}} \underline{J}_\ell^{(k)}$$

$$i = N, N-1, \ \ldots \ , \ 1; \ k = 1, 2, \ \ldots \tag{1.2}$$

where $k$ is the index of iteration, $\underline{I}_{Li}$ is the complex current of load at bus $i$, $\underline{I}_{Ci}$ is the complex current of a var compensator at bus $i$, $\underline{J}_\ell$ is the complex current in branch $\ell$ emanating from bus $i$, $\underline{Y}_i^o$ is the shunt admittance at bus $i$, $\alpha_{\ell i}$ is the set of buses connected with bus $i$, and $\underline{V}_i$ is the complex voltage at bus $i$.

For a specified power of load, $\underline{S}_{Li} = P_{Li} + jQ_{Li}$, the current of load $\underline{I}_{Li}$ is calculated in accordance with load model, as follows:

Constant power model (PQ):

$$\underline{I}_{Li}^{(k)} = \left( \frac{\underline{S}_{Li}}{\underline{V}_i^{(k-1)}} \right)^* \tag{1.3}$$

Constant current model (I):

$$\underline{I}_{Li}^{(k)} = \left| \frac{\underline{S}_{Li}}{V_{nom}} \right| \cdot e^{j\left(\theta^{(k)} - \phi\right)} \tag{1.4}$$

Constant impedance model (Z):

$$\underline{I}_{Li}^{(k)} = \underline{V}_i^{(k-1)} \cdot \frac{\underline{S}_{Li}^*}{V_{nom}^2} \tag{1.5}$$

where $V_{nom}$ is the nominal voltage, $\delta^{(k)}$ is the voltage angle at bus $i$ in the $k$th iteration, and $\varphi$ is the phase angle of the complex power $\underline{S}_{Li}$.

The current $\underline{I}_{Ci}^{(k)}$ in (1.2) is calculated in an analogous way.

*Step 4. Forward sweep:* Bus voltages are updated in a forward sweep starting from branches connected to the root bus toward those in the last. For each branch, $i$, the voltage at bus $i$ is calculated using the updated voltage at bus $m$

(Figure 1.2) and the branch current calculated in the preceding backward sweep:

$$\underline{V}_i^{(k)} = \underline{V}_m^{(k)} - \underline{Z}_i \underline{J}_i^{(k)} \tag{1.6}$$

where $\underline{Z}_i$ is the series impedance of branch $i$.

*Step 5. Check convergence criteria*: Steps 3 and 4 are repeated until convergence is achieved. For example, a convergence criterion can be adopted as follows:

$$\max\left|V_i^{(k)} - V_i^{(k-1)}\right| \leq \varepsilon_V = (10^{-3} - 10^{-5}) \tag{1.7}$$

### *1.2.1 DG units treatment*

Depending on the control, the DG may be set to output power at either a constant power factor for small DG or constant voltage for large DG. Therefore, the bus to which the DG is connected should be modeled as a PQ bus for the first case and a PV bus for the second.

Figure 1.3 shows a part of the DN with DG connected at bus $i$. In accordance with Figure 1.3, the current in line section $i$ at iteration $k$ is

$$\underline{J}_i^{(k)} = \underline{I}_{Li}^{(k)} + \underline{I}_{Ci}^{(k)} - \underline{I}_{DGi}^{(k)} + \underline{Y}_i^o \cdot \underline{V}_i^{(k-1)} + \sum_{\substack{\ell \in \alpha_{\ell i} \\ \ell \neq i}} \underline{J}_\ell^{(k)} \quad i = N, N-1, \ldots, 1;\ k = 1, 2, \ldots \tag{1.8}$$

where $\underline{I}_{DGi}^{(k)}$ is the current injections at bus $i$ corresponding to DG.

In case bus $i$ is modeled as a PQ bus, the DG unit should be treated as a negative PQ load. Accordingly, the current injection of the DG at bus $i$ can be calculated as follows:

$$\underline{I}_{DGi}^{(k)} = \left(\frac{P_{DGi}^{sp} + jQ_{DGi}^{sp}}{\underline{V}_i^{(k-1)}}\right)^* \quad P_{DGi}^{sp} = const;\ Q_{DGi}^{sp} = const; \tag{1.9}$$

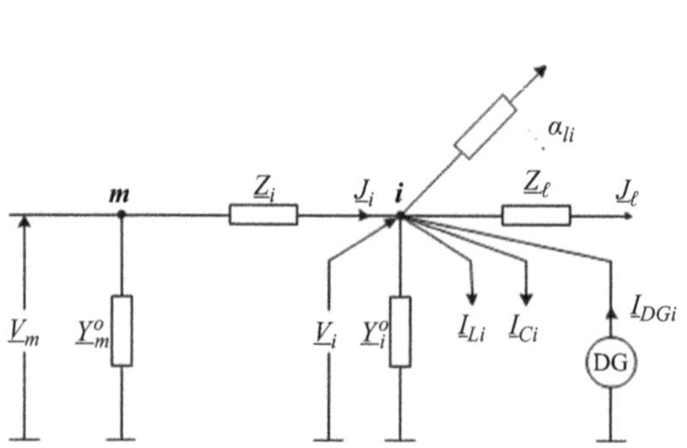


*Figure 1.3 A part of distribution network with DG*

where $P_{DGi}^{sp}$ and $Q_{DGi}^{sp}$ are specified active and reactive power outputs of the DG connected at bus $i$, respectively.

In case bus $i$ in Figure 1.3 is modeled as a PV bus, handling the PV buses in the power flow algorithm requires some additional processes, which are as follows:

1. Initially, the DG active power outputs, reactive power generation limits, and voltage magnitudes at PV buses are specified. The reactive power of the DGs is initialized to zero. Perform power flow calculation considering the DGs as negative PQ loads.
2. Calculate the voltage at each PV bus

$$\underline{V}_i^{cal(k)} = V_i^{cal(k)}\left[\cos\left(\theta_i^{(k)}\right) + j\sin\left(\theta_i^{(k)}\right)\right] \quad i = 1,\ldots,NPV \tag{1.10}$$

$$\underline{V}_i^{sp(k)} = V_i^{sp}\left[\cos\left(\theta_i^{(k)}\right) + j\sin\left(\theta_i^{(k)}\right)\right] \quad i = 1,\ldots,NPV \tag{1.11}$$

Calculate the complex value of the specified voltage for all PV buses:

3. Calculate the vector of the voltage mismatches for PV buses

$$\Delta\underline{\mathbf{V}}^{(k)} = \underline{\mathbf{V}}^{sp(k)} - \underline{\mathbf{V}}^{cal(k)} \tag{1.12}$$

If any of the mismatches is greater than a threshold, i.e. $\left|\Delta V_i^{(k)}\right| > \varepsilon$, then perform the next step.

4. Calculate the required reactive power generation of the DGs for all PV buses,

$$\mathbf{Q}_{DG}^{(k)} = \mathbf{Q}^{(k)} + \mathbf{Q}_L \tag{1.13}$$

where $\mathbf{Q}_L$ is the vector of reactive power loads and $\mathbf{Q}^{(k)}$ is the vector of the new reactive power injections at PV buses. The vector of the new reactive power injections at PV buses is

$$\mathbf{Q}^{(k)} = \mathbf{Q}^{(k-1)} + \Delta\mathbf{Q}^{(k)} \tag{1.14}$$

where $\Delta\mathbf{Q}^{(k)}$ is the vector of the reactive power increments at PV nodes

$$\Delta\mathbf{Q}^{(k)} = \mathrm{Im}\left\{\underline{\mathbf{V}}^{sp(k)} \cdot \left(\underline{\mathbf{Z}}_{PV}^{-1} \cdot \Delta\underline{\mathbf{V}}^{(k)}\right)^*\right\} \tag{1.15}$$

where $\underline{\mathbf{Z}}_{PV}$ is the PV bus sensitivity matrix. The dimension of $\underline{\mathbf{Z}}_{PV}$ is equal to the number of PV buses. The diagonal entry, $\underline{Z}_{PVii}$, in $\underline{\mathbf{Z}}_{PV}$ is equal to the sum positive sequence impedance of all branches between PV bus $i$ and the root bus (substation bus). If two PV buses, $i$ and $j$, have completely different paths to the root bus, then the off-diagonal entry $\underline{Z}_{PVij}$ is zero. If $i$ and $j$ share a piece of common path to the root bus, then $\underline{Z}_{PVij}$ is equal to the positive sequence impedances of all branches on this common path.

5. $Q_{DGi}^{(k)}$ then is compared with the reactive power generation limits. If $Q_{DGi}^{(k)}$ is within the limits, i.e. $Q_{DGi}^{\min} < Q_{DGi}^{(k)} < Q_{DGi}^{\max}$, then the DG current injected to PV bus $i$ is calculated as

$$\underline{\mathbf{I}}_{DGi}^{(k)} = \left( \frac{P_{DGi}^{sp} + jQ_{DGi}^{(k)}}{\underline{V}_i^{(k-1)}} \right)^* \tag{1.16}$$

In subsequent iterations, these currents will be combined with other bus current injections according to (1.8). Otherwise, if $Q_{DGi}^{(k)}$ violates any reactive power generation limit, it will be set to that limit, and combined with the reactive load at this bus. Subsequently, the rows and columns in the PV node sensitivity matrix, $\underline{\mathbf{Z}}_{PV}$, corresponding to this bus are removed and the LU factors of $\underline{\mathbf{Z}}_{PV}$ are updated.

Now, the backward/forward sweep algorithm is performed again. The power flow solution will be terminated when the voltage magnitudes of all PV buses converge to their specified values.

## 1.3 Power flow calculation in unbalanced DNs

The application of the backward/forward sweep power flow algorithm for unbalanced DNs implies a calculation in phase coordinates. Consequently, proper three-phase modeling of the DN components is crucial for the efficiency and accuracy of the power flow calculations.

### *1.3.1 Line modeling*

Since unbalanced DNs consists of un-transposed three-phase, two-phase, and single-phase lines with a mainly unbalanced load, it is necessary to take into account the magnetic coupling between the phases of the lines and the influence of the earth as a return path of the unbalance current. In addition, due to the uneven geometric position of the phases, the mutual phase capacities and the influence of the earth on the capacitance of the lines should be taken into account [3]. Therefore, the lines in DNs should be modeled by the (3 × 3)-dimensional three-phase series impedance and shunt admittance matrices, as illustrated in Figure 1.4. The calculation of these two matrices is based on the Carson equations and Kron reductions [4].

### *1.3.2 Transformer modeling*

The first problem in applying the backward/forward sweep-based power flow method for the unbalanced radial distribution networks with multiple voltage levels

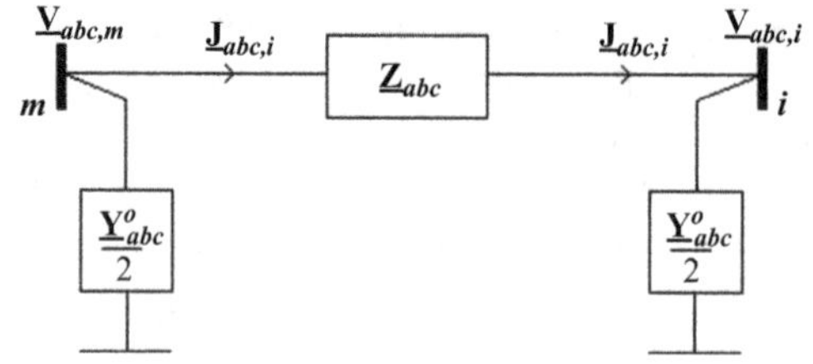


*Figure 1.4 Model of a three-phase unbalanced line*

is reflected in the inability to direct summarize branch currents (backward sweep) and calculate the bus voltages (forward sweep) for three-phase transformer whose secondary windings connections are delta or ungrounded wye. In the literature, three approaches can be identified for modeling the distribution transformers in the power flow methods for unbalanced DNs. The first approach is based on the modeling of three-phase distribution transformers over nodal admittance matrices. Although it is considered a classic way of modeling the transformer, this approach has a significant disadvantage which is reflected in the problem of the singularity of the nodal admittance matrices for some transformer connections such as $Y_nD$ or Dd [5]. Another approach to modeling the distribution transformer is based on using the equations in the phase domain for the currents at the primary side as a function of currents at the secondary side of the transformer (used in a backward sweep), and the equations for the voltages at the secondary side as a function of voltages at the primary side and currents at the secondary side of the transformer (used in a forward sweep). In [4], a matrix form of these equations is given for several transformer connections, but not for all connections, for example, $Y_nd$. Finally, the third approach is based on the symmetrical components model of three-phase transformers.

In this section, the usage of symmetrical component models of three-phase distribution transformers for common transformer connections in the backward/forward sweep-based power flow algorithm is demonstrated [6]. A matrix block model of a three-phase transformer is presented in Figure 1.5. As shown in Figure 1.5, a series block represents the transfer matrix of the transformer for currents $\underline{\mathbf{M}}_J$, the transfer matrix of the transformer for voltages $\underline{\mathbf{M}}_V$, and the matrix of phase impedances of transformer $\underline{\mathbf{Z}}_{Tabc}$. A shunt block $\underline{\mathbf{Y}}^m_{abc}$ models magnetizing admittance of the transformer. The shunt block should be placed on the side to which windings are connected in the grounded way ($Y_n$). In all other cases the shunt block can be connected on either side of the transformer (arbitrarily selected).

The backward/forward method involves two sweeps of calculations. In the backward sweep, starting from the branches connected to the far ends and moving toward a certain swing bus, the currents at the sending end of the branch are calculated as a function of the currents at its receiving end. In the forward sweep, starting from the branches connected to the source bus and moving toward the branches connected to the far ends, the voltages at the receiving end of the branch

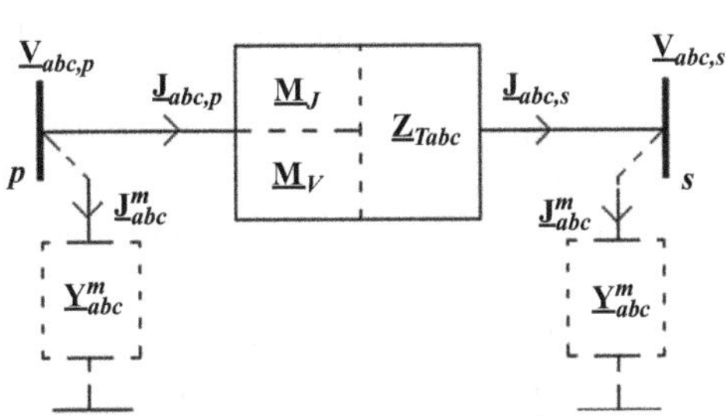

*Figure 1.5 Model of a three-phase transformer*

are calculated as a function of the voltages at its sending end and the currents at its receiving end. Therefore, when the branch represents a three-phase transformer, the currents at the primary ($p$) side of the transformer are calculated as a function of the currents at its secondary ($s$) side. The voltages at the secondary side of the transformer are calculated as a function of the voltages at its primary side and the currents at its secondary side. The relations between the symmetrical components of the voltages and currents of a three-phase transformer can be expressed in matrix form as follows:

$$\underline{\mathbf{J}}_{di0,p} = \underline{\mathbf{m}}_{J} \cdot \underline{\mathbf{J}}_{di0,s} \tag{1.17}$$

$$\underline{\mathbf{V}}_{di0,s} = \underline{\mathbf{m}}_{V} \cdot \underline{\mathbf{V}}_{di0,p} - \underline{\mathbf{Z}}_{Tdi0} \cdot \underline{\mathbf{J}}_{di0,s} \tag{1.18}$$

where

$\underline{\mathbf{J}}_{di0,p} = \begin{bmatrix} \underline{J}_{d,p} & \underline{J}_{i,p} & \underline{J}_{0,p} \end{bmatrix}^T$ is the vector of symmetrical components ($d$-positive sequence, $i$-negative sequence, and 0-zero sequence) of three-phase currents at the primary side of the transformer;
$\underline{\mathbf{J}}_{di0,s} = \begin{bmatrix} \underline{J}_{d,s} & \underline{J}_{i,s} & \underline{J}_{0,s} \end{bmatrix}^T$ is the vector of symmetrical components of three-phase currents at the secondary side of the transformer;
$\underline{\mathbf{V}}_{di0,p} = \begin{bmatrix} \underline{V}_{d,p} & \underline{V}_{i,p} & \underline{V}_{0,p} \end{bmatrix}^T$ is the vector of symmetrical components of three-phase voltages at the primary side of the transformer;
$\underline{\mathbf{V}}_{di0,s} = \begin{bmatrix} \underline{V}_{d,s} & \underline{V}_{i,s} & \underline{V}_{0,s} \end{bmatrix}^T$ is the vector of symmetrical components of three-phase voltages at the secondary side of the transformer.

The transfer matrices $\underline{\mathbf{m}}_{J}$ and $\underline{\mathbf{m}}_{V}$, as well as the matrix of symmetrical components of the transformer impedance $\underline{\mathbf{Z}}_{Tdi0}$, are given in Table 1.1.

As can be seen from Table 1.1, the transfer of the positive and negative sequences of the voltages and currents does not depend on the type of transformer connections. Transfer of the zero sequences of the currents and voltages, as well as the form of the matrix of symmetrical components of the transformer impedance $\underline{\mathbf{Z}}_{Tdi0}$, depends on the type of transformer connections and the groundedness of the neutral point.

### *1.3.3 Load modeling*

The loads in DNs can be three-phase, two-phase, or single-phase, connected in wye (Y) or delta (Δ) configuration [4]. In the power flow calculation, the currents per network phases are relevant. In addition to the load configuration (Y or Δ), the calculation of these currents also depends on the load model [3]. There are three load models: constant power model (PQ), constant impedance model (Z), and constant current model (I), or any combination of the three. Formulas for calculating line currents for various combinations of three-phase load models are listed in Table 1.2, according to the labels in Figure 1.6.

Single-phase and two-phase loads are modeled by setting the currents of the missing phases to zero, regardless of the connection type (wye or delta). Although

*Table 1.1 Transfer matrices for different connection types of the transformer*

| Connection | $\mathbf{m}_J$ | $\mathbf{m}_V$ | $\underline{\mathbf{Z}}_{Tdi0}$ |
|---|---|---|---|
| $Y_ny_n$ ($k$=0,4,8) | $\begin{bmatrix} 1/\underline{m}_T^* & 0 & 0 \\ 0 & 1/\underline{m}_T & 0 \\ 0 & 0 & 1/m_T \end{bmatrix}$ | $\begin{bmatrix} 1/\underline{m}_T & 0 & 0 \\ 0 & 1/\underline{m}_T^* & 0 \\ 0 & 0 & 1/m_T \end{bmatrix}$ | $\begin{bmatrix} \underline{Z}_T & 0 & 0 \\ 0 & \underline{Z}_T & 0 \\ 0 & 0 & \underline{Z}_T \end{bmatrix}$ |
| $Y_ny_n$ ($k$=2,6,10) | $\begin{bmatrix} 1/\underline{m}_T^* & 0 & 0 \\ 0 & 1/\underline{m}_T & 0 \\ 0 & 0 & -1/m_T \end{bmatrix}$ | $\begin{bmatrix} 1/\underline{m}_T & 0 & 0 \\ 0 & 1/\underline{m}_T^* & 0 \\ 0 & 0 & -1/m_T \end{bmatrix}$ | $\begin{bmatrix} \underline{Z}_T & 0 & 0 \\ 0 & \underline{Z}_T & 0 \\ 0 & 0 & \underline{Z}_T \end{bmatrix}$ |
| $Yy_n$ ($k$=0,2,4,6,8,10)<br>$Dy_n$ ($k$=1,3,5,7,9,11) | $\begin{bmatrix} 1/\underline{m}_T^* & 0 & 0 \\ 0 & 1/\underline{m}_T & 0 \\ 0 & 0 & 0 \end{bmatrix}$ | $\begin{bmatrix} 1/\underline{m}_T & 0 & 0 \\ 0 & 1/\underline{m}_T^* & 0 \\ 0 & 0 & 0 \end{bmatrix}$ | $\begin{bmatrix} \underline{Z}_T & 0 & 0 \\ 0 & \underline{Z}_T & 0 \\ 0 & 0 & \underline{Z}_T \end{bmatrix}$ |
| $Y_ny$ ($k$=0,2,4,6,8,10)<br>$Yy$ ($k$=0,2,4,6,8,10)<br>$Y_nd$ ($k$=1,3,5,7,9,11)<br>$Yd$ ($k$=1,3,5,7,9,11)<br>$Dy$ ($k$=1,3,5,7,9,11)<br>$Dd$ ($k$=0,2,4,6,8,10) | $\begin{bmatrix} 1/\underline{m}_T^* & 0 & 0 \\ 0 & 1/\underline{m}_T & 0 \\ 0 & 0 & 0 \end{bmatrix}$ | $\begin{bmatrix} 1/\underline{m}_T & 0 & 0 \\ 0 & 1/\underline{m}_T^* & 0 \\ 0 & 0 & 0 \end{bmatrix}$ | $\begin{bmatrix} \underline{Z}_T & 0 & 0 \\ 0 & \underline{Z}_T & 0 \\ 0 & 0 & \underline{Z}_T \end{bmatrix}$ |

*Note*: where $\underline{m}_T = m_T \cdot e^{jk30^\circ}$ is the complex transmission ratio and $\underline{Z}_T$ is the transformer impedance.

it seems trivial, it is still important to take care of it because when calculating unbalanced power flows, the quantities must be defined in all three phases, in each branch and in each bus of the network. In the software implementation of the algorithm for unbalanced power flows, currents, and voltages are presented in matrix form, that is, in the form of vectors with dimensions 3 × 1, so that each element of these vectors must be defined.

### *1.3.4 Backward/forward power flow algorithm for unbalanced DNs*

In principle, the algorithm consists of five main steps, as described for balanced DNs.

*Step 1. Read input data*: Define network configuration, parameters of network elements, load data, DG units data, and convergence criteria;

*Step 2. Initialization*: Set the phase voltages at the root bus and assume the initial phase voltages at all other network buses. In the per-unit system, the initial voltage values are usually adopted as follows:

$$\underline{\mathbf{V}}_{abc,i}^{(0)} = \begin{bmatrix} \underline{V}_{a,i}^{(0)} \\ \underline{V}_{b,i}^{(0)} \\ \underline{V}_{c,i}^{(0)} \end{bmatrix} = \begin{bmatrix} 1 \cdot e^{0^\circ} \\ 1 \cdot e^{-j120^\circ} \\ 1 \cdot e^{j120^\circ} \end{bmatrix} \quad i = 1, \ \ldots, \ N \tag{1.19}$$

*Table 1.2 Load models in three-phase distribution networks*

| Load type | Connection: Wye | Connection: Delta |
|---|---|---|
| Constan PQ | $\underline{I}_{La} = \left(\frac{\underline{S}_{La}}{\underline{V}_a}\right)^*$ | $\underline{I}_{Lab} = \left(\frac{\underline{S}_{Lab}}{\underline{V}_a - \underline{V}_b}\right)^* = \left(\frac{\underline{S}_{Lab}}{\underline{V}_{ab}}\right)^*$ |
| | $\underline{I}_{Lb} = \left(\frac{\underline{S}_{Lb}}{\underline{V}_b}\right)^*$ | $\underline{I}_{Lbc} = \left(\frac{\underline{S}_{Lbc}}{\underline{V}_b - \underline{V}_c}\right)^* = \left(\frac{\underline{S}_{Lbc}}{\underline{V}_{bc}}\right)^*$ |
| | $\underline{I}_{Lc} = \left(\frac{\underline{S}_{Lc}}{\underline{V}_c}\right)^*$ | $\underline{I}_{Lca} = \left(\frac{\underline{S}_{Lca}}{\underline{V}_c - \underline{V}_a}\right)^* = \left(\frac{\underline{S}_{Lca}}{\underline{V}_{ca}}\right)^*$ |
| | | $\underline{I}_{La} = \underline{I}_{Lab} - \underline{I}_{Lca};\ \underline{I}_{Lb} = \underline{I}_{Lbc} - \underline{I}_{Lab};\ \underline{I}_{Lc} = \underline{I}_{Lca} - \underline{I}_{Lbc}$ |
| Constant I | $\underline{I}_{La} = \left\lvert\frac{\underline{S}_{La}}{\underline{V}_{a,nom}}\right\rvert e^{j(\delta_a - \varphi_{La})}$ | $\underline{I}_{Lab} = \left\lvert\frac{\underline{S}_{Lab}}{\underline{V}_{ab,nom}}\right\rvert e^{j(\delta_{ab} - \varphi_{Lab})}$ |
| | $\underline{I}_{Lb} = \left\lvert\frac{\underline{S}_{Lb}}{\underline{V}_{b,nom}}\right\rvert e^{j(\delta_b - \varphi_{Lb})}$ | $\underline{I}_{Lbc} = \left\lvert\frac{\underline{S}_{Lbc}}{\underline{V}_{bc,nom}}\right\rvert e^{j(\delta_{bc} - \varphi_{Lbc})}$ |
| | $\underline{I}_{Lc} = \left\lvert\frac{\underline{S}_{Lc}}{\underline{V}_{c,nom}}\right\rvert e^{j(\delta_c - \varphi_{Lc})}$ | $\underline{I}_{Lca} = \left\lvert\frac{\underline{S}_{Lca}}{\underline{V}_{ca,nom}}\right\rvert e^{j(\delta_{ca} - \varphi_{Lca})}$ |
| | | $\underline{I}_{La} = \underline{I}_{Lab} - \underline{I}_{Lca};\ \underline{I}_{Lb} = \underline{I}_{Lbc} - \underline{I}_{Lab};\ \underline{I}_{Lc} = \underline{I}_{Lca} - \underline{I}_{Lbc}$ |
| Constant Z | $\underline{I}_{La} = \underline{V}_a \frac{\underline{S}^*_{La}}{\lvert\underline{V}_{a,nom}\rvert^2}$ | $\underline{I}_{Lab} = \underline{V}_{ab} \frac{\underline{S}^*_{Lab}}{\lvert\underline{V}_{ab,nom}\rvert^2}$ |
| | $\underline{I}_{Lb} = \underline{V}_b \frac{\underline{S}^*_{Lb}}{\lvert\underline{V}_{b,nom}\rvert^2}$ | $\underline{I}_{Lbc} = \underline{V}_{bc} \frac{\underline{S}^*_{Lbc}}{\lvert\underline{V}_{bc,nom}\rvert^2}$ |
| | $\underline{I}_{Lc} = \underline{V}_c \frac{\underline{S}^*_{Lc}}{\lvert\underline{V}_{c,nom}\rvert^2}$ | $\underline{I}_{Lca} = \underline{V}_{ca} \frac{\underline{S}^*_{Lca}}{\lvert\underline{V}_{ca,nom}\rvert^2}$ |
| | | $\underline{I}_{La} = \underline{I}_{Lab} - \underline{I}_{Lca};\ \underline{I}_{Lb} = \underline{I}_{Lbc} - \underline{I}_{Lab};\ \underline{I}_{Lc} = \underline{I}_{Lca} - \underline{I}_{Lbc}$ |

*Notes*: where $\underline{I}_{La}$, $\underline{I}_{Lb}$, and $\underline{I}_{Lc}$ are line (per phase) currents of load; $\underline{S}_{La}$, $\underline{S}_{Lb}$, and $\underline{S}_{Lc}$ are specified complex powers of load per phases; $\underline{S}_{Lab}$, $\underline{S}_{Lbc}$, and $\underline{S}_{Lca}$ are specified complex powers of load phase-to-phase; $\underline{V}_a$, $\underline{V}_b$, and $\underline{V}_c$ are complex phase voltages; $\underline{V}_{a,nom}$, $\underline{V}_{b,nom}$, and $\underline{V}_{c,nom}$ are nominal phase voltages; $\delta_a$, $\delta_b$, and $\delta_c$ are phase voltage angles; and $\varphi_{La}$, $\varphi_{Lb}$, and $\varphi_{Lc}$ are load power factor angles.

where $\underline{\mathbf{V}}^{(0)}_{abc,i}$ is the vector of initial guesses of the voltage phasors at bus $i$ and $N$ is the total number of buses.

*Step 3. Backward sweep*: At iteration $k$, starting from the branch in the last layer (branch corresponding to last bus $N$) and moving toward the branches connected to the root bus, the line current in branch $i$, $\underline{\mathbf{J}}_{abc,i}$, according to

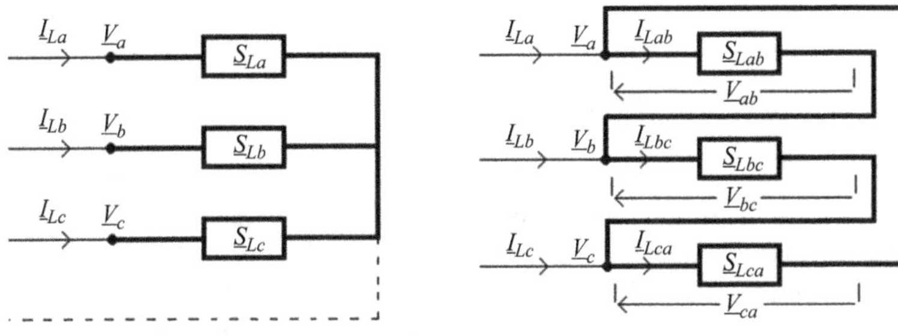


*Figure 1.6 Load connections: (a) wye and (b) delta*

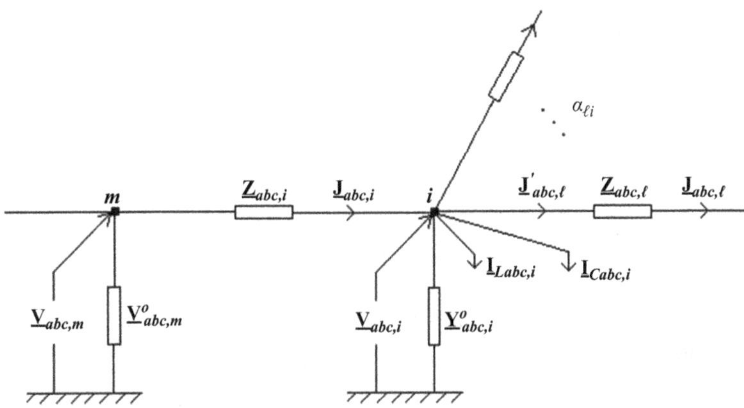

*Figure 1.7 A part of an unbalanced distribution network*

Figure 1.7, is calculated as follows:

$$\underline{\mathbf{J}}_{abc,i}^{(k)} = \underline{\mathbf{I}}_{Labc,i}^{(k)} + \underline{\mathbf{I}}_{Cabc,i}^{(k)} + \underline{\mathbf{Y}}_{abc,i}^{o} \cdot \underline{\mathbf{V}}_{abc,i}^{(k-1)} + \sum_{\substack{\ell \in \alpha_{\ell i} \\ \ell \neq i}} \underline{\mathbf{J}}_{abc,\ell}^{\prime(k)}$$

$$i = N, N-1, \ \ldots\ , 1; \ k = 1, 2, \ \ldots \tag{1.20}$$

where $k$ is the index of iteration, $\underline{\mathbf{J}}_{abc,i}^{(k)} = \left[ \underline{J}_{a,i}^{(k)} \ \ \underline{J}_{b,i}^{(k)} \ \ \underline{J}_{c,i}^{(k)} \right]^T$ is the current phasor vector at the end of the branch $i$ in the $k$th iteration, $\underline{\mathbf{I}}_{Labc,i}^{(k)} = \left[ \underline{I}_{La,i}^{(k)} \ \ \underline{I}_{Lb,i}^{(k)} \ \ \underline{I}_{Lc,i}^{(k)} \right]^T$ is the load current phasor vector in the bus $i$ in the $k$th iteration, $\underline{\mathbf{I}}_{Cabc,i}^{(k)} = \left[ \underline{I}_{Ca,i}^{(k)} \ \ \underline{I}_{Cb,i}^{(k)} \ \ \underline{I}_{Cc,i}^{(k)} \right]^T$ is the capacitor current phasor vector in the bus $i$ in the $k$th iteration, $\underline{\mathbf{J}}'_{abc,\ell}(k) = \left[ \underline{J}'_{a,\ell}(k) \ \ \underline{J}'_{b,\ell}(k) \ \ \underline{J}'_{c,\ell}(k) \right]^T$ is the current phasor vector at the beginning of branch $\ell$ that starts from bus $i$ in the $k$th iteration, $\underline{\mathbf{V}}_{abc,i}^{(k-1)} = \begin{bmatrix} \underline{V}_{a,i}^{(k-1)} & \underline{V}_{b,i}^{(k-1)} & \underline{V}_{c,i}^{(k-1)} \end{bmatrix}^T$ is the voltage phasor vector at the bus $i$ in iteration $(k-1)$, and $\underline{\mathbf{Y}}_{abc,i}^{o}$ is the summary phase shunt admittance matrix in bus $i$. This matrix includes the phase shunt admittances of the lines and/or the magnetizing admittances of the transformers connected to bus $i$ (in accordance with the modeling of the lines and transformers in Figures 1.4 and 1.5, respectively); $\alpha_{\ell i}$ is the set of buses connected with bus $i$.

Note that the load and capacitor currents are calculated according to the equations listed in Table 1.2, using the voltage phasors from the previous iteration $(k-1)$.

If the branch $\ell$ is a line segment (feeder), the phasor vector of the current at the beginning of the branch is equal to the phasor vector of the current at the end of the branch, i.e. $\underline{\mathbf{J}}'_{abc,\ell}(k) = \underline{\mathbf{J}}_{abc,\ell}^{(k)}$.

When branch $\ell$ is a transformer, the current at the beginning of the branch (current from the primary side of the transformer) is calculated in accordance with the way of modeling the transformer (Figure 1.5), i.e.

$$\underline{\mathbf{J}}'_{abc,\ell}(k) = \underline{\mathbf{M}}_J \cdot \underline{\mathbf{J}}_{abc,\ell}^{(k)} \tag{1.21}$$

where $\underline{\mathbf{M}}_J = \mathbf{T} \cdot \underline{\mathbf{m}}_J \cdot \mathbf{T}^{-1}$ is the transformation matrix of the transformer for currents, $\mathbf{T}$ is the Fortescue's transformation matrix, and $\underline{\mathbf{m}}_J$ is the transfer matrix for currents as given in Table 1.1.

*Step 4. Forward sweep*: Bus voltages are updated in a forward sweep starting from branches connected to the root bus toward those in the last. For each branch, $i$, the voltage at bus $i$ is calculated using the updated voltage at bus $m$ (Figure 1.7), and the branch current calculated in the preceding backward sweep:

$$\underline{\mathbf{V}}_{abc,i}^{(k)} = \underline{\mathbf{V}}_{abc,m}^{(k)} - \underline{\mathbf{Z}}_{abc,i} \cdot \underline{\mathbf{J}}_{abc,i}^{(k)} \tag{1.22}$$

In case the branch between nodes $m$ and $i$ is a transformer, the following expression applies:

$$\underline{\mathbf{V}}_{abc,i}^{(k)} = \underline{\mathbf{M}}_{Vi} \cdot \underline{\mathbf{V}}_{abc,m}^{(k)} - \underline{\mathbf{Z}}_{abc,i} \cdot \underline{\mathbf{J}}_{abc,i}^{(k)} \tag{1.23}$$

where $\underline{\mathbf{V}}_{abc,i}^{(k)} = \begin{bmatrix} \underline{V}_{a,i}^{(k)} & \underline{V}_{b,i}^{(k)} & \underline{V}_{c,i}^{(k)} \end{bmatrix}^T$ is the voltage phasor vector at bus $i$ in iteration $(k)$, $\underline{\mathbf{V}}_{abc,m}^{(k)} = \begin{bmatrix} \underline{V}_{a,m}^{(k)} & \underline{V}_{b,m}^{(k)} & \underline{V}_{c,m}^{(k)} \end{bmatrix}^T$ is the voltage phasor vector at the bus $m$ in the $k$th iteration, $\underline{\mathbf{Z}}_{abc,i}$ is the phase impedance matrix of branch $i$, $\underline{\mathbf{M}}_V = \mathbf{T} \cdot \underline{\mathbf{m}}_V \cdot \mathbf{T}^{-1}$ is the transformation matrix of the transformer for voltages, $\mathbf{T}$ is Fortescue's transformation matrix, and $\underline{\mathbf{m}}_V$ is the transfer matrix for voltages as given in Table 1.1.

*Step 5. Check convergence criteria*: Steps 3 and 4 are repeated until convergence is achieved. For example, a convergence criteria can be adopted as follows:

$$\max\left|\mathbf{V}_{abc,i}^{(k)} - \mathbf{V}_{abc,i}^{(k-1)}\right| \le \varepsilon_V = \left(10^{-3} - 10^{-5}\right) = \left(10^{-3} - 10^{-5}\right) \tag{1.24}$$

#### 1.3.4.1 DG modeling

The primary energy of DG units may be injected into the DN via either a synchronous or an asynchronous electric machine, which is directly connected to the grid, a combination of an electric machine and a power electronic interface, or only via a power electronic interface [7]. If the electric machine is directly connected to the grid, its operation determines the model of DG for power flow studies. In other cases, the characteristics of the interface control circuit determine the DG model. The node with DG can be represented as a PQ or PV bus [1]. Figure 1.8 shows a part of the DN with DG connected at bus $i$. In accordance with Figure 1.8, the current in line section $i$ at iteration $k$ is

$$\underline{\mathbf{J}}_{abc,i}^{(k)} = \underline{\mathbf{I}}_{Labc,i}^{(k)} - \underline{\mathbf{I}}_{DGabc,i}^{(k)} + \underline{\mathbf{Y}}_{abc,i}^{o} \cdot \underline{\mathbf{V}}_{abc,i}^{(k-1)} + \sum_{\substack{\ell \in \alpha_{\ell i} \\ \ell \neq i}} \underline{\mathbf{J}}_{abc,\ell}^{'(k)}$$

$$i = N, N-1, \ldots, 1; k = 1, 2, \ldots \tag{1.25}$$

where $\underline{I}_{DGabc,i}^{(k)}$ are phase current injections at bus $i$ corresponding to DG.

In case node $i$ is modeled as a PQ bus, the DG unit should be treated as a negative PQ load. Accordingly, the current injection of the DG at bus $i$ can be calculated as follows:

$$\underline{\mathbf{I}}_{DGabc,i}^{(k)} = \begin{bmatrix} \left(\dfrac{P_{DGa,i}^{sp} + jQ_{DGa,i}^{sp}}{\underline{V}_{a,i}^{(k-1)}}\right)^* \\ \left(\dfrac{P_{DGb,i}^{sp} + jQ_{DGb,i}^{sp}}{\underline{V}_{b,i}^{(k-1)}}\right)^* \\ \left(\dfrac{P_{DGc,i}^{sp} + jQ_{DGc,i}^{sp}}{\underline{V}_{c,i}^{(k-1)}}\right)^* \end{bmatrix} \tag{1.26}$$

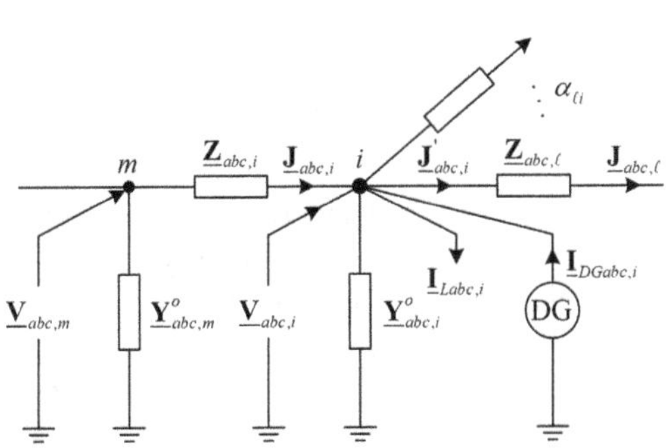


*Figure 1.8 A part of unbalanced DN with DG*

where $P^{sp}_{DGabc,i}$, $Q^{sp}_{DGabc,i}$ are specified active and reactive power outputs of the DG connected at bus $i$.

In case bus $i$ from Figure 1.8 is modeled as a PV bus, handling the PV buses in the power flow algorithm requires some additional processes, as follows:

(i) Initially, the DG active power outputs, reactive power generation limits, and positive sequence voltage magnitudes at PV buses are specified. The reactive powers of the DGs are initialized to zero. The backward/forward sweeps are performed considering the DGs as negative PQ loads. When the power flow is converged the next step is done.

(ii) Calculate the positive sequence voltage at each PV bus

$$\underline{V}^{cal(k)}_{d,i} = V^{cal(k)}_{d,i}\left[\cos\left(\theta^{(k)}_{d,i}\right) + j\sin\left(\theta^{(k)}_{d,i}\right)\right] \quad i = 1, 2, \ldots, N_{PV} \tag{1.27}$$

(iii) Calculate the complex value of the specified voltage for all PV buses

$$\underline{V}^{sp(k)}_{d,i} = V^{sp}_{d,i}\left[\cos\left(\theta^{(k)}_{d,i}\right) + j\sin\left(\theta^{(k)}_{d,i}\right)\right] \quad i = 1, 2, \ldots, N_{PV} \tag{1.28}$$

(iv) Calculate the vector of the positive sequence voltage mismatches for PV nodes

$$\Delta\underline{\mathbf{V}}^{(k)} = \underline{\mathbf{V}}^{sp(k)}_{d} - \underline{\mathbf{V}}^{cal(k)}_{d} \tag{1.29}$$

If any of the mismatches is greater than a threshold, i.e. $\left|\Delta V^{(k)}_{i}\right| > \varepsilon$, then perform the next step.

(v) Calculate the required reactive power generation of the DGs for all PV buses

$$\mathbf{Q}^{(k)}_{DGabc} = \mathbf{Q}^{(k)}_{abc} + \mathbf{Q}_{Labc} \tag{1.30}$$

where $\mathbf{Q}_{Labc}$ is the vector of reactive power loads at PV nodes and $\mathbf{Q}^{(k)}_{abc}$ is the vector of the new reactive power injections at PV buses.

The vector of the new reactive power injections at PV buses is

$$\mathbf{Q}^{(k)}_{abc} = \mathbf{Q}^{(k-1)}_{abc} + \Delta\mathbf{Q}^{(k)} \tag{1.31}$$

$\Delta\mathbf{Q}^{(k)}_{abc}$ is the vector of the reactive power increments at PV nodes

$$\Delta\mathbf{Q}^{(k)} = \mathrm{Im}\left\{\underline{\mathbf{V}}^{sp(k)}_{d} \cdot \left(\underline{\mathbf{Z}}^{-1}_{PV} \cdot \Delta\underline{\mathbf{V}}^{(k)}\right)^{*}\right\} \tag{1.32}$$

where $\underline{\mathbf{Z}}_{PV}$ is the PV node sensitivity matrix. The dimension of $\underline{\mathbf{Z}}_{PV}$ is equal to the number of PV nodes. The diagonal entry, $\underline{Z}_{PVii}$, in $\underline{\mathbf{Z}}_{PV}$ is equal to the sum positive sequence impedance of all line sections between PV bus $i$ and the root bus (substation bus). If two PV buses, $i$ and $j$, have completely different paths to the root bus, then the off-diagonal entry $\underline{Z}_{PVij}$ is zero. If $i$ and $j$ share a piece of common path to the root node, then $\underline{Z}_{PVij}$ is equal to the positive sequence impedances of all line sections on this common path.

(vi) $Q_{DGabc,i}^{(k)}$ then is compared with the reactive power generation limits. If $Q_{DGabc,i}^{(k)}$ is within the limits, i.e. $Q_{DGabc,i}^{\min} < Q_{DGabc,i}^{(k)} < Q_{DGabc,i}^{\max}$, then the corresponding DG currents are injected to PV node $i$ according to

$$\underline{\mathbf{I}}_{DGabc,i}^{(k)} = \begin{bmatrix} \left( \dfrac{P_{DGa,i}^{sp} + jQ_{DGa,i}^{(k)}}{\underline{V}_{a,i}^{(k-1)}} \right)^{*} \\ \left( \dfrac{P_{DGb,i}^{sp} + jQ_{DGb,i}^{(k)}}{\underline{V}_{b,i}^{(k-1)}} \right)^{*} \\ \left( \dfrac{P_{DGc,i}^{sp} + jQ_{DGc,i}^{(k)}}{\underline{V}_{c,i}^{(k-1)}} \right)^{*} \end{bmatrix} \tag{1.33}$$

In subsequent iterations, these currents will be combined with other nodal current injections according to (1.25). Otherwise, if $Q_{DGabc,i}$ violate any reactive power generation limit, it will be set to that limit and combined with the reactive load at this bus. Subsequently, the rows and columns in the PV node sensitivity matrix, $\underline{\mathbf{Z}}_{PV}$, corresponding to this node are removed, and the LU factors of $\underline{\mathbf{Z}}_{PV}$ are updated.

Now, the backward/forward sweeps are performed again. After the convergence, Steps (ii)–(vi) are repeated. The power flow solution will be terminated when the positive sequence voltages of all PV buses converge to their specified value.

## 1.4 MATLAB® programs

The above-described backward/forward power flow algorithms were implemented in MATLAB. The power flow programs, both for balanced and unbalanced DNs, are available as supplementary files of this chapter on the book's homepage at the IET digital library.

Users can modify and customize these files according to their needs.

### *1.4.1 Power flow program for balanced DNs*

The power flow program for balanced DNs is developed in the form of an m-files package *pfdnm*. The *pfdnm* is an educational program package of MATLAB m-files for power flow computation in balanced DNs. The iterative backward/forward power flow method is implemented in the solver *pfdnm.m*. It is available in six test systems, two of which are standard IEEE test systems.

Figure 1.9 shows an example of the application of *pfdnm.m* solver for power flow calculation in the IEEE 33-bus test feeder.

### *1.4.2 Power flow program for unbalanced DNs*

The iterative backward/forward power flow algorithm for unbalanced DNs is implemented in the *pfudnm* program package. It is available in four test systems, including IEEE 4, 13, 37, and 123-bus test feeders. Figure 1.10 shows an example of running this program to calculate the power flow in the IEEE 13-bus test feeder.

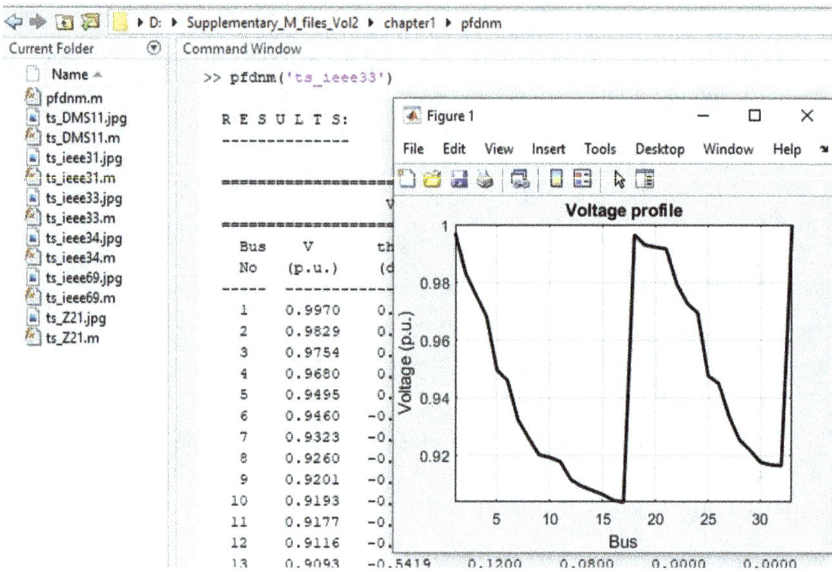


*Figure 1.9 Example of using the pfdnm program package*

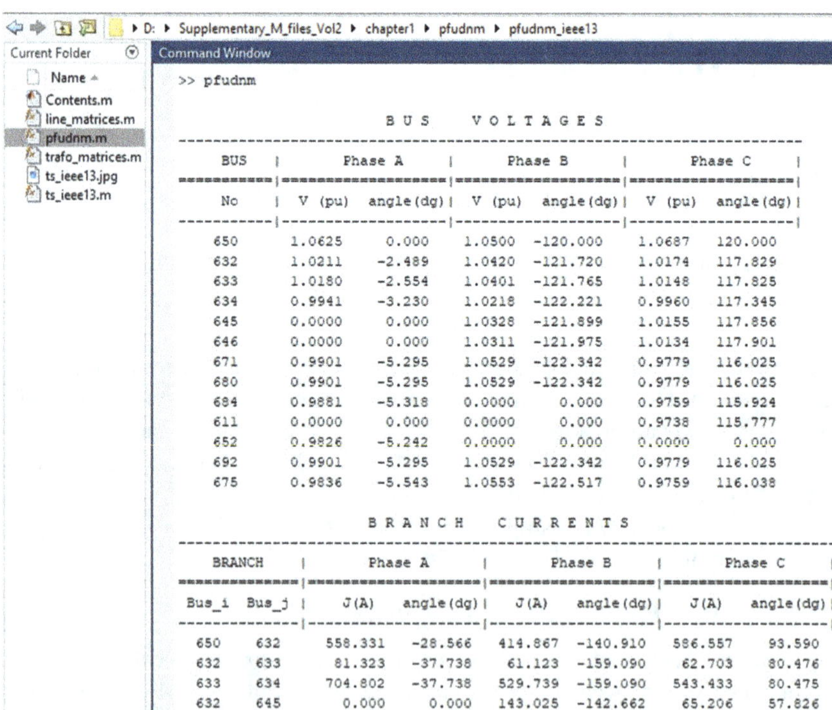


*Figure 1.10 Example of using the pfudnm program package*

## 1.5 Conclusion

This chapter describes the backward/forward sweep algorithm for power flow computation in balanced and unbalanced DN. Power flow calculation is one of the most important and most used functions in the analysis of DNs. In that context, this chapter presents an introduction to the next chapters dealing with different optimization problems in DNs such as the OPF, Volt/Var control, optimal integration of DG units, distribution system state estimation, etc.

## References


[1] Shirmohammadi D., Hong H.W., Semlyen A., and Luo G.X. 'A Compensation-based power flow method for weakly meshed distribution and transmission networks'. *IEEE Transactions on Power Systems*. 1988;**3**(2): 753–62.

[2] Khushalani S., Solanki J.M., and Shulz N.N. 'Development of three-phase unbalanced power flow using PV and PQ models for distributed generation and study of the impact of DG models'. *IEEE Transactions on Power Systems*. 2007; **22**(3):1019–25.

[3] Ullah Z., Elkadeem M.R., Wang S., and Radosavljević J. 'A novel PSOS-CGSA method for state estimation in unbalanced DG-integrated distribution systems'. *IEEE Access*. 2020; **8**:113219–29.

[4] Kersting W.H. *Distribution System Modeling and Analysis*. Boca Raton, FL: CRC Press; 2002. pp. 125–39.

[5] Xiao P., Yu D.C., and Yan W. 'A unified three-phase transformer model for distribution load flow calculations'. *IEEE Transactions on Power Systems*. 2006; **21**(1):153–59.

[6] Radosavljević J., Jevtić M., and Klimenta D. 'Modeling of the transformers and distributed generation for power flow calculation in unbalanced distribution networks'. *VIII Symposium on Serbian Distribution Networks with Regional Participation (CIRED Serbia)*, Vrnjačka Banja, September 23–28, 2012, R-6.17 (in Serbian).

[7] Moghaddas S.M., and Mashhour E. 'Distributed generation modeling for power flow studies and a three-phase unbalanced power flow solution for radial distribution systems considering distribution generation'. *Electric Power Systems Research*. 2009;**79**:680–86.

# Chapter 2
# Optimal power flow in distribution networks

## 2.1 Introduction

Recently, due to the growth of distributed generation (DG) penetration, the nature of the distribution network has been altered from a passive network to an active one and by a distribution company market, which is operated by a distribution system operator (DSO). DSO is responsible for the operation of the distribution network in an effective manner, maintaining the required reliability of electricity supply and the quality of supply as well as coordinating the operation of the coordinated high voltage (HV) grid in cooperation with the transmission system operator.

Electricity distribution companies tend to integrate different types of distributed DG units into the distribution network. Integration of DG units with renewable energy such as photovoltaic (PV) systems, small wind turbines ($WT_s$), and mini hydropower plants (MHP) is influenced by geographical and meteorological conditions. The DG units with non-renewable energy, including fuel cells (FCs), microturbines (MTs), and diesel-electric generators (DEGs), can be connected to any point of the distribution network, but the generating cost should be taken into account. Depending on the DG unit technologies, the primary energy of DG units may be injected into the distribution network via either a synchronous or an asynchronous electric machine that is directly connected to the grid, a combination of an electric machine and a power electronic interface, or only via a power electronic interface. If the electric machine is directly connected to the grid, its operation determines the model of the DG unit (constant active and reactive power – PQ bus or constant active power and voltage magnitude – PV bus) for power flow studies. In other cases, the characteristics of the interface control circuit determine the DG unit model. As a general rule, in case when the control circuit of the converter is designed to control active power and voltage independently, the DG unit model shall be as a PV bus, and when it is designed to control active and reactive power independently, the DG unit model shall be as a PQ bus [1,2].

If properly planned and controlled, the DG units may offer improved voltage profile and power loss reduction of the distribution network, better economics, and a reduced dependence on the local utility [3]. Safe, reliable, and economic operation of the active distribution network can be achieved only by efficiently coordinating the DG unit operation, voltage regulation (voltage regulators, ULTC), and reactive power compensation (VAR compensators); that is, by solving an optimal power flow (OPF) problem. Accordingly, the OPF calculation is one of the important tools for DSO to perform this duty. The main objective of the OPF for a distribution network is to minimize the fuel cost of DG units, voltage profile improvement, var/volt coordination, and minimization of power losses through optimal settings of the control variables while at the same time satisfying various distribution system operating constraints.

The traditional OPF analysis is addressed as a *deterministic OPF problem* based on the premise that the input variables and network parameters are accurately known. In other words, it is supposed that the network parameters, power of load, and output power of DG units are deterministic known. Therefore, the deterministic OPF approaches are highly dependent on the accuracy of the input data. However, many random disturbances or uncertain factors, such as the variation of load, the change in network configuration, and especially the stochastic nature of the output power of the DG with renewable energy such as WT and PV units, really exist in distribution network operation. In these cases, it is necessary to incorporate these uncertainties in OPF modeling to overcome inaccuracy in the OPF results. Therefore, the OPF problem is transformed into a probabilistic OPF problem [4].

In its most general mathematical formulation, the OPF is a non-linear, non-convex, large-scale, static optimization problem with both continuous and discrete control variables. Several different approaches, such as methods of linear and quadratic programming (QP) and recently very popular metaheuristic optimization techniques, have been applied to solve the OPF problem in distribution networks. The author in [5] proposed a mixed integer linear programming model for distribution network optimization and analyzed various operating conditions. In [6], an optimal distribution power flow strategy based on the decomposition of the overall system problem into two components: economic dispatch for energy at the system level and loss minimization at the distribution level is proposed. The optimization problem was solved by a combined QP and sectioning algorithm. Gan *et al.* [7] proposed conic programming for solving the OPF problem in radial distribution networks with the objective of minimizing the generation cost of DG units and system power loss. The authors in [8] used the sequential quadratic programming method to solve a multi-objective OPF problem for distribution networks with objectives such as minimal power loss, minimal voltage deviation, and maximum DG active power output.

In recent years, many population-based optimization techniques have been used to solve complex constrained optimization problems. Some of the population-based methods such as genetic algorithm (GA), particle swarm optimization (PSO), and gravitational search algorithm (GSA), have been successfully applied to find the optimal solution for OPF problem in distribution networks. A brief overview of these and other metaheuristic methods for solving the problem of OPF in distribution networks is given in the first part of the book (*Metaheuristic Optimization in Power Engineering: Vol. 1: Algorithms and Power Dispatch – Using MATLAB®-Based software*, Chapter 6, Section 6.4).

In this chapter, a generalized approach for the OPF analysis in distribution networks is presented. This implies the inclusion of different types of DG units, taking into account the uncertainties of the input variables, the multi-objective optimization, and the application of efficient metaheuristic methods for solving the OPF problem.

The rest of the chapter is organized as follows. In Section 2.2, the deterministic OPF as a non-linear multi-objective optimization problem with equality and inequality constraints is mathematically formulated. This formulation includes the DG units with renewable and non-renewable energy sources. Optimization models of the DG units are given in Section 2.3. Section 2.4 briefly describes two metaheuristic optimization methods proposed to solve the OPF problems, namely the GA and GSA methods. In addition to the deterministic, a probabilistic OPF approach based on the $2m + 1$ point estimate method is presented in Section 2.5. The Weibull and normal distributions are employed to model the input random variables, namely the wind speed, solar irradiance, and load power. The Gram Charlier expansion series is used to obtain PDFs of the OPF results. Simulation results are presented in Section 2.6. Section 2.7 describes a MATLAB®-based program package for OPF in distribution networks by using different metaheuristic optimization methods. Finally, the main conclusions are summarized in Section 2.8.

## 2.2 Deterministic OPF

The goal of the OPF is to minimize a selected objective function via optimal settings of the control variables, subjected to various equality and inequality constraints. Generally, the OPF problem can be formulated as follows:

$$\min F(\mathbf{x}, \mathbf{y}) \tag{2.1}$$

Subject to:

$$g(\mathbf{x}, \mathbf{y}) = 0 \tag{2.2}$$

$$h(\mathbf{x}, \mathbf{y}) \leq 0 \tag{2.3}$$

$$\mathbf{x} \in \mathbf{X} \tag{2.4}$$

where $F$ is the objective function to be minimized and $\mathbf{x}$ and $\mathbf{y}$ are vectors of control and dependent variables, respectively.

For the distribution network with DG units, the vector of control variables ($\mathbf{x}$) consisting of

- The active power output of the DG units with nonrenewable energy sources $P_{DG}$;
- Root bus voltage $V_0$;
- The terminal voltage at PV buses $V_{PV}$;
- Transformer (voltage regulation) tap setting $T$; and

- The output of shunt VAR compensators $Q_C$.

Therefore, the vector of control variables can be expressed as

$$\mathbf{x} = [P_{DG1}\ldots P_{DGNNR}, V_0, V_{PV1}\ldots V_{PVNPV}, T_1\ldots T_{NT}, Q_{C1}\ldots Q_{CNC}] \tag{2.5}$$

where *NNR*, *NPV*, *NT*, and *NC* are the number of the nonrenewable DG units, the number of PV buses (i.e. DG units modeled as PV buses), the number of regulating transformers, and the number of VAR compensators, respectively.

The vector of dependent variables (**y**) consisting of

- The active power bought/sold from/to the utility grid $P_{gr}$. This is active power that is delivered through the root bus of the distribution network.
- Load bus voltage, including DG units, which are modeled as PQ buses, $V_L$.
- The reactive power outputs of the DG units, which are modeled as PV buses $Q_{DG}$.
- Branch flow $S_l$.

Therefore, **y** can be expressed as

$$\mathbf{y} = \left[P_{gr}V_{L1}\ldots V_{LNL}, Q_{DG1}\ldots Q_{DGNPV}, S_{l1}\ldots S_{lN}\right] \tag{2.6}$$

where *NL*, *NPV*, and *NB* are the number of load buses, the number of PV buses, and the total number of buses (namely branches) in the distribution network, respectively.

### 2.2.1 Objective function

The objective function can take different forms. Several cases have been considered herein.

*Case 1*: Generating cost minimization

$$F = f_{gr}\left(P_{gr}\right) + \sum_{i=1}^{NNR} f_i(P_{DGi}) \tag{2.7}$$

where $f_{gr}\left(P_{gr}\right)$ is the cost characteristics for the active power bought/sold from/to the utility grid and $f_i(P_{DGi})$ is the generating cost characteristic of the $i$th DG unit.

Usually, the generating cost characteristics are defined as quadratic cost function of active power as follows:

$$f_i(P_i) = a_i + b_iP_i + c_iP_i^2 \tag{2.8}$$

where $P_i$ stands for $P_{gr}$ or $P_{DGi}$ and $a_i$, $b_i$, and $c_i$ are the appropriate cost coefficients.

*Case 2*: Power loss minimization

$$F = \sum_{i=1}^{NB} P_{lossi} \tag{2.9}$$

where $P_{lossi}$ is the power loss in branch $i$.

*Case 3*: Multi-objective OPF as simultaneous minimization of the generating cost and power losses:

$$F = f_{gr}(P_{gr}) + \sum_{i=1}^{NNR} f_i(P_{DGi}) + w_P \sum_{i=1}^{NB} P_{lossi} \tag{2.10}$$

where $w_P$ is the weighting factor for power losses.

*Case 4*: Multi-objective OPF as simultaneous minimization of the generating cost and voltage deviation at load buses:

$$F = f_{gr}(P_{gr}) + \sum_{i=1}^{NNR} f_i(P_{DGi}) + w_V \sum_{i=1}^{NL} |1 - V_i| \tag{2.11}$$

where $V_i$ and $w_V$ are the voltage magnitude at load bus $i$ and the weighting factor for voltage deviation, respectively.

*Case 5*: Multi-objective OPF as simultaneous minimization of the generating cost, power losses, and voltage deviation at load buses:

$$F = f_{gr}(P_{gr}) + \sum_{i=1}^{NNR} f_i(P_{DGi}) + w_P \sum_{i=1}^{NB} P_{lossi} + w_V \sum_{i=1}^{NL} |1 - V_i| \tag{2.12}$$

### 2.2.2 Constraints

The equality constraints (2.2) represent a typical power balance and power flow model for distribution networks. The power balance equation in the distribution network with DG units with renewable and nonrenewable energy sources can be expressed as follows:

$$\sum_{i=1}^{NNR} P_{DGi} + P_{gr} = \sum_{i=1}^{NL} P_{Li} + \sum_{i=1}^{NB} P_{lossi} - \sum_{i=1}^{NR} P_{DGi} \tag{2.13}$$

where $NR$ is the number of DG units with renewable energy sources.

The power flow model for distribution networks is represented by the backward/forward sweep power flow equations given in Chapter 1 of this book.

The inequality constraints (2.3) are the functional operating constraints that contain load bus voltage magnitude limits, DG unit reactive power capabilities, and branch flow limits:

$$V_i^{\min} \leq V_i \leq V_i^{\max} \quad i = 1, \ldots, NL \tag{2.14}$$

$$Q_{DGi}^{\min} \leq Q_{DGi} \leq Q_{DGi}^{\max} \quad i = 1, \ldots, NPV \tag{2.15}$$

$$S_i \leq S_i^{\max} \quad i = 1, \ldots, NB \tag{2.16}$$

Constraints (2.4) define the feasibility region of the problem control variables such as DG unit active power output limits, root bus voltage limits, PV bus voltage limits, transformer tap setting limits, and shunt VAR compensation limits:

$$P_{DGi}^{\min} \le P_{DGi} \le P_{DGi}^{\max} \quad i = 1, \ldots, NNR \tag{2.17}$$

$$V_0^{\min} \le V_0 \le V_0^{\max} \tag{2.18}$$

$$V_{DGi}^{\min} \le V_{DGi} \le V_{DGi}^{\max} \quad i = 1, \ldots, NPV \tag{2.19}$$

$$T_i^{\min} \le T_i \le T_i^{\max} \quad i = 1, \ldots, NT \tag{2.20}$$

$$Q_{Ci}^{\min} \le Q_{Ci} \le Q_{Ci}^{\max} \quad i = 1, \ldots, NC \tag{2.21}$$

It should be noted that in the optimization process, the control variables are self-constrained. The inequality constraints of dependent variables, such as $V_L$, $Q_{DG}$, and $S_l$, can be incorporated in the objective function $F$ as quadratic penalty terms. The new expanded objective function to be minimized becomes

$$F_p = F + \lambda_V \sum_{i=1}^{NL} \left(V_i - V_i^{\lim}\right)^2 + \lambda_{QDG} \sum_{i=1}^{NPV} \left(Q_{DGi} - Q_{DGi}^{\lim}\right)^2 + \lambda_S \sum_{i=1}^{NB} \left(S_i - S_i^{\lim}\right)^2 \tag{2.22}$$

where $\lambda_V$, $\lambda_{QDG}$, and $\lambda_S$ are defined as penalty factors and $x^{\lim}$ is the limit value of the dependent variable $x$, and given as follows:

if $x > x^{\max}$, then $x^{\lim} = x^{\max}$
else if $x < x^{\min}$, then $x^{\lim} = x^{\min}$
else
$x^{\lim} = x$; end

## 2.3 DG units modeling for OPF

This section describes the models of different types of DG units used for the OPF computation in distribution networks. These models are related to the generating cost characteristics of DGs.

### 2.3.1 Diesel generator

Usually, the diesel fuel consumption data in L/h at 25%, 50%, 70%, and 100% of the DEG power rating (kW) are given by the manufacturer. Based on these data, the fuel consumption characteristic of the DEG can be estimated as a quadratic function of its active power output:

$$Fuel_{DEG} = a + bP_{DEG} + cP_{DEG}^2 \tag{2.23}$$

where $Fuel_{DEG}$ is the DEG fuel consumption (*L/h*); $P_{DEG}$ is the DEG power output (kW); and $a$, $b$, and $c$ are coefficients of the fuel consumption characteristic. Accordingly, the total DEG fuel cost $f_{DEG}$ can be calculated as

$$f_{DEG} = c_{DEG} \cdot Fuel_{DEG} = c_{DEG}\left(a + bP_{DEG} + cP_{DEG}^2\right) \tag{2.24}$$

where $c_{DEG}$ is the diesel fuel price to supply the DEG ($/L). For example, let us consider the HDJW 410 T6, 369 kW DEG. The diesel fuel consumption data of the DEG are given in Table 2.1 [9]. Based on these data and the fuel price adopted $c_{DEG} = 0.6$ $/L, the fuel costs characteristic shown in Figure 2.1 is obtained.

### *2.3.2 Fuel cell*

Fuel cost for the FC is dependent on the active power output and the FC efficiency [10]:

$$f_{FC} = c_{FC} P_{FC} / \eta_{FC} \tag{2.25}$$

where $f_{FC}$ is the fuel cost for the FC ($/h), $P_{FC}$ is the FC output (electrical) power (kW), $\eta_{FC}$ is the FC efficiency, and $c_{FC}$ is the fuel (natural gas) price to supply the FC ($/kWh). The FC efficiency can be expressed as the ratio of actual operating voltage $V$ and 1.482 (V):

$$\eta_{fc} = V / 1.482 \tag{2.26}$$

As an example, the proton exchange membrane (PEM) FC has been considered. The actual operating voltage of the FC is a function of power level and can

*Table 2.1 Fuel consumption data for the HDJW 410 T6, 369 kW DEG*

| **Load level (%)** | **100** | **75** | **50** | **25** |
|---|---|---|---|---|
| Fuel consumption (L/h) | 100.9 | 74.8 | 51.9 | 29.5 |

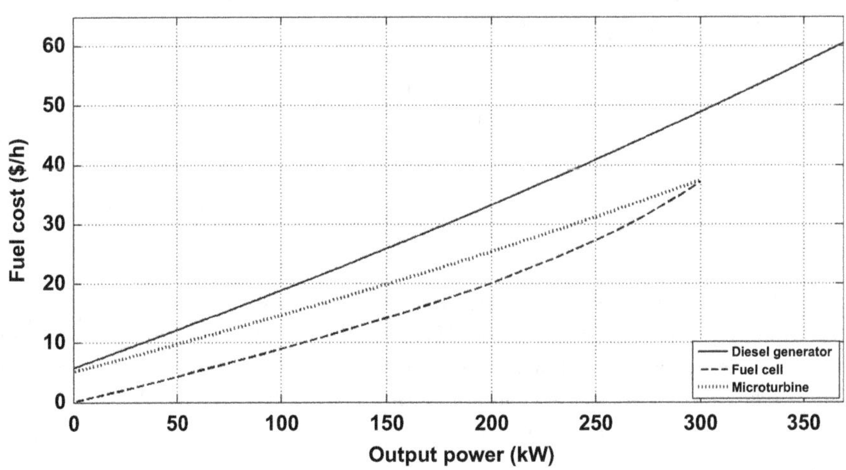


*Figure 2.1 Fuel cost characteristics of the DG units*

be calculated using the following formula [11]:

$$V = \frac{1}{2}\left[V_0 + \sqrt{V_0^2 - 4\left(V_0 V_{Pnom} - V_{Pnom}^2\right)\cdot \xi}\right] \tag{2.27}$$

where $V_0$ is the theoretical or thermodynamic FC potential (0.9 V), $V_{Pnom}$ is the selected cell potential at nominal power (0.45–0.75 V), $\xi = P_{fc}/P_{nom}$ is the power level, and $P_{nom}$ is the nominal power of the FC. Let us assume the PEM FC efficiency characteristic for $V_0 = 0.9$ V and $V_{Pnom} = 0.6$ V. Figure 2.1 shows the fuel cost curve of the FC estimated for $P_{nom} = 300$ kW and $c_{fc} = 0.05$ \$/kWh.

### 2.3.3 Microturbine

The MT fuel costs, $f_{MT}$, can be calculated as follows:

$$f_{MT} = c_{MT} P_{MT} / \eta_{MT} \tag{2.28}$$

where $P_{MT}$ is the MT electrical power output (kW), $\eta_{MT}$ is the MT electric efficiency, and $c_{MT}$ is the fuel price to supply the MT (\$/kWh). The efficiency of the MT increases with the increase of the output power [12]. The electric efficiency characteristic of the MT can be estimated as a quadratic function of its power output:

$$\eta_{MT} = a + bP_{MT} + cP_{MT}^2 \tag{2.29}$$

From a typical electrical efficiency curve of a 300 kW CGT301-302 MT given in [12], the MT fuel costs curve is obtained, as shown in Figure 2.1. It is assumed that the fuel price to supply the MT is $c_{MT} = 0.05$ \$/kWh.

### 2.3.4 Wind turbine

To calculate power output of a WT, two main factors must be known: the wind speed at a certain location and the power curve of the WT. According to Atwa *et al.* [13], the power curve of a WT can be modeled by means of a function split into four different parts:

$$P_{WT} = \begin{cases} 0, & v \le v_{ci} \\ \dfrac{v^2 - v_{ci}^2}{v_{nom}^2 - v_{ci}^2}\cdot P_{nom}, & v_{ci} < v \le v_{nom} \\ P_{nom}, & v_{nom} < v \le v_{co} \\ 0, & v > v_{co} \end{cases} \tag{2.30}$$

where $P_{nom}$, $v_{nom}$, $v_{ci}$, and $v_{co}$ are nominal power, nominal wind speed, cut-in wind speed, and cut-out wind speed of the WT, respectively; and $P_{WT}$ and $v$ are denoted power output of the WT and wind speed. These data, as well as the experimentally determined power curve (2.30), are given by the manufacturers of WTs. Fuel costs at the wind farm are equal to zero, so they usually operate with the maximum available power.

For the OPF simulation tests presented in Section 2.6, Vestas V44/600 kW WT model is adopted [14]. The parameters used to model the power curve obtained

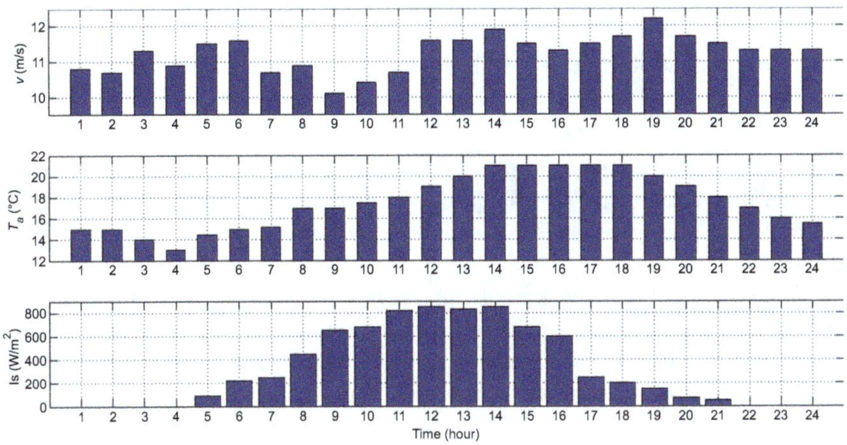


*Figure 2.2 The wind speed, ambient temperature, and solar irradiance data*

from the owner's manual are as follows: $P_{nom}$ = 600 kW, $v_{ci}$ = 4 m/s, $v_{nom}$ = 16 m/s, and $v_{co}$ = 20 m/s. Figure 2.2 shows the wind speed data used to calculate the power generated by the WT generator in the deterministic OPF algorithm.

## *2.3.5 Photovoltaic*

The power output of the PV module is dependent on the solar irradiance and ambient temperature of the site as well as the characteristics of the module itself. The following equation can be used to calculate the power output of the PV module $P_{PV}$ [15]:

$$P_{PV} = P_{STC}\frac{Is}{1000}[1 + \gamma(T_c - 25)] \tag{2.31}$$

where $P_{STC}$ is the PV module maximum power at standard test condition (STC) (W), $I_S$ is the solar irradiance on the PV module surface (W/m$^2$), $\gamma$ is the PV module temperature coefficient for power (°C$^{-1}$), and $T_c$ is the PV cell (module) temperature (°C).

The PV module temperature can be calculated as a function of solar irradiance and ambient temperature based on the module's nominal operating cell temperature (NOCT). The NOCT model equation is [16]

$$T_c = T_a + \frac{Is}{800}\cdot(T_{NOCT} - 20) \tag{2.32}$$

where $T_a$ is the ambient temperature (°C) and $T_{NOCT}$ is the nominal operating cell temperature (NOCT) (°C), of the module.

Sunmodule SW 250 monomodules are used in the OPF simulation tests performed in Section 2.6. Their performance characteristics are $P_{STC}$ = 250(W), $\gamma$ = −0.0045, $T_{NOCT}$ = 46 °C. Figure 2.2 shows the ambient temperature and solar irradiance data used to calculate the output power generated by the PV modules.

### 2.3.6 Mini hydropower plants

The power output of MHP is calculated according to the expression:

$$P_{MHE} = \rho g Q H \eta_T \eta_G \tag{2.33}$$

where $\rho$ is the water density (1000 kg/m$^3$), $g$ is the gravitation constant (9.81 m/s$^2$), $Q$ is the flow rate (m$^3$/s), $H$ is the head (the difference in height between the inlet and outlet surfaces) (m), $\eta_T$ is the water turbine efficiency, and $\eta_G$ is the generator efficiency.

The turbine efficiency depends on the flow rate, $\eta_T = f(Q)$. It can be determined experimentally or by calculation based on data provided by turbine manufacturers. Fuel costs for MHP are equal to zero, so they always operate with the maximum available power [according to (2.33)].

### 2.3.7 Electric grid

The market energy costs $f_{gr}$ (\$/h) can be represented by quadratic function as

$$f_{gr} = a + bP_{gr} + cP_{gr}^2 \tag{2.34}$$

where $P_{gr}$ is the electrical power (kW) from the electric grid and $a$, $b$, and $c$ are cost coefficients. In the OPF test cases considered in Section 2.6, the coefficient values were adopted as $a = 0$, $b = 0.1$ \$/kWh, and $c = 5 \times 10^{-5}$ \$/kW$^2$h.

## 2.4 Solution methods

This book deals with the application of metaheuristic methods for solving problems in the power industry. Accordingly, this chapter describes the application of two representative methods for solving the OPF problem in distribution networks.

The first method is a GA, which was proposed in [17] for the solution of OPF in distribution networks with DG units, including FCs, MTs, diesel generators, PV systems, and WTs. In addition, the application of GSA to the OPF problem as proposed in [18] is also described here. The GSA approach was successfully applied and tested on two distribution systems and the simulation results were compared with the GA method.

### 2.4.1 Genetic algorithm

The GA is one of the techniques, that is, one optimization procedure based on the natural evolution process imitation [19]. They belong to the methods of a directed random search of the solution domain with the aim of finding a global optimum. The classical optimization methods start from a single possible initial solution, and they reach the optimum by applying the heuristic rules iteratively. The GA starts from the population, which is a group of individuals. Each individual represents a potential solution to the optimization problem. The individuals are presented in the same way, usually through a column or a string of data. The quality of each solution or individual is determined based on the fitness function values. Through a series of GA operations, a new population is obtained, and its individuals are engendered by the individuals from the previous population according to the natural evolution principles: the choice of parents, crossover, and mutation. The basic operations of the GA are

1. Representation of individuals: All data (variables) that make an individual are written in a string. A string is composed of substrings. Each substring represents a binary encoded variable on which the process of optimization is carried out. The number of substrings, therefore, the size of a string, depends on the number of variables that are optimized.
2. Initialization: Individuals with random strings are generated that set up the initial population.
3. Fitness function calculation: It is used to rate the quality of an individual, and it represents an equivalent of the function that should be optimized, that is, objective function.
4. Selection: During the selection process, the individuals who will participate in the reproduction (parents) are selected. The point of the selection is to store and transfer good individuals to the next generation.
5. Crossover: The way in which coded column parts (substrings) are crossed over actually makes a GA. Crossover is an exchange process of column parts between two individuals, that is, "parents." One or two new individuals engendered by the crossover, that is, a "child." The possibility of inheriting the first parent's characteristics by a child is introduced during this process.
6. Mutation: Mutation is a way to give a new piece of information to an individual. Mutation represents an accidental bit variation of an individual, generally with a constant probability for each bit within a population. The mutation probability can further vary depending on the size of the population, application and preferences of the explorer. A fixed value, which is often kept during the whole GA, is used for each generation.
7. Ending conditions: The process of finding the optimal solution by the use of the GA is an iterative process that ends when a maximum number of generations is achieved or when another criterion is fulfilled, such as a minimum offset from the best fitness value and medium fitness value of all individuals in a current population. If end conditions are fulfilled, the best individual thus obtained is the semi-optimal solution in question. Otherwise, return to 3.

### *2.4.2 GA implementation to the deterministic OPF problem*

A general flowchart that briefly describes the application of GA in solving the OPF in distribution networks is presented in Figure 2.3. The MATLAB program realization of the GA within the toolbox/gads module is applied herein. The power flow program to the evaluation of the objective function was based on the backward/forward power flow method.

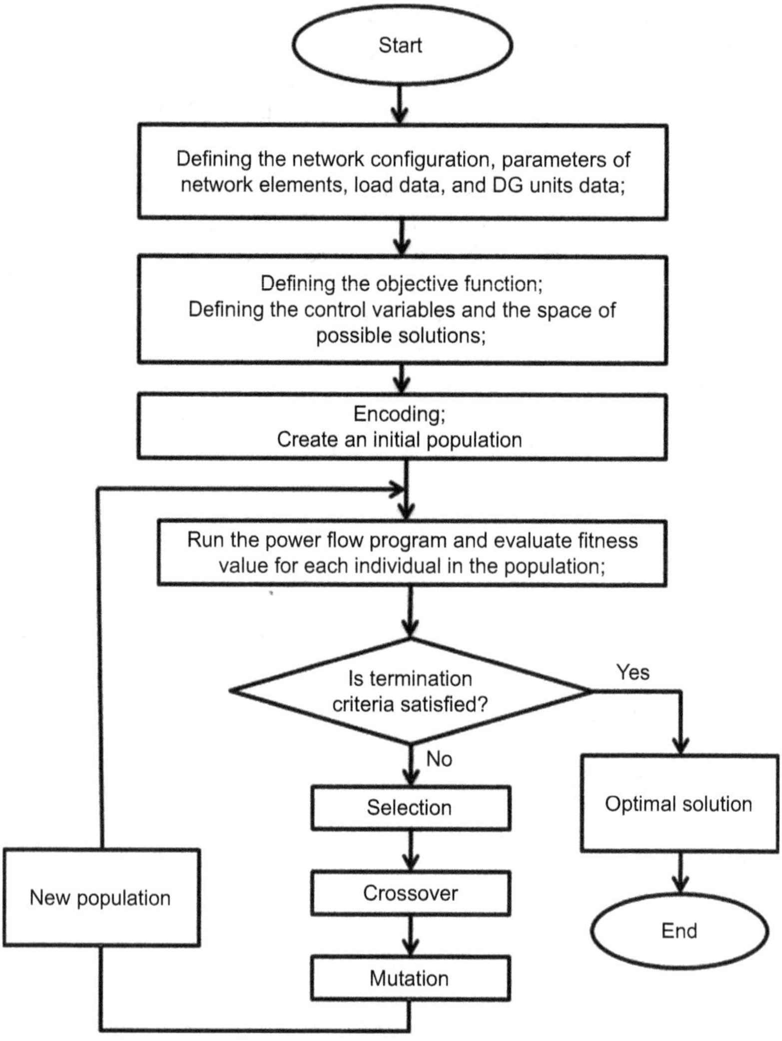


*Figure 2.3 Flowchart of GA for OPF in distribution networks*

### *2.4.3 Gravitational search algorithm*

The GSA is a metaheuristic optimization algorithm where the search agents are considered as a collection of masses that interact with each other based on the Newtonian gravity and the laws of motion [20]. The position of the mass corresponds to the solution of the problem, and its gravitational and inertial masses are determined using a fitness function. In other words, each mass presents a potential solution. The algorithm is navigated by properly adjusting the gravitational and inertial masses. In a system with $N$ agents (masses), the position of the $i$th agent is defined by

$$\mathbf{x}_i = \left[x_i^1, \ldots, x_i^d, \ldots, x_i^n\right] \; i = 1, \ldots, N \tag{2.35}$$

where $n$ is the number of control variables and $x_i^k$ defines the position of the $i$th agent in the $k$th dimension, i.e. the value of the $k$th control variable in the $i$th potential solution.

After evaluating the current population fitness, the mass of each agent is calculated as follows:

$$M_i(t) = \frac{m_i(t)}{\sum_{j=1}^{N} m_j(t)} \tag{2.36}$$

where

$$m_i(t) = \frac{fit_i(t) - worst(t)}{best(t) - worst(t)} \tag{2.37}$$

where $fit_i(t)$ represents the fitness value of agent $i$ at iteration $t$ and $best(t)$ and $worst(t)$ are the best and worst fitness of all agents, respectively.

According to Newton's gravitation theory, the total force that acts on the $i$th agent at iteration $t$ is specified as follows:

$$\mathbf{F}_i(t) = \sum_{j \in Kbest, j \neq i} r \cdot G(t) \frac{M_j(t) \cdot M_i(t)}{R_{i,j}(t) + \varepsilon} \left(\mathbf{x}_j(t) - \mathbf{x}_i(t)\right) \tag{2.38}$$

where $r$ is a random number within the interval [0, 1]. $G(t)$ is the gravitational constant at iteration $t$, $M_i(t)$ and $M_j(t)$ are masses of agents $i$ and $j$, $\varepsilon$ is a small constant, and $R_{ij}(t)$ is the Euclidian distance between the two agents $i$ and $j$, given by the following equation:

$$R_{ij}(t) = \left\|\mathbf{x}_i(t), \mathbf{x}_j(t)\right\|_2 \tag{2.39}$$

*Kbest* is the set of first $K$ agents with the best fitness value and biggest mass, which is a function of time, initialized to $K_0$ at the beginning and decreased with time. In such a way, at the beginning, all agents apply the force, and as time passes, *Kbest* is decreased linearly, and at the end, there will be just one agent applying force to the others.

According to the law of motion, the acceleration of the $i$th agent, at iteration $t$, is given by the following equation:

$$\mathbf{a}_i(t) = \frac{\mathbf{F}_i(t)}{M_i(t)} \tag{2.40}$$

The searching strategy on this notion can be defined as finding the next velocity and next position of an agent. Next velocity of an agent is defined as a function of its current velocity added to its current acceleration. Hence, the next position and next velocity of an agent can be computed as follows:

$$\mathbf{v}_i(t+1) = r_i \cdot \mathbf{v}_i(t) + \mathbf{a}_i(t) \tag{2.41}$$

$$\mathbf{x}_i(t+1) = \mathbf{x}_i(t) + \mathbf{v}_i(t+1) \tag{2.42}$$

where $r_i$ is a uniform random number in the interval [0, 1]; $\mathbf{x}_i(t)$, $\mathbf{v}_i(t)$, and $\mathbf{a}_i(t)$ are the position, velocity, and acceleration of agent $i$ at iteration $t$, respectively.

The gravitational constant $G(t)$ in (2.38) is a function of the initial value $G_0$ and time (iteration) $t$:

$$G(t) = G_0 \cdot e^{-\alpha \frac{t}{t_{\max}}} \tag{2.43}$$

The parameters of maximum iteration $t_{\max}$, population size $N$, initial gravitational constant $G_0$ and constant $\alpha$ control the performance of GSA.

### 2.4.4 GSA implementation to the deterministic OPF problem

The described GSA approach has been applied to solve the OPF problem in the following steps [18]:

*Step 1*: Load the distribution network configuration, line data, transformer data, shunt VAR compensator data, load data, and DG unit data. Use the per-unit system.

*Step 2*: Specify the vector of control variables $\mathbf{x}$ (2.5) and their lower and upper limits (2.17)–(2.21); specify the dependent variables $\mathbf{y}$ (2.6), and their lower and upper limits (2.14)–(2.16); specify the objective function to be optimized $F(\mathbf{x},\mathbf{y})$ (2.7)–(2.12), and the penalty factors ($\lambda_{V,QDG,S}$) to form the expanded objective function $F_p(\mathbf{x},\mathbf{y})$ (2.22);

*Step 3*: Initialize GSA parameters, such as the population size ($N$), maximum iteration number ($t_{\max}$), initial constant $K_0$, initial gravitational constant $G_0$, and constant $\alpha$.

*Step 4*: Generate an initial random population of $N$ agents. The initial positions of each agent are randomly selected between minimum and maximum values of the control variables. Initialize the velocity of agents by a zero matrix of dimension $N \times n$, where $N$ is the number of agents (population size) and $n$ is the number of control variables.

*Step 5*: Run the power flow program based on the backward/forward power flow method for each agent $\mathbf{x}_i(t)$ from the current population $\mathbf{POP}(t)$ and calculate the corresponding values of the objective function (fitness values).

*Step 6*: Calculate the gravitational constant $G(t)$ (2.43), the best and worst fitness values, and the mass of each agent $M_i(t)$ (2.36), $i = 1, \ldots, N$.

*Step 7*: Calculate the total force in different directions using (2.38), and the acceleration of each agent using (2.40).

*Step 8*: Update the velocity of each agent using (2.41).

*Step 9*: Update the position of each agent using (2.42).

*Step 10*: Repeat Steps 5–9 until the stop criteria are reached, that is the maximum number of iterations $t_{\max}$.

*Step 11*: Return the best solution obtained in the last iteration; stop.

## 2.5 Probabilistic OPF

The results of deterministic OPF for a distribution network are highly dependent on the accuracy of the input variables such as the power of load and the power output of DG units. When some of the input variables are uncertain, the OPF problem becomes probabilistic. Because of the probabilistic nature of wind speed and solar irradiance, the power output of the WT and the PV units are random variables as well. Moreover, it is very hard to expect that the forecast of the load demand is exactly correct due to unexpected disturbance, forecast error, or load variation. Such randomly occurring factors would be the main source of uncertainties. Every probabilistic formulation requires statistical characterization of the input random variables and a method for evaluating the statistical features of the output variables [21].

The probabilistic approach to solve the OPF problem has two key steps. The first step is the statistical characterization of the input random variables. The uncertainty of independent input variables is the source of the uncertainty of dependent variables in the system. In other words, due to the random character of input variables such as WT's power outputs and power of loads in the distribution network, the obtained OPF results such as bus voltages, power by branches, or fuel costs of DG units should also be considered as random variables. Evaluating the statistical characteristics of the OPF results is the second key step in the probabilistic OPF approach.

### *2.5.1 Statistical characterization of the input random variables*

#### 2.5.1.1 Wind speed modeling

The wind speed probability density function at a certain location is generally described by a Weibull distribution:

$$f_v(v) = \frac{k}{C} \cdot \left(\frac{v}{C}\right)^{k-1} \cdot e^{-\left(\frac{v}{C}\right)^k} \tag{2.44}$$

The cumulative density function (CDF) for the Weibull distribution is

$$F_v(v) = 1 - \cdot e^{-\left(\frac{v}{C}\right)^k} \tag{2.45}$$

The CDF with its inverse has been utilized to calculate the wind speed:

$$v = C \cdot (-\ln(r))^{\frac{1}{k}} \tag{2.46}$$

where $r$ are the random numbers uniformly distributed on [0, 1]. Constants $C$ and $k$ are the scale and shape parameters of the Weibull distribution. Different methods can be used to calculate the Weibull parameters. Usually, parameters $k$ and $C$ are calculated, approximately, using mean wind speed $v_m$ and standard deviation $\sigma$ as follows [13]:

$$k = \left(\frac{\sigma}{v_m}\right)^{-1.086} \tag{2.47}$$

$$C = \frac{v_m}{\Gamma(1 + 1/k)} \tag{2.48}$$

where $\Gamma(x)$ is the gamma function defined as

$$\Gamma(x) = \int_0^{\infty} t^{x-1} e^{-t} dt \text{ за } x > 0 \tag{2.49}$$

#### 2.5.1.2 Solar irradiance modeling

Basically, solar irradiance has a stochastic nature. Therefore, the probability density function should be adopted. The solar irradiance uncertainty may be characterized by normal probability density function [22]:

$$f_{I_S}(I_S) = \frac{1}{\sigma\sqrt{2\pi}} \cdot e^{-\frac{(I_S-\mu)^2}{2\sigma^2}} \tag{2.50}$$

CDF for the normal distribution is

$$F_{I_S}(I_S) = \frac{1}{2}\left[1 + \text{erf}\left(\frac{I_S - \mu}{\sqrt{2}\sigma}\right)\right] \tag{2.51}$$

CDF with its inverse has been utilized to calculate the solar irradiance:

$$I_S = \mu + \sqrt{2} \cdot \sigma \cdot \text{erf}^{-1}(2r - 1) \tag{2.52}$$

where $r$ is a random variable having uniform distribution in [0, 1]; $\mu$ is the hourly mean value of the solar irradiance and $\sigma$ is the standard deviation of the $I_S$; erf and $\text{erf}^{-1}$ are the error function and the inverse error function, respectively, defined as follows:

$$\text{erf}(x) = \frac{2}{\sqrt{\pi}}\int_0^{x} e^{-t^2} dt; \ \text{erf}^{-1}(x) = 1 - \text{erf}(x) = \frac{2}{\sqrt{\pi}}\int_x^{\infty} e^{-t^2} dt \tag{2.53}$$

The standard deviation of the hourly mean value of the solar irradiance can be pre-determined for the morning period, afternoon period, and hours around the solar noon. For the purpose of the OPF simulation tests in Section 2.6, it was assumed that the hourly means of solar irradiance $\mu$ are equal to their forecasted values from Figure 2.3, and values of $\sigma$ are pre-determined with $\frac{\sigma}{\mu} \cdot 100 = 8\%$ at morning (5 h–8 h) and evening hours (17 h–21 h), and 3% around the solar noon (9 h–16 h).

#### 2.5.1.3 Probabilistic load model

The load is assumed to be a random variable ($L$) following the same probability density function within each hour of a given daily load diagram. The PDF of the active/reactive power load will be assumed to be under the normal distribution [23]:

$$f_L(L) = \frac{1}{\sigma\sqrt{2\pi}} \cdot e^{-\frac{(L-\mu)^2}{2\sigma^2}} \tag{2.54}$$

where $\mu$ is the mean value given as a fixed load level on the daily load diagram and $\sigma$ is the standard deviation of $L$.

Normal CDF (2.55) with its inverse (2.56) has been used to generate the load value:

$$F_L(L) = \frac{1}{2}\left[1 + \text{erf}\left(\frac{L-\mu}{\sqrt{2}\sigma}\right)\right] \tag{2.55}$$

$$L = \mu + \sqrt{2} \cdot \sigma \cdot \text{erf}^{-1}(2r - 1) \tag{2.56}$$

where $r$ is a random variable having uniform distribution in [0, 1] and erf and $\text{erf}^{-1}$ are the error function and the inverse error function, respectively.

In the simulation tests, it is assumed that the mean values of the active/reactive power load at each load bus are given in Figure 2.3 with a standard deviation of 5%.

### *2.5.2 Statistical evaluation of the output variables*

Generally, the probabilistic OPF can be expressed as

$$\mathbf{Y} = F(\mathbf{X}) \tag{2.57}$$

where $\mathbf{X}$ is the vector of input random variables and $\mathbf{Y}$ is the vector of output random variables.

In the technical literature, there are several methods for evaluating the statistical features of the output variables. These methods may be classified into three main categories: Monte Carlo simulation, analytical methods, and approximate methods.

Monte Carlo simulation [23] randomly generates values for uncertain input variables, and these values are taken into account to solve a deterministic problem. The simulation is repeated many times (several hundred to several thousand) in order to determine with sufficient accuracy the statistical characteristics of the results (i.e. $\mathbf{Y}$). The main drawback of the Monte Carlo method is the great number of simulations required to attain convergence.

Analytical methods [25,26] are based on certain mathematical simplifications of the problem, in order to obtain analytically the statistical characteristics of random output variable $\mathbf{Y}$ based on the statistical characteristics of random input variables $\mathbf{X}$.

Approximate methods [21,24,27] provide an approximate description of the statistical properties of output random variables. One of these methods has been explained and applied in this chapter. It is the point estimate method. This method, like Monte Carlo, uses a deterministic approach to solving a probabilistic problem, but with a drastically lower number of simulations.

The point estimate method can guarantee a great reduction of the computational efforts compared to the Monte Carlo simulation procedure [24]. This method concentrates the statistical information provided by the first few central moments of the $m$ input random variables on $K = 2, 3$, or 5 points for each variable, named concentrations. By using these points and function $F$, which relates input and output variables, statistical moments of the output variables can be obtained. To obtain these moments, function $F$ has to be calculated $2m$, $2m+1$, or $4m+1$ times depending on the adopted scheme. In this work, $2m+1$ scheme was used.

The $2m+1$ scheme is more accurate than the $2m$ scheme, and performances of the $2m+1$ and $4m+1$ schemes are practically the same. The $2m+1$ scheme requires $2m+1$ evaluations of function $F$ and uses only a two-point concentration for each input random variable [21].

### *2.5.3 Procedure for solving probabilistic OPF*

Procedure for computing the moments of the output variables for the probabilistic OPF problem can be summarized by the following steps:

1. Determine number $m$ of input random variables.
2. Set the vector of the $j$th moment of the output variable equal to zero: $E(\mathbf{Y}^j) = 0$
3. Set $t = 1$ ($t = 1,2, \ldots, m$)
4. Determine two standard locations:

$$\xi_{t,i} = \frac{\lambda_{t,3}}{2} + (-1)^{3-i} \cdot \sqrt{\lambda_{t,4} - \frac{3}{4}\lambda_{t,3}^2}\; i = 1,2 \tag{2.58}$$

   where $\lambda_{t,3}$ is the skewness and $\lambda_{t,4}$ is the kurtosis of the input random variable $x_t$.

5. Determine two locations $x_{t,i}$

$$x_{t,i} = \mu_{x_t} + \xi_{t,i} \cdot \sigma_{x_t} i = 1,2 \tag{2.59}$$

   where $\mu_{x_t}$ and $\sigma_{x_t}$ are the mean and standard deviation of $x_t$, respectively.

6. Run *the deterministic OPF algorithm* for both locations $x_{t,i}$ using the two input variable vectors:

$$\mathbf{X}_i = \left[\mu_{x1}, \mu_{x2}, \ldots, x_{t,i}, \ldots, \mu_{xm}\right]\; i = 1,2 \tag{2.60}$$

   where $\mu_{xk}$ ($k = 1,2,\ldots,m$; and $k \neq t$) is the mean value of the remaining random input variables.

7. Determine the weight factors:

$$w_{t,i} = \frac{(-1)^{3-i}}{\xi_{t,i}(\xi_{t,1} - \xi_{t,2})}\; i = 1,2 \tag{2.61}$$

8. Update $E(\mathbf{Y}^j)$:

$$E(\mathbf{Y}^j) = E(\mathbf{Y}^j) + \sum_{i=1}^{2} w_{t,i}[F(\mathbf{X}_i)]^j \tag{2.62}$$

9. Repeat Steps 4–8 for $t = t + 1$ until the list of random input variables is exhausted.
10. Run *the deterministic OPF algorithm* using as input variable vector:

$$\mathbf{X}_\mu = \left[\mu_{x1}, \mu_{x2}, \ldots, \mu_{x,t}, \ldots, \mu_{xm}\right] \tag{2.63}$$

11. Determine the weight factor of OPF solution of Step 10:

$$w_0 = 1 - \sum_{t=1}^{m} \frac{1}{\lambda_{t,4} - \lambda_{t,3}^2} \tag{2.64}$$

12. Update $E(\mathbf{Y}^j)$:

$$\begin{aligned} E(\mathbf{Y}^j) &= E(\mathbf{Y}^j) + w_0\left[F(\mathbf{X}_\mu)\right]^j \\ E(\mathbf{Y}^j) &= \sum_{t=1}^{m}\sum_{i=1}^{2} w_{t,i}\left[F\left(\mu_{x_1}, \mu_{x_1}, \ldots, x_{t,i}, \ldots, \mu_{x_m}\right)\right]^j + w_0\left[F(\mathbf{X}_\mu)\right]^j \end{aligned} \tag{2.65}$$

Knowing statistical moments of the output random variable, the mean and standard deviation can be computed as

$$\mu_Y = E(\mathbf{Y}) \tag{2.66}$$

$$\sigma_Y = \sqrt{E(\mathbf{Y}^2) - \mu_Y^2} \tag{2.67}$$

Based on statistical moments, it is possible to approximate the probability density functions (PDFs) of the output random variables of interest using the Gram–Charlier series approach. For example, let us consider $Y$ as any random variable with a continuous distribution, whose mean value is $\mu_Y$ and standard deviation $\sigma_Y$. For the standardized variable $x = (Y - \mu_Y)/\sigma_Y$, its cumulative function and density function are denoted as $F(x)$ and $f(x)$, respectively. According to Gram–Charlier expansion, the cumulative and the density functions can be written as [25]

$$F(x) = \Phi(x) + \frac{c_1}{1!}\Phi'(x) + \frac{c_2}{2!}\Phi''(x) + \frac{c_3}{3!}\Phi'''(x) + \ldots \tag{2.68}$$

$$f(x) = \varphi(x) + \frac{c_1}{1!}\varphi'(x) + \frac{c_2}{2!}\varphi''(x) + \frac{c_3}{3!}\varphi'''(x) + \ldots \tag{2.69}$$

where $\Phi(x)$ and $\varphi(x)$ represent the cumulative distribution function (CDF) and probabilistic density function (PDF) of the normal distribution with $\mu = 0$ and

$\sigma = 1$. The constant coefficients $c_k$ are calculated as follows:

$$c_1 = c_2 = 0 \tag{2.70}$$

$$c_3 = -\frac{E(Y^3) - 3\mu_Y E(Y^2) + 2\mu_Y^3}{\sigma_Y^3} \tag{2.71}$$

$$c_4 = \frac{E(Y^4) - 3[E(Y^2)]^2 - 4\mu_Y E(Y^3) + 12\mu_Y^2 E(Y^2) - 6\mu_Y^4 + 3\sigma_Y^4}{\sigma_Y^4} - 3 \tag{2.72}$$

*Note*: $E(Y^1)$, $E(Y^2)$, $E(Y^3)$, $E(Y^4)$, ... are the $j$th statistical moments of the output random variable $Y$, $E(Y^j)$, $= 1, 2, 3, 4, \ldots$.

## 2.6 Simulation results

The proposed OPF approach was tested on a modified IEEE 34-bus test feeder shown in Figure 2.4. The original IEEE 34-bus test feeder [28] is modified using the following assumptions [29]:

- Only the three-phase sections of the original test feeder are included. The single-phase lateral sections were removed, and their loads were connected at their respective sending end nodes. The three-phase sections are assumed to be transposed and therefore considered as only positive sequence impedances. The shunt admittances of the sections are neglected.

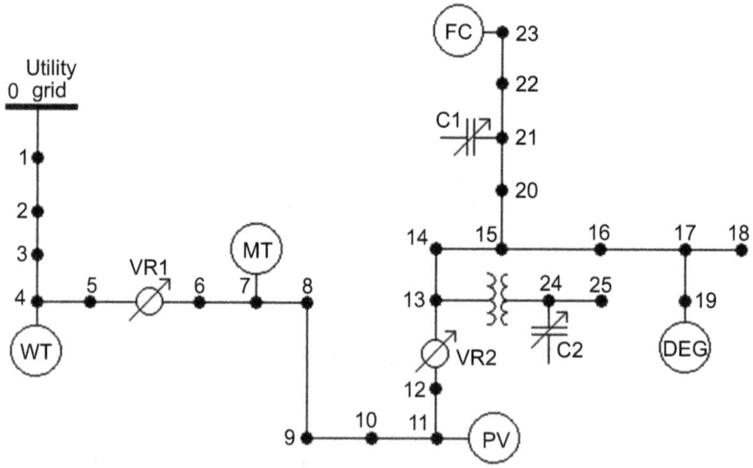


*Figure 2.4 Single-line diagram of the modified IEEE 34-bus test system*

- The unbalanced phase loads in each three-phase section are summed up and taken as total three-phase balanced loads. The loads on laterals have already been "absorbed" into the corresponding three-phase "root" sections. The distributed loads in all sections are represented as spot loads. One-half of the total distributed load is placed at the beginning of the section, while the remaining half of the total distributed load is placed at the end of the section. The branch impedances and the nominal active and reactive power of the load are given in Table 2.2. It has been assumed that the active and reactive power of load at each bus varies based on the daily load diagram shown in Figure 2.5 relative to nominal values.
- Five different DG units were introduced in the modified test feeder. It is adopted that the DG units connected at buses 4 (WT), 7 (MT), 11 (PV), and 19 (DEG) operate in PQ mode with 0.9 power factor (produce reactive power). The DG unit at bus 23 (FC) is capable of controlling active power and voltage independently and therefore operates in PV mode. The DG unit data are listed

*Table 2.2 Line data and load data of the modified IEEE 34-bus test feeder*

| **Send. bus** | **Rec. bus** | **Section parameters** | | **Load at receiving bus** | | ***Load model*** |
|---|---|---|---|---|---|---|
| | | $R$ **(Ω)** | $X$ **(Ω)** | $P_L$ **(kW)** | $Q_L$ **(kVAr)** | |
| 0 | 1 | 0.5473 | 0.4072 | 27.5 | 14.5 | PQ |
| 1 | 2 | 0.3670 | 0.2730 | 27.5 | 14.5 | PQ |
| 2 | 3 | 6.8373 | 5.0864 | 16 | 8 | I |
| 3 | 4 | 7.9553 | 5.9181 | 0 | 0 | PQ |
| 4 | 5 | 6.3069 | 4.6919 | 0 | 0 | PQ |
| 5 | 6 | 0.0032 | 0.0016 | 0 | 0 | PQ |
| 6 | 7 | 0.0992 | 0.0494 | 171.5 | 88 | PQ |
| 7 | 8 | 3.2681 | 1.6265 | 44.5 | 22 | I |
| 8 | 9 | 0.2689 | 0.1338 | 5.5 | 2.5 | PQ |
| 9 | 10 | 6.5426 | 3.2562 | 48.5 | 21.5 | Z |
| 10 | 11 | 0.1664 | 0.0828 | 4 | 2 | PQ |
| 11 | 12 | 11.7889 | 5.8672 | 0 | 0 | PQ |
| 12 | 13 | 0.0032 | 0.0016 | 7.5 | 3.5 | Z |
| 13 | 14 | 1.5684 | 0.7806 | 25.5 | 13 | PQ |
| 14 | 15 | 1.8661 | 0.9288 | 89 | 45 | Z |
| 15 | 16 | 0.6466 | 0.3218 | 174 | 106 | PQ |
| 16 | 17 | 0.8578 | 0.4269 | 61 | 31.5 | PQ |
| 17 | 18 | 0.2753 | 0.1370 | 47 | 31 | I |
| 17 | 19 | 0.0896 | 0.0446 | 28 | 14 | PQ |
| 15 | 20 | 0.0896 | 0.0446 | 4.5 | 2.5 | PQ |
| 20 | 21 | 0.4321 | 0.2151 | 432 | 329 | Z |
| 21 | 22 | 1.1651 | 0.5799 | 34 | 17 | PQ |
| 22 | 23 | 0.1696 | 0.0844 | 71.5 | 53.5 | PQ |
| 13 | 24 | 11.7800 | 25.2864 | 0 | 0 | PQ |
| 24 | 25 | 80.2600 | 59.7059 | 450 | 225 | I |

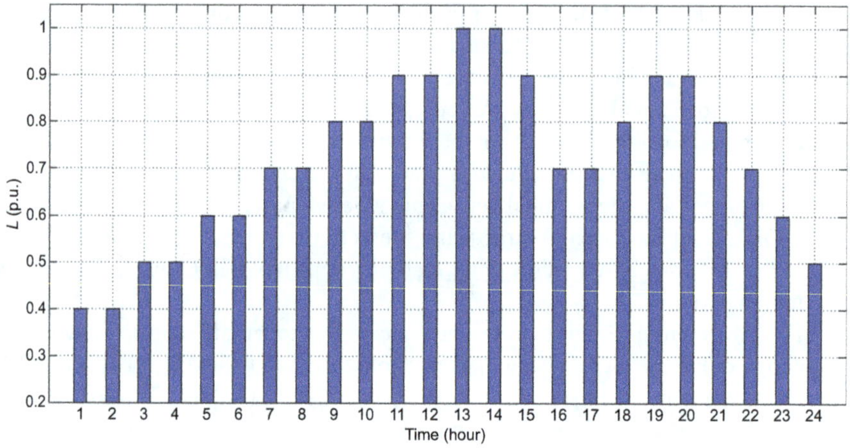


*Figure 2.5 Daily load diagram*

*Table 2.3 DG unit's data*

| Location | Type | Mode | $P_{Dgmin}$ (kW) | $P_{DGnom}$ (kW) | $Q_{DG}$ (kVAr) | Cost coefficients | | |
|---|---|---|---|---|---|---|---|---|
| | | | | | | *a* ($/h) | *b* ($/kWh) | *c* ($/kW$^2$h) |
| 4 | WT | PQ | 0 | 600 | $\cos\varphi = 0.9$ | 0 | 0 | 0 |
| 7 | MT | PQ | 90 | 300 | $\cos\varphi = 0.9$ | 5.3 | $88 \times 10^{-3}$ | $61 \times 10^{-6}$ |
| 11 | PV | PQ | 0 | 250 | $\cos\varphi = 0.9$ | 0 | 0 | 0 |
| 19 | DEG | PQ | 0 | 369 | $\cos\varphi = 0.9$ | 5.8 | $124 \times 10^{-3}$ | $65 \times 10^{-6}$ |
| 23 | FC | PV | 0 | 300 | −200–225 | 0.57 | $62 \times 10^{-3}$ | $190 \times 10^{-6}$ |
| 0 | Electric grid | Slack | – | – | – | 0 | $100 \times 10^{-3}$ | $50 \times 10^{-6}$ |

in Table 2.3. The fuel cost coefficients (*a, b, c*) are determined according to the fuel cost and output power characteristics of the DG units given in Section 2.4. The power outputs of WT and PV are calculated in accordance with the forecasted values of wind speed, ambient temperature, and solar irradiance given in Figure 2.3.

- It is assumed there are two shunt VAR compensators connected at nodes 21 and 24.

There are a total of nine control variables for the OPF problem as follows: three DG unit active power outputs, one root bus voltage magnitude, one PV bus voltage magnitude, two voltage regulation tap settings, and two shunt VAR compensator reactive power outputs. The lower and upper limits for the control variables are given in Table 2.4. The lower and upper limits of the load bus voltages, including the DG units, which are modeled as PQ buses, are 0.95 and 1.05 p.u., respectively.

### 2.6.1 Deterministic OPF analysis

In the deterministic OPF, the output powers of WT and PV units as well as the load level are equal to their specified (forecasted) values. Three optimization cases were considered.

*Case 1*: fuel cost minimization for DG units (MT, FC, and DEG);

*Case 2*: simultaneous minimization of the fuel cost and power losses;

*Case 3*: simultaneous minimization of the fuel cost, power losses, and voltage deviation.

Table 2.5 shows the optimal settings of the control variables and appropriate values of the fuel cost, power losses, and voltage deviation for the maximum load

*Table 2.4 The limits of control variables*

| Control variables | Min | Max |
|---|---|---|
| $P_{MT}$ (kW) | 0 | 300 |
| $P_{FC}$ (kW) | 0 | 300 |
| $P_{DEG}$ (kW) | 0 | 369 |
| $V_0$ (p.u.) | 0.97 | 1.05 |
| $V_{23}$ (p.u.) | 0.98 | 1.05 |
| $t_{VR1}$ (p.u.) | 0.90 | 1.10 |
| $t_{VR2}$ (p.u.) | 0.90 | 1.10 |
| $Q_{C1}$ (kVAr) | 0 | 300 |
| $Q_{C2}$ (kVAr) | 0 | 300 |

*Table 2.5 Optimal settings of control variables for max load level (T = 14 h)*

| | GA | | | GSA | | |
|---|---|---|---|---|---|---|
| | **Case 1** | **Case 2** | **Case 3** | **Case 1** | **Case 2** | **Case 3** |
| $P_{MT}$ (kW) | 300.0 | 300.0 | 292.3 | 300 | 300 | 300 |
| $P_{FC}$ (kW) | 261.2 | 291.8 | 220.7 | 253.9 | 300 | 298.2 |
| $P_{DEG}$ (kW) | 297.1 | 369.0 | 367.3 | 305.2 | 369 | 369 |
| $V_0$ (p.u.) | 0.9995 | 1.0500 | 1.0113 | 1.0017 | 1.0500 | 1.0027 |
| $V_{23}$ (p.u.) | 1.0498 | 1.0490 | 0.9891 | 1.0328 | 1.0301 | 1.0297 |
| $t_{VR1}$ (p.u.) | 0.9444 | 0.9888 | 0.9650 | 0.9793 | 0.9889 | 0.9870 |
| $t_{VR2}$ (p.u.) | 0.9667 | 0.9865 | 0.9528 | 0.9469 | 0.9927 | 0.9462 |
| $Q_{C1}$ (kVAr) | 148.4 | 115.1 | 296.6 | 270.2 | 191.5 | 113.6 |
| $Q_{C2}$ (kVAr) | 215.4 | 300.0 | 267.7 | 300 | 227 | 300 |
| *Fcost* ($/h) | **178.36** | 179.75 | 179.15 | **177.18** | 179.19 | 178.47 |
| *Ploss* (kW) | 76.36 | **64.35** | 79.12 | 77.95 | **62.80** | 65.47 |
| *VD* (p.u.) | 0.7252 | 0.9802 | 0.6185 | 0.5183 | 0.9416 | 0.4719 |

*Note*: Bold indicates the objective function that was optimized.

level ($T$ = 14 h) obtained by using GA and GSA. The bold values indicate the optimum value of the fuel cost, power losses, and voltage deviation in accordance with the optimization cases. Figure 2.6 shows voltage profiles of the above cases. As can be seen, the values of voltage at all buses are within the allowable limits.

The optimal settings of the control variables for other load levels were also determined. Figures 2.7–2.10 show the results obtained by applying GA for Case 3.

Daily values of the fuel cost, energy loss, and max voltage deviation in accordance with these optimal settings of control variables are shown in Figure 2.11. Minimization of the fuel cost in Case 1 causes the maximum energy

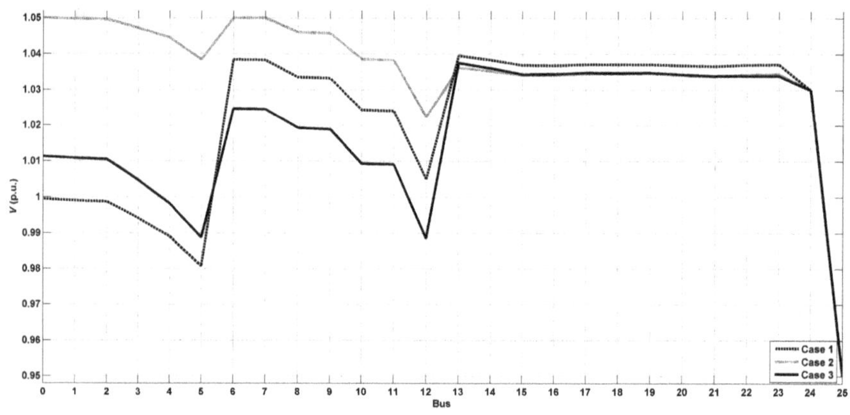


*Figure 2.6 Voltage profile for the maximum load level (T = 14 h)*

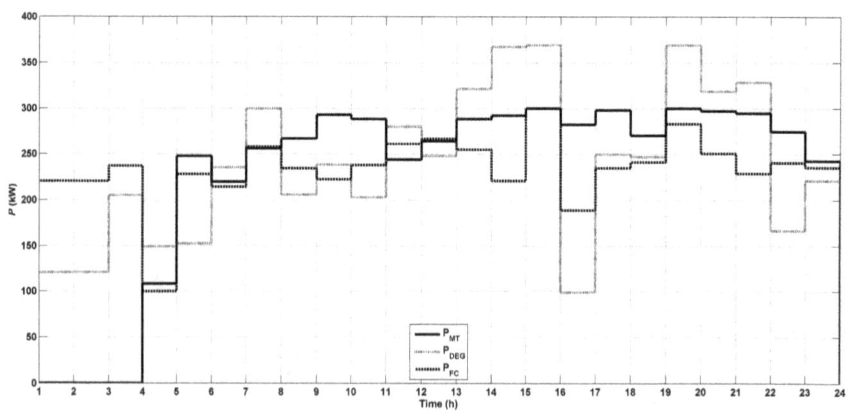


*Figure 2.7 Optimal power outputs of DG units for Case 3*

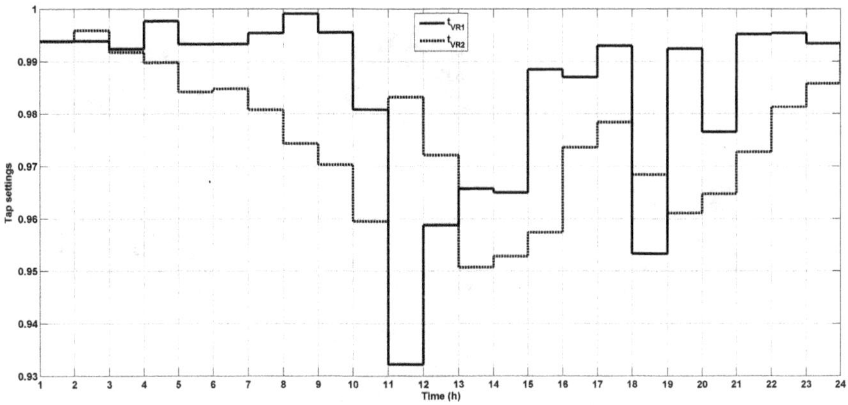


*Figure 2.8 Optimal tap settings of voltage regulation transformers for Case 3*

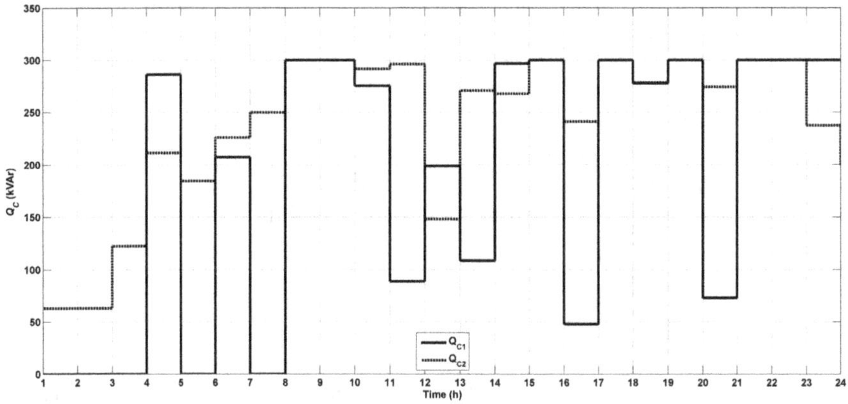


*Figure 2.9 Optimal reactive powers of shunt VAR compensators for Case 3*

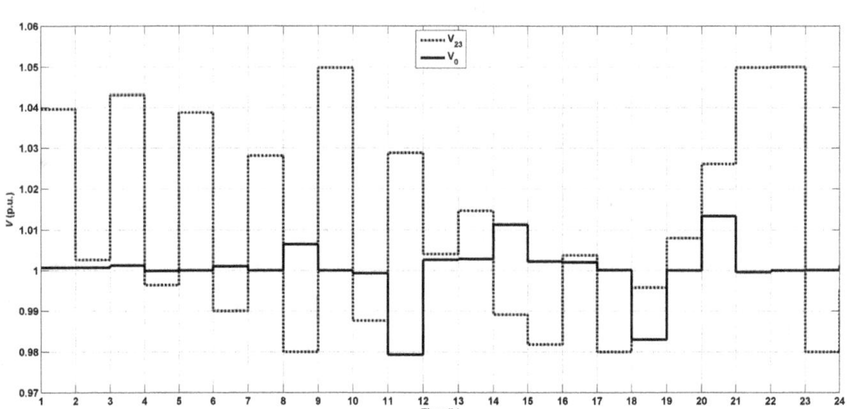


*Figure 2.10 Optimal settings of voltage magnitude at the root bus and PV bus for Case 3*

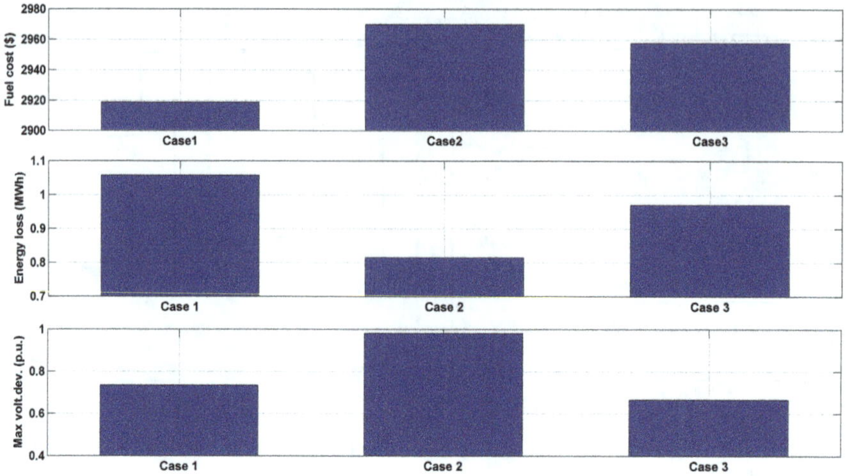


*Figure 2.11 Daily values of energy costs, energy losses, and maximum voltage deviation*

loss in relation to other cases. In Case 2, the minimum energy loss is obtained, but the fuel cost and voltage deviation are increased. The results obtained in Case 3 can be adopted as a compromise solution to the OPF problem.

### *2.6.2 Probabilistic OPF analysis*

In this section, the WT and PV unit power generation as well as the load level are adopted as uncorrelated input random variables. Using proper PDFs modeling for the hourly wind speed and solar irradiance given in Sections 2.5.1.1 and 2.5.1.2, the output power of WT and PV units are evaluated. All loads are fully correlated and follow the same PDF given in Section 2.5.1.3.

The $2m + 1$ point estimate method is implemented along with the GA to find the optimal solution of the probabilistic OPF. Table 2.6 reports the mean and standard deviation of the OPF results. Optimization Case 1 and Case 3 were considered for Time = 14 h, 8 h, and 2 h. Figure 2.12 shows the PDFs of the results for optimization Case 3 and Time = 14 h. To obtain these PDFs, the Gram–Charlier expansion was used.

In order to determine the impact of individual input random variables on the statistical characteristics of the results, the probabilistic OPF was performed for five different combinations of the input random variables. Results are given in Table 2.7. From these results, it is clear that input random variable $L$ has a higher impact on the standard deviation of the output variables than input random variables $P_{WT}$ and $P_{PV}$. The reason for this is that the total power load in the distribution network is much higher than the power outputs of WT and PV units.

*Table 2.6 Mean and standard deviation results of the probabilistic OPF*

| | | Time = 14 h | | Time = 8 h | | Time = 2 h | |
|---|---|---|---|---|---|---|---|
| | | **Case 1** | **Case 3** | **Case 1** | **Case 3** | **Case 1** | **Case 3** |
| $P_{mt}$ (kW) | $\mu$ | 297.1 | 289.3 | 271.0 | 244.7 | 58.7 | 38.7 |
| | $\sigma$ | 4.2 | 8.8 | 11.6 | 60.5 | 96.1 | 48.8 |
| $P_{fc}$ (kW) | $\mu$ | 250.4 | 240.0 | 209.3 | 184.6 | 199.7 | 183.9 |
| | $\sigma$ | 19.9 | 17.8 | 20.5 | 47.8 | 16.0 | 20.7 |
| $P_{dg}$ (kW) | $\mu$ | 318.9 | 317.7 | 159.5 | 145.6 | 49.2 | 101.5 |
| | $\sigma$ | 32.7 | 48.3 | 51.0 | 71.3 | 26.9 | 52.1 |
| $V_{800}$ (p.u.) | $\mu$ | 1.0083 | 1.0035 | 1.0050 | 0.9960 | 0.9938 | 0.9997 |
| | $\sigma$ | 0.0260 | 0.0226 | 0.0180 | 0.0118 | 0.0132 | 0.0032 |
| $V_{848}$ (p.u.) | $\mu$ | 1.0186 | 1.0073 | 1.0068 | 0.9980 | 0.9970 | 1.0046 |
| | $\sigma$ | 0.0166 | 0.0143 | 0.0163 | 0.0142 | 0.0036 | 0.0090 |
| $t_{\mathrm{VR1}}$ (p.u.) | $\mu$ | 0.9645 | 0.9621 | 0.9945 | 0.9675 | 1.0092 | 0.9975 |
| | $\sigma$ | 0.0273 | 0.0238 | 0.0277 | 0.0143 | 0.0111 | 0.0050 |
| $t_{\mathrm{VR2}}$ (p.u.) | $\mu$ | 0.9554 | 0.9548 | 0.9698 | 0.9779 | 0.9823 | 0.9952 |
| | $\sigma$ | 0.0152 | 0.0119 | 0.0190 | 0.0232 | 0.0116 | 0.0030 |
| $Q_{C1}$ (kVAr) | $\mu$ | 197.8 | 231.9 | 216.1 | 196.2 | 247.5 | 178.8 |
| | $\sigma$ | 81.4 | 20.3 | 33.2 | 30.8 | 53.1 | 70.3 |
| $Q_{C2}$ (kVAr) | $\mu$ | 236.8 | 254.3 | 209.8 | 209.9 | 197.3 | 181.6 |
| | $\sigma$ | 66.2 | 33.1 | 76.4 | 39.3 | 62.3 | 23.2 |
| *Fcost* (\$/h) | $\mu$ | 178.9161 | 179.1978 | 111.9691 | 115.0566 | 55.4705 | 59.2724 |
| | $\sigma$ | 16.5964 | 17.1368 | 9.7492 | 10.6029 | 5.5932 | 5.5934 |
| *Ploss* (kW) | $\mu$ | 77.7 | 79.2 | 38.1 | 45.4 | 13.6 | 17.5 |
| | $\sigma$ | 9.4 | 10.3 | 4.1 | 12.7 | 2.6 | 4.2 |
| *VD* (p.u.) | $\mu$ | 0.7302 | 0.6936 | 0.3603 | 0.3280 | 0.5550 | 0.0576 |
| | $\sigma$ | 0.1402 | 0.1380 | 0.0576 | 0.1148 | 0.0523 | 0.0075 |

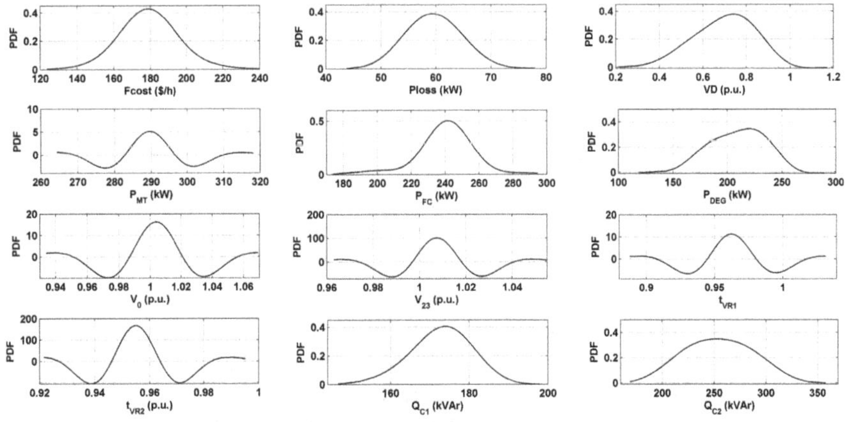


*Figure 2.12 PDF functions of the probabilistic OPF results for Case 3 in T = 14 h*

*Table 2.7 Mean and standard deviation results for different combinations of the input random variables*

| | | **Case C, time = 14 h Input random variables** | | | | |
|---|---|---|---|---|---|---|
| | | $P_{WT}$, $P_{PV}$, $L$ | $P_{WT}$, $P_{PV}$ | $P_{wt}$ | $P_{pv}$ | $L$ |
| *Fcost* ($/h) | $\mu$ | 179.1978 | 178.8243 | 178.8353 | 178.8601 | 179.2445 |
| | $\sigma$ | 17.1368 | 4.8933 | 4.6168 | 1.6218 | 16.4223 |
| *Ploss* (kW) | $\mu$ | 79.2 | 76.4 | 72.7 | 75.0 | 74.0 |
| | $\sigma$ | 10.3 | 5.0 | 2.8 | 5.2 | 10.4 |
| *VD* (p.u.) | $\mu$ | 0.6936 | 0.7144 | 0.7351 | 0.6956 | 0.6955 |
| | $\sigma$ | 0.1380 | 0.0672 | 0.0458 | 0.0406 | 0.1208 |

## 2.7 MATLAB® program *opfdm*

Computer program *opfdm* is an educational software tool for solving the OPF problems in distribution networks using metaheuristic optimization methods. The *opfdnm* has been developed to present the efficiency of different metaheuristic optimization methods in solving the OPF problem for distribution networks with DG units. The *opfdnm* software package is available as a supplementary file to this chapter on the book's home page in the IET Digital Library. Figure 2.13 shows a general flowchart for solving the OPF problem using metaheuristic optimization methods.

The aim of this program is to encompass the main steps in solving the OPF problem using metaheuristic methods. These steps include (i) selection of test system and edit system data; (ii) selection of objective function; (iii) selection of solution method, setting the algorithm parameters; (iv) program execution; and (v) display of the results.

The *opfdm* was developed in MATLAB (The MathWorks, Natick, Masachusetts, USA), because it integrates computation, programming, analyze data, and producing graphical displays in an easy-to-use environment where problems and solutions are expressed in familiar mathematical notation.

It is available 11-bus, 31-bus, 33-bus, 34-bus, and 69-bus test systems. For each test system, one jpg file with a single-line diagram of the system and one input m file are given. The input m file contains branch data, bus data, generator data, transformer data, and VAR compensation data for the selected test system. The data are given in matrix form. The user can change these data and save those modifications.

The user selects the objective function from six OPF objectives offered, such as minimization of fuel cost (*Fcost*), minimization of real power loss (*Ploss*), voltage profile improvement (*VD*), simultaneous minimization of fuel cost and real power loss (*Fcost_Ploss*), simultaneous minimization of fuel cost and voltage deviation (*Fcost_VD*), and simultaneous minimization of fuel cost, real power loss, and voltage deviation (*Fcost_Ploss_VD*).

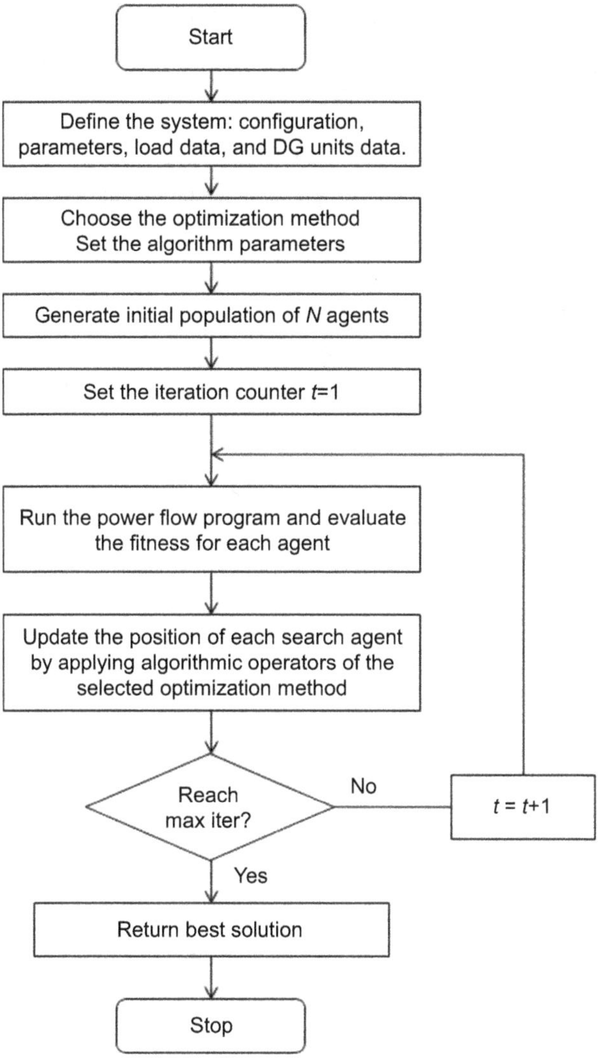


*Figure 2.13 OPF in distribution networks using metaheuristic optimization*

The *opfdm* program package offers more than 20 user-selectable metaheuristic algorithms for solving the OPF problem. In addition to well established methods such as PSO, GSA, ABC, ACO, TLBO, WDO, and GWO, there are several recently published algorithms such as PO, RUN, GBO, EO, and different hybrid variants of PSO and GSA (PSOGSA, PPSOGSA, PSOSCGSA), etc. These algorithms are given in the form of MATLAB m functions. The user can use the default (defined) algorithmic parameters, or to enter the desired values for algorithm parameters within the m file of the selected algorithm.

OPF calculation starts by running the main program, *opfdm*, as follows:

```
opfdm(method,testsystem,Case,N,max_it,nruns)
```

Various options can be selected for the input arguments, as follows.

```
method: e.g. 'ABC', 'PSO' 'GSA', 'WDO', 'CS', 'PSOGSA', etc.
testsystem: e.g. 'ts_dms11', 'ts_ieee31', 'ts_ieee33', etc.
Case - objective function: e.g. 'Fcost' 'Ploss', 'VD', etc.
N - Number of agents (e.g. 25).
max_it - Maximum number of iterations (e.g. 100).
nruns - Number of runs (e.g. 10).
```

Figure 2.14 shows the main window of *opfdm* while it operates on a Windows-based computer. An example of the execution of the program on the modified IEEE

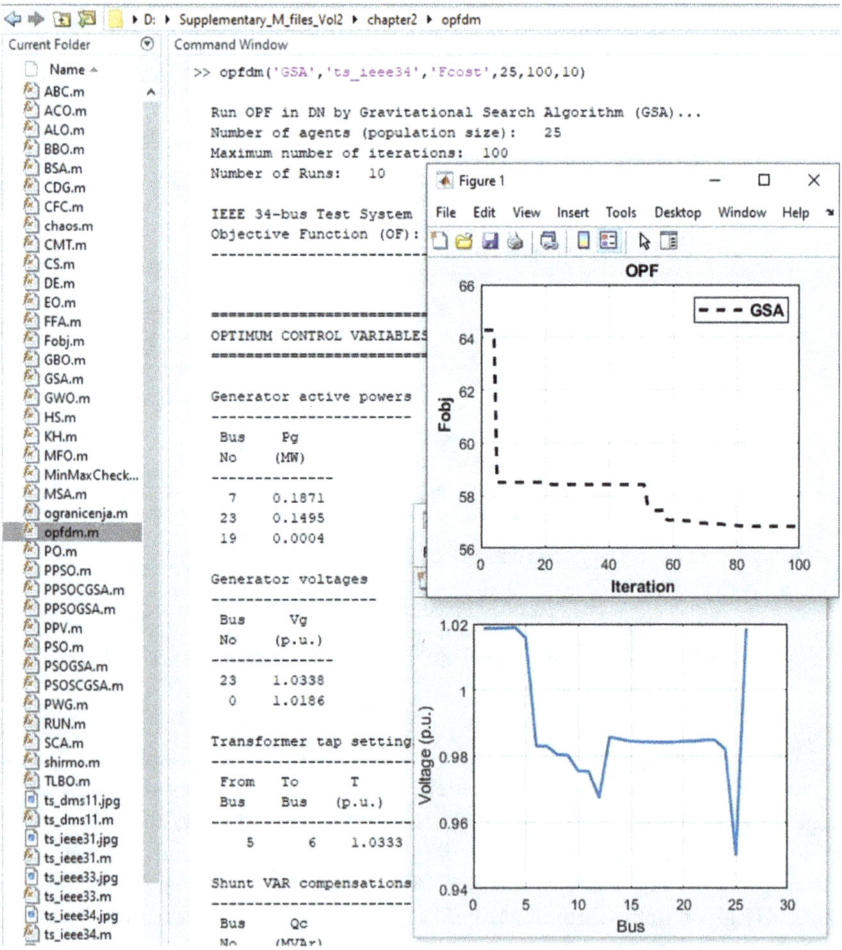


Figure 2.14 Main window of the opfdm program

34-bus test system is shown. Minimization of the fuel cost (*Fcost*) is selected as the objective function, and GSA is selected as the solution method. The OPF was performed for the $T$ = 10 h which determines load level, wind speed, ambient temperature, and solar irradiance on the daily diagrams of forecasted values.

The program package *opfdm* generates different segments of the OPF results, including

- The best results, that is objective function values in the last iteration.
- Convergence profile of the optimization algorithm.
- Optimal values of control variables, such as generator active power outputs, generator voltages, transformer tap settings, and shunt Var compensations.
- Bus voltages and bus powers in the system under optimum control variables.
- Branch power flows and losses in the system under optimum control variables.
- Checking if operating constraints are violated, as well as which operating constraints are violated.
- Statistical evaluation of the results, which includes the number of program execution, the minimum value of the objective function, the maximum value of the objective function, the mean value of the objective function, the standard deviation of the objective function, and the mean time of execution of the program. On the basis of these statistical indicators, a relevant comparison of different metaheuristic optimization methods can be achieved.

## 2.8 Conclusion

OPFs in distribution networks are achieved by the proper coordination of the operation of DG units, voltage regulators, shunt VAR compensators, and other regulation resources. This formulation includes the DG units with renewable and non-renewable energy sources. The OPF is very important tool for achieving economic, secure, and reliable operation of distribution networks.

To achieve a more realistic approach to the problem of optimizing distribution networks, many random disturbances or uncertain factors, such as the variation of load, the change in network configuration, and especially the stochastic nature of the output power of the DG with renewable energy such as WT and PV units, must be taken in consideration. This implies the transformation of the deterministic OPF in a probabilistic OPF approach. The OPF solution can serve as a useful decision-making supporting tool for distribution network operators and help to find out how the different uncertainties in input random variables affect the statistical indicators of the depending variables which define the operating state of the distribution network.

From a mathematical point of view, the OPF is a complex non-linear multi-objective optimization problem with equality and inequality constraints. Metaheuristic optimization methods, such as GA and GSA, have proved to be very effective in solving this problem.

Additionally, the *opfdm* program developed in MATLAB for solving OPF in distribution networks using different metaheuristic methods is presented in this chapter. This software package can be a very useful tool for researchers and students in the analysis of distribution networks.

## References


[1] Safigianni A.S., Koutroumpezis G.N., and Poulios V.C. 'Mixed distributed generation technologies in a medium voltage network'. *Electric Power Systems Research*. 2013;**96**:75–80.

[2] Moghaddas-Tafreshi S.M., and Mashhour E. 'Distributed generation modeling for power flow studies and a three-phase unbalanced power flow solution for radial distribution systems considering distributed generation'. *Electric Power Systems Research*. 2009;**79**:680–86.

[3] Li S., Tomsovic K., and Hiyama T. 'Load following functions using distributed energy resources'. *Proceedings of the IEEE/PES 2000 Power Engineering Society Summer Meeting*; Seattle, WA, USA, July 2000. IEEE; 2002. pp. 1756–61.

[4] Li X., Li Y., and Zhang S. 'Analysis of probabilistic optimal power flow taking account of the variation of load power'. *IEEE Transaction on Power Systems*. 2008;**23**(3):992–99.

[5] Borghetti A. 'Using mixed integer programming for the volt/var optimization in distribution feeders'. *Electric Power Systems Research*. 2013;**98**:39–50.

[6] Zhu Y., and Tomsovic K. 'Optimal distribution power flow for systems with distributed energy resources'. *Electrical Power Energy Systems*. 2007; **29**:260–67.

[7] Gan L., Li N., Topcu U., and Low S.H. 'Optimal power flow in tree networks'. *Proceedings of 52nd IEEE Conference on Decision and Control*; Florence, Italy, Dec 2013. IEEE; 2014. pp. 2313–18.

[8] Sheng W., Liu K.Y., and Cheng S. 'Optimal power flow algorithm and analysis in distribution system considering distributed generation'. *IET Generation, Transmission & Distribution*. 2014;**8**(2):261–72.

[9] http://www.marind.ca/medias/pdf/equipement_industriel/hipower/HJW410T6.pdf (Accessed August 2017).

[10] Azmy A. M., and Erlich I. 'Online optimal management of PEM fuel cells using neural networks'. *IEEE Transactions on Power Delivery*. 2005;**20**(2):1051–58.

[11] Barbir F., and Gomez T. 'Efficiency and economics of proton exchange membrane (PEM) fuel cells'. *International Journal of Hydrogen Energy*. 1997;**22**(10/11):1027–37.

[12] Campanari S., and Macchi E. 'Technical and tariff scenarios effect on microturbine trigenerative applications' *Journal of Engineering for Gas Turbines and Power*. 2004;**126**:581–89.

[13] Atwa Y. M., El-Saadany E.F., Salama M.M.A., Seethapathy R., Assam M., and Counti S. 'Adequacy evaluation of distribution system including wind/

solar DG during different modes of operation'. *IEEE Transactions on Power Systems*. 2001;**26**(4):1945–52.

[14] https://en.wind-turbine-models.com/turbines/272-vestas-v-44, (Accessed August 2017).

[15] Tiwari G. N., and Dubey S. *Fundamentals of Photovoltaic Modules and Their Applications*. Cambridge: RSC Energy Series; 2010. p. 422.

[16] Design qualification type approval of commercial PV Modules, IEC 61215 10.6 Performance at NOCT with IEC 60904-3 reference solar spectral irradiance distribution, 2005.

[17] Radosavljević J., Jevtić M., Klimenta D., and Arsić N. 'Optimal power flow for distribution networks with distributed generation'. *Serbian Journal of Electrical Engineering*. 2015;**12**(2):145–70.

[18] Radosavljević J., Jevtić M., Arsić N., and Klimenta D. 'Optimal power flow for distribution networks using gravitational search algorithm'. *Electrical Engineering*. 2014;**96**:335–45.

[19] Lee K.Y., and El-Sharkawi M.A. *Modern Heuristic Optimization Techniques with Applications to Power Systems*. Cambridge: IEEE Power Engineering Society (02TP160); 2002. p. 221.

[20] Rashedi E., Nezamabadi-pour H., and Saryazdi S. 'GSA: A gravitational search algorithm'. *Information Sciences*. 2009;**179**:2232–48.

[21] Caramia P., Carpinelli G., and Varilone P. 'Point estimate schemes for probabilistic three-phase load flow'. *Electric Power Systems Research*. 2010; **80**:168–75.

[22] Kaplanis S., and Kaplani E. 'A model to predict expected mean and stochastic hourly global solar radiation $I(h;n_j)$ values'. *Renewable Energy*. 2007;**32**:1414–25.

[23] Conti S., and Raiti S. 'Probabilistic load flow using Monte Carlo techniques for distribution network with photovoltaic generators'. *Solar Energy*. 2007; **81**:1473–81.

[24] Morales J.M., and Ruiz J.P. 'Point estimate schemes to solve the probabilistic power flow'. *IEEE Transactions on Power Systems*. 2007;**22** (4):1594–1601.

[25] Zhang P., and Lee S.T. 'Probabilistic load flow computation using the method of combined cumulants and Gram-Charlier expansion'. *IEEE Transactions on Power Systems*. 2004;**19** (1):676–82.

[26] Ruiz-Rodriguez F.J., Hernandez J.C., and Jurado F. 'Probabilistic load flow for photovoltaic distributed generation using the Cornish–Fisher expansion'. *Electric Power Systems Research*. 2012;**89**:129–38.

[27] Verbič G., and Canizares C.A. 'Probabilistic optimal power flow in electricity markets based on a two-point estimate method'. *IEEE Transactions on Power Systems*. 2006;**21**(4):1883–93.

[28] Distribution Test Feeders. Available https://ewh.ieee.org/soc/pes/dsacom/testfeeders/ (Accessed August 2017).

[29] Mwakabuta N., and Sekarv. 'Comparative study of the IEEE 34 node test feeder under practical simplifications; *39th North American Power Symposium*, Las Cruces, NM, USA, Oct 2007. IEEE; 2007. pp. 484–9.

# Chapter 3

# Multi-period optimal power flow in distribution networks

## 3.1 Introduction

In the last 20 years, we have witnessed the expansion of renewable energy sources (RES), primarily wind and solar photovoltaic (PV) power plants. The development and application of these technologies have several causes. First of all, it is the constant growth of electricity consumption and the need for new generation capacities. Another, no less important cause, is the ultimate request to limit the use of thermal power plants in order to reduce global warming due to greenhouse gas emissions. In addition, the deregulation and liberalization of the electricity market and the preferential prices for green energy have contributed rapid growth of RES in the last two decades.

Numerous researchers have intensively dealt with various aspects of the integration of RES into distribution networks (DNs). It has been shown that optimal planning and operation of RES can achieve positive effects, both in terms of technical performance of the system, and from the environmental and economic aspects. However, the uncontrollability and stochastic nature of wind speed and solar irradiance as the primary sources of RES remained their main shortcomings. The energy crisis that has been present lately imposes the need for much more flexible use of RES to fully exploit their energy potential. This means finding technically efficient and economically acceptable solutions for energy storage in order to fully exploit the potential of RES integrated into the system.

There is no way to directly store active electricity. There are more or less developed technologies and solutions for indirect storage of the electricity, i.e. converting the electricity into another type of energy that can be stored in a certain medium. The energy stored in this way is, when necessary, converted back into electricity, which is injected into the transmission or distribution system.

At the level of DNs, the battery energy storage systems (BESS) are imposed as a solution. For their application in DNs with significant penetration levels of RES, type WT and PV, it is necessary to solve two questions beforehand. The first one refers to the optimal planning of the BESS, which includes the choice of capacity, rated power, and location in the DN. The second one is the optimal exploitation, i.e. determination of the BESS operation mode in a given period of time, usually 24 h. This implies defining the charging/discharging regime, i.e. the time schedule and the charging/discharging power of the BESS in accordance with the adopted function that it should perform in the DN. These functions can be (i) providing energy management and optimal power flow (OPF) for reducing the cost of electricity, minimizing power losses, and improving the power quality; (ii) enabling/facilitating the realization of smart grids; (iii) meeting peak electrical load demands; and (iv) increasing the reliability of power supply. The mode of operation of the BESS is determined taking into account variable load, as well as the variable and stochastic power of RES (PV and WT) [1,2].

In essence, this is a problem of OPF in DNs. Given the specificity of the BESS, i.e. the limitation related to the state of charge (SoC), which directly affects the battery status in the next time segment (charging/discharging and power level), the problem of OPF must be observed in a certain (given) time horizon, usually a period of 24 h. Therefore, the problem of OPF becomes a multi-period OPF problem on a certain time horizon. The solution to this problem is performed by time segments (e.g. from 1 to 24 h), where the solution in a given segment ($t$) is conditioned by the solution obtained for the previous segment ($t-1$) [3].

In the last few years, a number of authors have addressed the issues of optimal power dispatch in DNs with integrated BESSs. These studies have been focused on several main aspects, such as modeling BESSs; defining the BESS function in DNs, i.e. objective functions of the optimization problem; defining control variables, dependent variables, and technical constraints; and the choice of solution methods.

A BESS is a combination of a voltage source converter and a battery pack such that the active and reactive power can be controlled independently [4–9]. In most research articles, BESSs are modeled as a set of equations defining that the stored energy at any given time segment is a function of the stored energy and charge/discharge power at the previous time segment. Internal losses of BESSs are accounted for through the charge/discharge efficiency coefficients or functions. The limits of the reactive power output to be provided by a BESS are also defined in these models.

Depending on the function that a BESS performs in the DN, the objectives for the OPF can be summarized in three groups: technical objectives, financial objectives, and multi-objectives. Financial objectives can be related to the maximization of revenues of an RES-BESS based on a hybrid power plant [10]; the reduction in the energy cost from the source grid [3,11]; or the minimization of distributed generator dispatch costs [12], power generation costs, and load-shedding costs [5]. Technical objectives are the base objectives and include static voltage stability improvement [6], transient voltage stability enhancement [8], energy loss minimization [9], voltage fluctuation mitigation [13], minimization of the voltage deviations at load buses [14], and reduction in power imbalances [15]. Researchers often adopt multi-objective functions with different combinations of financial and technical objectives, such as the simultaneous maximization of the total yield from an RES and minimization of the total costs of the energy loss in DNs [4] or the simultaneous minimization of power exchange cost between a DN and the transmission network and the penalty cost of the voltage deviation in the DN [7].

In its most general mathematical formulation, the multi-period OPF in DNs with integrated RES and BESS is a non-linear, non-convex, large-scale, dynamic optimization problem with constraints. There are two approaches to solving it, namely classical optimization methods and metaheuristic methods.

Non-linear optimization models are implemented in the GAMS environment and solved using CONOPT3 solver [4], CPLEX solver [5], and KNITRO solver [11]. The big M method was used in [7] to transform the model into a mixed-integer linear programming problem and solved using commercial software. The interior-point solver fmincon in MATLAB® R2021a [9] and the simplex solver through the docplex Python library [10] are also suggested for OPF in DNs with a BESS, whereas Giuntoli *et al.* [12] used the CaADdi framework to interface with the IPOPT solver for non-linear programming. In recent research, some population-based methods, such as PSO, have been proposed to find the optimal power dispatch solution in DNs with BESSs [6,14].

This chapter deals with multi-period OPF related to optimal day-ahead active–reactive power dispatching in unbalanced DNs with integrated single-phase PV generation and BESS [16]. The objectives are the minimization of cost for electricity, energy losses in the DN, and voltage unbalance at three-phase load buses by optimal management of active and reactive power flows. To solve this highly constrained non-linear optimization problem, a hybrid particle swarm optimization with sigmoid-based acceleration coefficients (PSOS) and a chaotic gravitational search algorithm (CGSA) called the PSOS–CGSA algorithm is proposed. A scenario-based approach encompassing the Monte Carlo simulation (MCS) method with a simultaneous backward reduction algorithm is used for the probabilistic assessment of the uncertainty of PV generation and power of loads. The effectiveness of the proposed procedure is evaluated through a series of test cases in a modified IEEE 13-bus feeder. The simulation results show that the proposed approach enables a large reduction in daily costs for electricity, as well as a reduction in expected daily energy losses in the DN by 22% compared to the base case without BESS while ensuring that the phase voltage unbalance rate (PVUR) is below the maximum limit of 2% for all three-phase buses in the DN.

The rest of this chapter is organized as follows. In Section 3.2, a scenario-based model is formulated for the optimal dispatching of active power in DNs by operating mode control of BESSs. The problem of optimal reactive power dispatch (ORPD) in unbalanced DNs with single-phase PV and BESS inverters is elaborated in Section 3.3. The solution method based on the hybrid PSOS–CGSA metaheuristic algorithm is briefly presented in Section 3.4. The simulation results are discussed in Section 3.5, whereas the main conclusions are listed in Section 3.6.

## 3.2 Problem formulation

The focus of this chapter is the problem of multi-period OPF in unbalanced DNs with single-phase PV generation and BESSs. The highlights of this study are as follows:

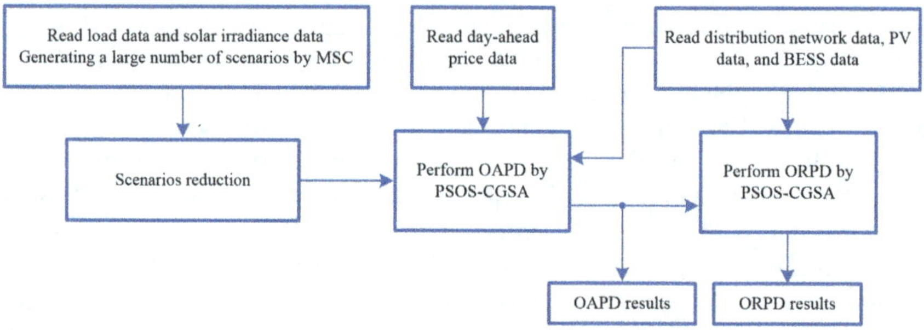


*Figure 3.1 Flow chart of the proposed methodology*

- A dynamic model of the optimal active power dispatch (OAPD) in unbalanced DNs with single-phase PV and BESSs is proposed to minimize energy costs from the source network.
- The ORPD in unbalanced DNs with single-phase PV and BESSs is considered. The objective functions consist of minimizing energy losses and voltage unbalance at three-phase load buses with the reactive powers of solar PV inverters and BESSs as control variables.
- A scenario-based approach encompassing the Monte Carlo simulation (MCS) method and the simultaneous backward reduction technique is proposed for modeling the uncertainty of PV generation and load in DNs.
- A novel hybrid PSOS–CGSA algorithm is proposed and applied to solve optimal active–reactive power dispatch problems.
- A modified IEEE 13-bus feeder for evaluating the solvability and applicability of the proposed approach is proposed and defined.

The proposed methodology is presented in a structured way through the flowchart in Figure 3.1.

## 3.3 Modeling of uncertainties

### 3.3.1 PV generation modeling

The power of the PV source as a function of solar irradiance can be expressed as follows:

$$P_{PV}(S) = \begin{cases} P_{pvn}\dfrac{S^2}{S_{stc}R_c} & \text{for } S < R_c \\ P_{pvn}\dfrac{S}{S_{stc}} & \text{for } S \geq R_c \end{cases} \tag{3.1}$$

where $P_{pvn}$ is the nominal power of the PV source, $S$ is the solar irradiance on the PV module surface, $S_{stc}$ is the solar irradiance under standard test conditions, and $R_c$ is a certain irradiance point.

To model the stochastic nature of the solar irradiance, it is most convenient to use the Beta probability density function (PDF):

$$f_b(S) = \begin{cases} \dfrac{\Gamma(\alpha+\beta)}{\Gamma(\alpha)\Gamma(\beta)} S^{(\alpha+1)}(1-S)^{(\beta-1)}, \\ \text{for } 0 \leq S \leq 1, \alpha \geq 0,\ \beta \geq 0 \\ \qquad 0, \text{ otherwise} \end{cases} \tag{3.2}$$

where $f_b(S)$ is the Beta function of the solar irradiance ($S$), $\alpha$ and $\beta$ are the shape parameters of the Beta distribution, and $\Gamma$ is the Gamma function.

Based on the data of the long-term measurement of the solar irradiance in a given area, characteristic daily diagrams of the solar radiation can be estimated with statistical indicators, i.e. mean value ($\mu_S$) and standard deviation of the solar irradiance ($\sigma_S$) in a given time interval $t$ (e.g. an hour). Using the mean value and standard deviation of the solar irradiance, the shape parameters of the Beta PDF can be calculated for a time interval ($t$) as follows:

$$\beta = (1-\mu_S) \cdot \left( \frac{\mu_S(1+\mu_S)}{(\sigma_S)^2} - 1 \right) \tag{3.3}$$

$$\alpha = \frac{\mu_S \cdot \beta}{1-\mu_S} \tag{3.4}$$

### 3.3.2 Load modeling

The normal (Gaussian) probability function is most commonly used to model load uncertainty in distribution systems. Generally, the load level is assumed to be a random variable ($L$) following the same PDF within each hour of a given daily load diagram.

$$f_n(L) = \frac{1}{\sqrt{2\pi}\sigma_L} e^{-\frac{(L-\mu_L)^2}{2\sigma_L^2}} \tag{3.5}$$

$$L = \mu_L + \sqrt{2}\sigma_L \mathrm{erf}^{-1}(2r-1) \tag{3.6}$$

where $\mu_L$ and $\sigma_L$ are the mean value and standard deviation of $L$, respectively; $r$ is a random number in the range of [0, 1]; and erf(·) and $\mathrm{erf}^{-1}(\cdot)$ are the error function and the inverse error function, respectively.

### 3.3.3 Scenario generation

The MCS method is suitable for modeling the uncertainties of solar irradiance and load level. Starting from probability density functions for solar irradiance and load, the parameters of which are determined based on historical data, e.g. previous years, the MCS method can generate a large number of scenarios for day-ahead profiles of solar irradiance and load levels. Although a larger number of scenarios (e.g. several thousand) provide a more faithful modeling of the uncertainty of the output power of the PV source and the load level; conversely, it means a higher computation burden. In distribution network operation control applications, such as OPF, methods with high accuracy and low computational burden are needed.

As a compromise between a good approximation (modeling) of the uncertainty of stochastic input variables (solar irradiance and load) on the one hand and a significant reduction in calculation time on the other hand, a scenario reduction strategy aggregating similar scenarios can be applied. In doing so, it must be ensured that the metric of the probability distribution of the reduced set is close enough to the original set of scenarios. There are several algorithms for reducing the number of scenarios, such as k-means clustering, backward reduction, and fast-forward selection. In this work, the simultaneous backward reduction algorithm [17–19] was employed to reduce a ternary scenario tree for the daily PV generations and load profiles.

## 3.4 Optimal active power dispatch

The battery operation mode is determined by two basic parameters: charge/discharge power ($P_{bat}$) and the level of stored energy, i.e. *SoC*. Since the battery capacity is limited, the *SoC* of the battery must be viewed as a dynamic quantity, as follows [10]:

$$SoC_{\phi,t} = SoC_{\phi,t-\Delta t} + \Delta SoC_{\phi} \tag{3.7}$$

$$\Delta SoC_{\phi} = \left[ P_{bat,ch,\phi,t-\Delta t} \cdot \eta_{ch,\phi} - \frac{P_{bat,dch,\phi,t-\Delta t}}{\eta_{dch,\phi}} \right] \frac{\Delta t}{C_{bat,\phi}} \tag{3.8}$$

where $C_{bat,\phi}$ is the total capacity of the battery; $P_{bat,ch}$ and $P_{bat,dch}$ are the charging and discharging power of the battery, respectively; $\eta_{ch}$ and $\eta_{dch}$ are the charging and discharging efficiency, respectively; $\Delta t$ is the time segment (e.g. 1 h); and $\phi$ denotes the phase, where $\phi \in (a,b,c)$.

Battery power and *SoC* must always be within defined limits, as follows:

$$0 \leq P_{bat,ch,\phi,t} \leq P_{bat,ch\ \max} \tag{3.9}$$

$$0 \leq P_{bat,dch,\phi,t} \leq P_{bat,dch\ \max} \tag{3.10}$$

$$SoC_{\min} \leq SoC_{\phi,t} \leq SoC_{\max} \tag{3.11}$$

$$SoC_{\phi,t=0} = SoC_{\phi,t=T} \tag{3.12}$$

$$P_{bat,ch,\phi,t} \cdot P_{bat,dch,\phi,t} = 0 \tag{3.13}$$

where $P_{bat,ch\ \max}$ and $P_{bat,dch\ \max}$ are the maximum charging and discharging powers of the battery, respectively; $SOC_{\min}$ and $SOC_{\max}$ are the predefined minimum and maximum charge levels, respectively; and $T$ is a time horizon, e.g. 24 h.

Battery charge/discharge power ($P_{bat}$) and $SoC$ are mutually dependent variables. Taking into account the variability of the electricity price (EP), their relationship can be expressed as follows:

$$P_{bat,\phi,t} = \begin{cases} 0 & \text{if} \quad c_{g,t} \leq c_{gav} \\ & \text{and} \quad SoC_{\phi,t} = SoC_{\max} \\ a_{\phi,t} \cdot P_{bat,ch\,\max} & \text{if} \quad c_{g,t} \leq c_{gav} \\ & \text{and} \quad SoC_{\min} \leq SoC_{\phi,t} < SoC_{\max} \\ 0 & \text{if} \quad c_{g,t} > c_{gav} \\ & \text{and} \quad SoC_{\phi,t} = SoC_{\min} \\ -a_{\phi,t} \cdot P_{bat,dch\,\max} & \text{if} \quad c_{g,t} > c_{gav} \\ & \text{and} \quad SoC_{\min} < SoC_{\phi,t} \leq SoC_{\max} \end{cases} \tag{3.14}$$

where $c_{g,t}$ is the day-ahead EP from the source grid at time $t$, $c_{gav}$ is the average EP during the day, and $a_{\phi,t}$ is a coefficient within the range [0–1].

The coefficient $a_{\phi,t}$, which defines the power of the BESS, can be determined in different ways. In [3], the coefficient $a_{\phi,t}$ was calculated based on the relation of the EP at the time $t$ with the mean EP during the day. A similar approach was adopted in [20], where the coefficient $a_{\phi,t}$ was calculated as a function of the difference in the EP from the predefined lower price limit for discharging and the upper price limit for charging. However, the charging/discharging powers of the BESS determined in this way do not guarantee the optimal dispatch of the active power according to the adopted objective function. Therefore, in this study, the coefficient $a_{\phi,t}$ is considered as a control variable, the value of which is optimized to ensure the minimization of a given objective function.

The basic equation of the active power dispatch problem is the power balance constraint (3.15), which must be satisfied at each time segment ($t$):

$$\sum_{\phi \in (a,b,c)} P_{g,\phi,t} = \sum_{\phi \in (a,b,c)} \left[ P_{L,\phi,t} - P_{PV,\phi,t} + P_{bat,\phi,t} \right] \tag{3.15}$$

where $P_{g,\phi,t}$ is the active power from/to the source grid at time $t$ in phase $\phi$; $P_{L,\phi,t}$ is the total active power of the load at time $t$ in phase $\phi$; $P_{PV,\phi,t}$ is the total active power generation of PV units at time $t$ in phase $\phi$; and $P_{bat,\phi,t}$ is the active power of the battery at time $t$ in phase $\phi$, where $\phi \in (a,b,c)$.

### 3.4.1 Objective function for OAPD

By controlling the operating mode of the battery, i.e. the charging/discharging power and SoC, the active power flows in the distribution network can be significantly influenced. In this way, the exchange power with the source power grid is directly affected, as well as other parameters of the distribution networks, such as power losses, voltage profile, etc. The operating costs of the distribution network mostly depend on the value of the energy from/to the source grid. Considering various PV generation and load demand scenarios with their probabilities, the objective function for OAPD can be formulated as the minimization of the expected cost for electricity, as follows:

$$\min Cost = \min\left[\sum_{\xi=1}^{\Xi}\left(C_g^{\xi} + C_{bat}^{\xi}\right)p(\xi)\right] \tag{3.16}$$

$$C_g^{\xi} = \sum_{t=1}^{T}\sum_{\phi \in a,b,c} c_{g,t}P_{g,\phi,t}^{\xi} \tag{3.17}$$

$$C_{bat}^{\xi} = \sum_{t=1}^{T}\sum_{\phi \in a,b,c} c_{bat}P_{bat,\phi,t}^{\xi} \tag{3.18}$$

where $\xi$ and $\Xi$ are the current scenario and the total number of scenarios, respectively; $p(\xi)$ is the probability of scenario $\xi$; $C_g^{\xi}$ and $C_{bat}^{\xi}$ are the total cost for electricity from the source grid and operating cost of the battery in scenario $\xi$, respectively; $c_g$ and $C_{bat}$ are the EP from the source grid and the unit operating cost of the battery, respectively; and $T$ is the considered period (i.e. 24 h).

Two assumptions should be kept in mind here, namely:

- The retail EPs have the same daily profile as the wholesale EPs and fluctuate hourly;
- The DN transacts the wholesale markets through some energy agent such as an aggregator.

The power output of a PV source is determined by the solar irradiation at a given moment, and its value cannot be influenced. Therefore, the cost of PV generation is not taken into account in the objective function (3.16) because it cannot be optimized by controlling the charge/discharge power of the battery.

For stable operation, constraints (3.9)–(3.13) and (3.15) must be satisfied. The output power of the PV generator is limited by the maximum output power, i.e. the nominal power. Additionally, $P_{g,\phi}(t)$ must be restricted by minimum and maximum values.

## 3.5 Optimal reactive power dispatch

In this study, it is assumed that PV arrays and batteries are connected to unbalanced DNs via single-phase converters. Depending on the nominal apparent power of the converter and the active power level of the corresponding PV source or battery, they have the ability to control the generation/absorption of reactive power, that is, they can be used as an additional regulation resource in DNs [21], as illustrated in Figure 3.2.

In this work, we do not specifically deal with power converter topologies. For practical realization, several power converter topologies can be applied to connect BESSs to the grid [22]. The conventional structure of a power converter for BESSs consists of a bidirectional DC/DC stage and a DC/AC stage. The basic requirement that they must fulfill for the ORPD problem is that their rated power be equal to or

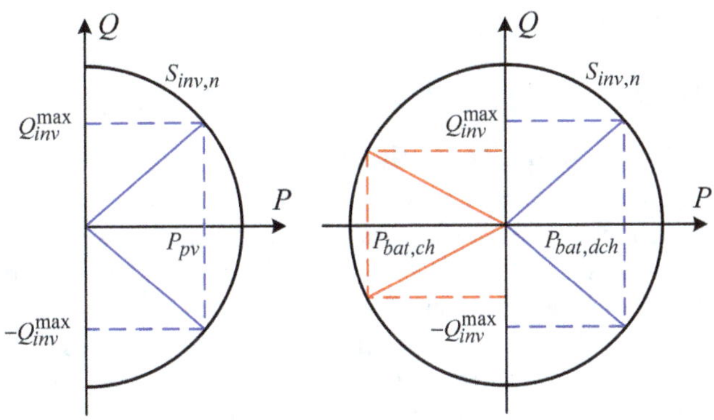


*Figure 3.2 Inverter capability curve: (a) diagram of the two-quadrant regulating capacity of a PV model; and (b) diagram of the four-quadrant regulating capacity of a BESS model. The red line indicates "consumer mode" and the blue line "generator mode" considering active power from/to the distribution network.*

greater than the rated power of the battery and that they have the ability to control the reactive power through a DC bus capacitor.

Therefore, for DNs with PV units and BESSs, the problem of ORPD can be posed as a problem of determining the optimal reference values of the available reactive powers of the inverters and the slack bus voltage magnitudes with a certain objective function under specified active power outputs of all PV and BESS inverters representing the solution of the active power dispatch, as explained in Section 2.

## *3.5.1 Objective function for ORPD*

### 3.5.1.1 Active power loss minimization

The level of active power losses is one of the most important indicators of the efficiency of distribution network management because they represent the direct costs for a distribution company. Therefore, the minimization of active power losses is one of the most common and important objective functions in OPF studies used by distribution system operators. Considering different (reduced) scenarios for PV generation and load profiles with their respective probabilities, the objective function for ORPD can be formulated as a minimization of the expected daily energy loss as follows:

$$Fobj1 = \min\left[\sum_{\xi=1}^{\Xi} W_{loss}^{\xi} p(\xi)\right] \tag{3.19}$$

$$W_{loss}^{\xi} = \sum_{t=1}^{T} \sum_{\substack{i=1 \\ \phi\in a,b,c}}^{NB} R_{i,\phi}\left(I_{i,\phi,t}^{\xi}\right)^2 \tag{3.20}$$

where $W_{loss}^{\xi}$ is the total energy loss in scenario $\xi$, $NB$ is the number of branches in the DN, $R_{i,\phi}$ is the resistance of branch $i$ in phase $\phi$, and $I_{i,\phi}^{\xi}$ is the current magnitude through the $i$th branch in phase $\phi$ for scenario $\xi$.

#### 3.5.1.2 Voltage unbalance minimization

High penetration of single-phase PV systems and relative variations in per-phase loading lead to an increase in voltage imbalance in distribution networks. It is well known that significant voltage unbalance may lead to higher network losses, increased heating of network components such as transformers, and costly damage or derating of three-phase motors. In this study, the PVUR is used as the voltage unbalance metric. The PVUR is calculated at each three-phase bus using the line-to-ground voltage magnitudes ($V_a$, $V_b$, and $V_c$) [23]:

$$\begin{aligned} PVUR(\%) &= \frac{\Delta V_P^{\max}}{V_P^{avg}} \cdot 100, \\ \text{where } V_P^{avg} &= \frac{V_a + V_b + V_c}{3}, \\ \Delta V_P^{\max} &= \max\{|V_a - V_P^{avg}|, |V_b - V_P^{avg}|, |V_c - V_P^{avg}|\} \end{aligned} \tag{3.21}$$

According to IEEE 141-1993 [24] the maximum value of $PVUR$ must not exceed 2% for distribution networks. To ensure that the $PVUR$ value at any three-phase bus does not exceed this limit, the minimization of the expected maximum of $PVUR$ in the DN can be defined as an objective function:

$$Fobj2 = \min\left[\sum_{\xi=1}^{\Xi} PVUR_{\max}^{\xi}\, p(\xi)\right] \tag{3.22}$$

where $PVUR_{\max}^{\xi}$ is the maximum value of the PVUR in the DN for scenario $\xi$.

### 3.5.2 Constraints

#### 3.5.2.1 Equality constraints

The equality constraints for the ORPD problem are typical non-linear power flow equations for three-phase unbalanced DNs [25,26], as given in Chapter 1 (Section 1.3).

#### 3.5.2.2 Inequality constraints

There are two types of inequality constraints: functional operational constraints, such as slack bus active power limits (3.23), load bus voltage limits (3.24), and branch flow limits (3.25); and feasibility region constraints defined by inverter reactive power limits (3.26) and slack bus voltage limits (27)

$$P_{grid,\min}^{\phi} \le P_{grid,t}^{\phi} \le P_{grid,\max}^{\phi} \tag{3.23}$$

$$V_{i,\min}^{\phi} \le V_{i,t}^{\phi} \le V_{i,\max}^{\phi} \quad i = 1,\ldots,NL \tag{3.24}$$

$$S_{i,t}^{\phi} \le S_{i,\max}^{\phi} \quad i = 1,\ldots,NB \tag{3.25}$$

$$-\sqrt{\left(S^{\phi}_{in,i}\right)^2 - \left(P^{\phi}_{in,i,t}\right)^2} \le Q^{\phi}_{in,i,t} \le \sqrt{\left(S^{\phi}_{in,i}\right)^2 - \left(P^{\phi}_{in,i,t}\right)^2} \tag{3.26}$$

$$V^{\phi}_{0,\,\min} \le V^{\phi}_{0,t} \le V^{\phi}_{0,\,\max} \tag{3.27}$$

where $P^{\phi}_{grid,t}$ is the active power from/to the source grid at time $t$ in phase $\phi$; $P^{f}_{grid,\,\min}$ and $P^{\phi}_{grid,\,\max}$ are the minimum and maximum power from/to the source grid, respectively; $V^{\phi}_{i,t}$ is the voltage magnitude at load bus $i$ in phase $\phi$; $V^{\phi}_{i,\,\min}$ and $V^{\phi}_{i,\,\max}$ are minimum and maximum limits of the voltage magnitude at load buses, respectively; $S^{\phi}_{i,t}$ and $S^{\phi}_{i,\,\max}$ are the apparent power value and maximum apparent power value of the $i$th branch in phase $\phi$, respectively; $Q^{\phi}_{in,i}$ is the reactive power injection of the inverter connected to phase $\phi$ at the $i$th bus; $S^{\phi}_{in,i}$ and $P^{\phi}_{in,i,t}$ are the maximum (rated) apparent power and the active power at time $t$ of the PV (or BESS) inverter connected to phase $\phi$ at the $i$th bus, respectively; and $V^{\phi}_{0,t}$, $V^{\phi}_{0,\,\min}$, and $V^{\phi}_{0,\,\max}$ are the value, minimum, and maximum limits of the voltage magnitude at the slack bus in phase $\phi$, respectively.

## 3.6 Solution method

The hybrid PSOS–CGSA algorithm introduced in [26] was applied to effectively solve the above-defined optimal active/reactive power dispatch. This is a meta-heuristic population-based optimization algorithm constructed by the hybridization of PSOS [27] and the CGSA [28]. Its performance in terms of simplicity, efficiency, and robustness has been demonstrated in solving various complex optimization problems, such as distribution system state estimation [26], dynamic economic dispatch [29], and OPF in DNs [3].

### *3.6.1 Application of PSOS–CGSA for active power dispatch*

In the OAPD problem presented in Section 3.4, the objective function is defined by (3.16). The control variables are coefficients $a_{\phi}(t)$, which define the power of the BESS according to (3.14). Therefore, a search agent can be written as

$$\mathbf{x}_{i,t} = \left[a_{a,i,t}, a_{b,i,t}, a_{c,i,t}\right]\ i = 1, \ldots, N \tag{3.28}$$

where $N$ is the total number of search agents (population size) and $a_{\phi,i,t}$ is the value of control variable $a$ within search agent $i$ for phase $\phi$ at iteration $t$.

A summary of the algorithm proposed to solve the active power dispatch is represented by the flowchart in Figure 3.3.

### *3.6.2 Application of PSOS–CGSA for reactive power dispatch*

The vector of control variables in the problem of ORPD consists of the reactive powers of the PV and BESS inverters and the slack bus voltage magnitudes and can

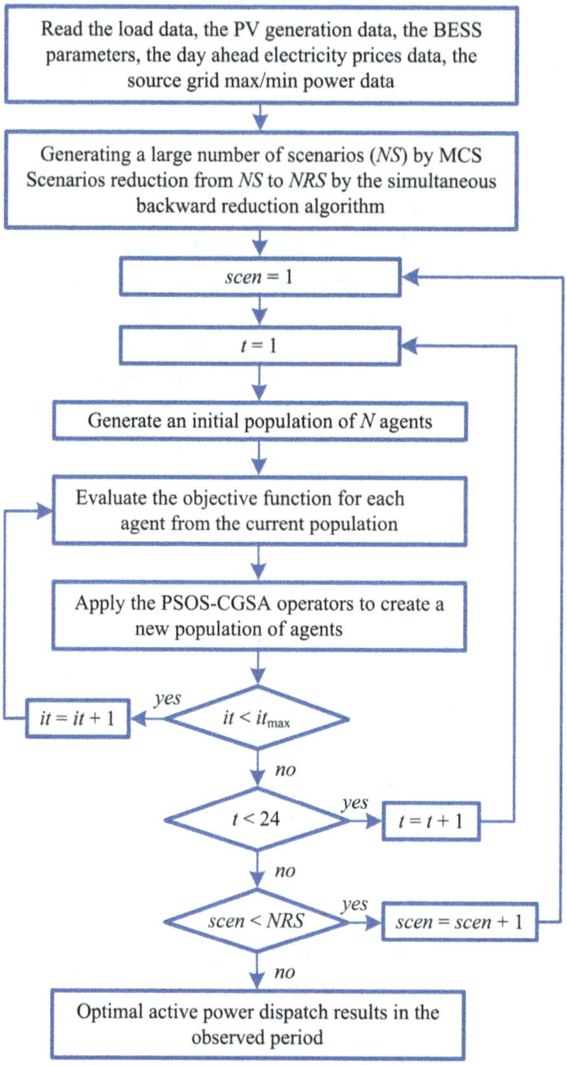


*Figure 3.3 Flowchart for the active power dispatch using PSOS–CGSA*

be expressed as follows:

$$\mathbf{x}_{i,t} = \left[Q^{\phi}_{i,1,t}, \ldots Q^{\phi}_{i,j,t}, \ldots, Q^{\phi}_{i,N_i,t}, V^{\phi}_{0,t}\right] \quad i = 1, \ldots, N \tag{3.29}$$

where $Q^{\phi}_{i,j}$ is the reactive power of the inverter at the $j$th bus in phase $\phi$, $N_i$ is the total number of buses in which inverters are connected, and $V^{\phi}_{0}$ is the voltage magnitude at the root bus of phase $\phi \in (a,b,c)$.

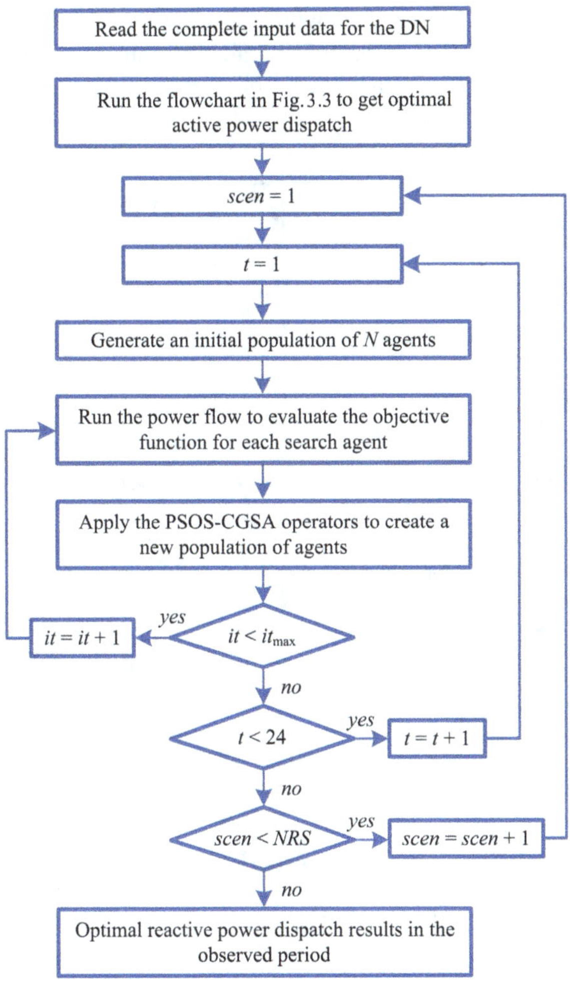


*Figure 3.4 Flowchart for the reactive power dispatch using PSOS–CGSA*

The application of PSOS–CGSA in solving the problem of ORPD is presented in the flowchart in Figure 3.4.

## 3.7 Simulation results

The proposed approach for the day-ahead optimal dispatching of active and reactive powers was simulated on a modified IEEE 13-bus feeder [30]. In the modified IEEE 13-bus feeder, the single-phase PV sources are connected to buses 634, 645, 646, 611, 675, 692, and 652, as shown in Figure 3.5. Single-phase PV sources, each with the same rated power of 250 kW, are connected to the grid via single-phase inverters. Single-phase BESSs, each with a rated power of 0.9 MW and a capacity of 3.6 MWh, are connected at bus 680 in all three phases via single-phase converters with a rated apparent power of 1.1 of the rated power of the batteries. Data on PV and BESS parameters are given in Tables 3.1 and 3.2, respectively.

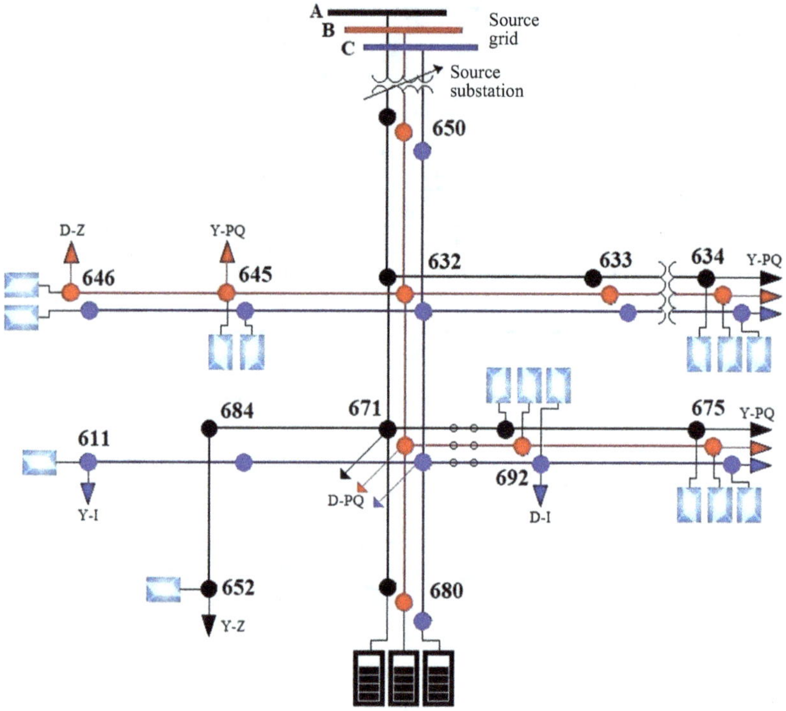


*Figure 3.5 Single-line diagram of the modified IEEE 13-bus test system*

*Table 3.1 Parameters of PV units*

| $P_{pvn}$ (kW) | $S_{stc}$ (kW/m$^2$) | $R_c$ (kW/m$^2$) |
|---|---|---|
| 250 | 1 | 0.12 |

*Table 3.2 Parameters of BESS*

| $P_{bat,ch\max}$ (MW) | $P_{bat,dch\max}$ (MW) | $SoC_{\max}$ (%) | $SoC_{\min}$ (%) | $\eta_{ch} = \eta_{dch}$ | $c_{bat}$ (EUR/MWh) |
|---|---|---|---|---|---|
| 0.9 | 0.9 | 80 | 20 | 1 | 10 |

The daily load profile is represented by the mean hourly values shown in Table 3.3, with a standard deviation of 10%. The daily solar irradiance profile used in the simulation is represented by the mean hourly values and standard deviations given in Table 3.4. The day-ahead EP data used are shown in Table 3.5. The

*Table 3.3 Load profile data*

| Hour | $\mu_L$ (p.u.) | Hour | $\mu_L$ (p.u.) | Hour | $\mu_L$ (p.u.) |
|---|---|---|---|---|---|
| 1 | 0.1591 | 9 | 0.5644 | 17 | 0.6715 |
| 2 | 0.2045 | 10 | 0.5372 | 18 | 1.0000 |
| 3 | 0.1948 | 11 | 0.6026 | 19 | 0.9310 |
| 4 | 0.1858 | 12 | 0.5496 | 20 | 0.7588 |
| 5 | 0.1963 | 13 | 0.4893 | 21 | 0.8244 |
| 6 | 0.1866 | 14 | 0.4847 | 22 | 0.9924 |
| 7 | 0.3154 | 15 | 0.4493 | 23 | 0.7869 |
| 8 | 0.5499 | 16 | 0.4390 | 24 | 0.5222 |

*Table 3.4 Solar irradiance data*

| Hour | $\mu_S$ (W/m$^2$) | $\sigma_S$ (W/m$^2$) |
|---|---|---|
| 8 | 158 | 120 |
| 9 | 386 | 248 |
| 10 | 538 | 277 |
| 11 | 669 | 261 |
| 12 | 748 | 242 |
| 13 | 819 | 221 |
| 14 | 862 | 207 |
| 15 | 870 | 197 |
| 16 | 853 | 205 |
| 17 | 779 | 237 |
| 18 | 700 | 258 |
| 19 | 572 | 263 |
| 20 | 357 | 206 |

*Table 3.5 EP data*

| Hour | $c_g$ (EUR/MWh) | Hour | $c_g$ (EUR/MWh) | Hour | $c_g$ (EUR/MWh) |
|---|---|---|---|---|---|
| 1 | 67.59 | 9 | 83.53 | 17 | 82.02 |
| 2 | 65.31 | 10 | 81.26 | 18 | 82.77 |
| 3 | 52.40 | 11 | 72.14 | 19 | 85.05 |
| 4 | 45.56 | 12 | 68.35 | 20 | 88.85 |
| 5 | 50.12 | 13 | 65.31 | 21 | 101.00 |
| 6 | 59.23 | 14 | 65.00 | 22 | 91.13 |
| 7 | 71.38 | 15 | 68.04 | 23 | 80.50 |
| 8 | 82.02 | 16 | 72.90 | 24 | 73.66 |

maximum value is estimated according to the EP for Serbia on the spot market in January–February 2023.

### 3.7.1 OAPD results

As explained in Section 3.3, using MCS, there are 2000 solar irradiance and load scenarios generated, as shown in Figure 3.6. By applying the simultaneous backward reduction technique, these 2000 scenarios were reduced to the ten scenarios shown in Figure 3.7.

By applying the proposed approach for the OAPD, the expected total operating costs of each scenario were determined; the results are shown in Table 3.6.

The optimal value of the expected daily cost for electricity is EUR 453.16. According to (3.16), this value is obtained as the sum of the costs in each reduced scenario multiplied by the corresponding probability. In view of this, the expected day-ahead charging/discharging schedule of the BESS was obtained, as shown in Figures 3.8 and 3.9.

For comparison, Figure 3.10 shows the daily costs for electricity in the following cases: (i) without a BESS integrated into the DN (base case), (ii) with a BESS in the DN applying the charge/discharge schedule approach presented in [3],

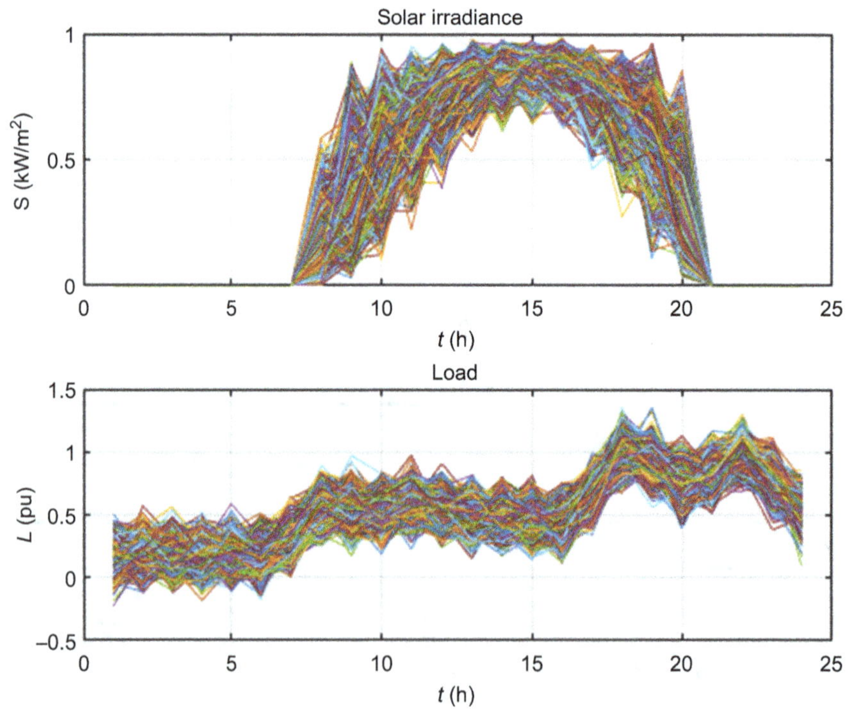


*Figure 3.6 Scenarios generated by MCS*

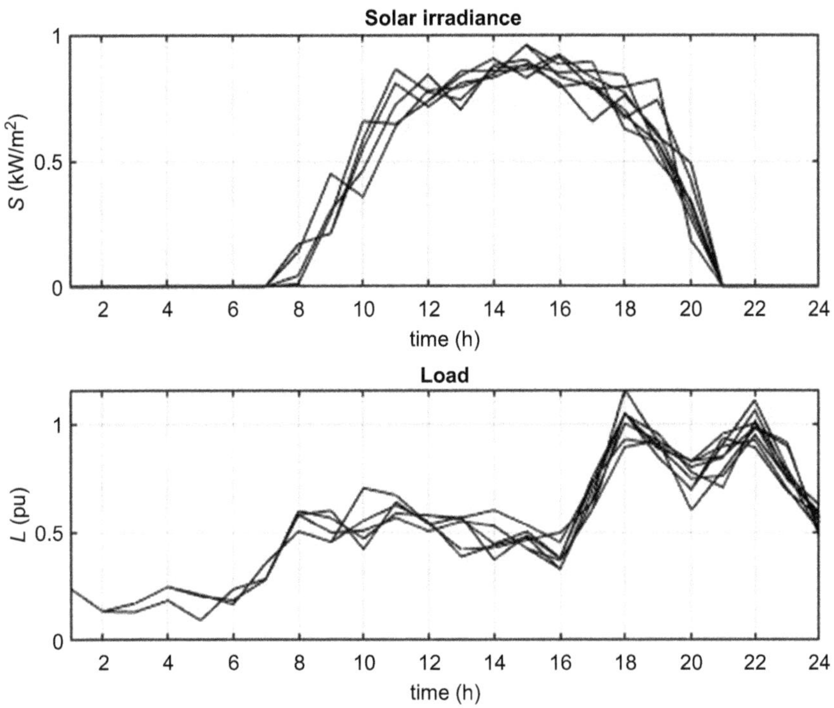


*Figure 3.7 Reduced scenarios of solar irradiance and load*

*Table 3.6 Expected costs obtained by OAPD*

| Scenario | Probability | Cost (€) |
|---|---|---|
| 1 | 0.0420 | 440.22 |
| 2 | 0.0760 | 590.36 |
| 3 | 0.0565 | 483.70 |
| 4 | 0.0760 | 552.52 |
| 5 | 0.0580 | 361.21 |
| 6 | 0.1100 | 462.06 |
| 7 | 0.2250 | 416.95 |
| 8 | 0.1265 | **339.61** |
| 9 | 0.1205 | 524.65 |
| 10 | 0.1095 | 444.86 |
| Expected | | **453.16** |

*Note*: The bold values indicate the best results.

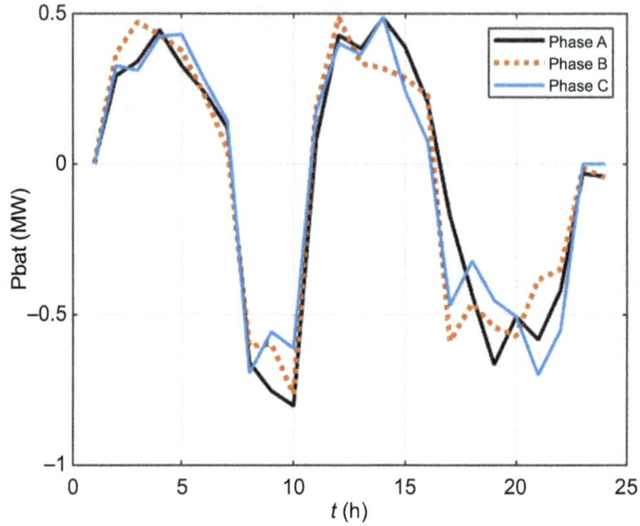


*Figure 3.8 Expected battery power during the day*

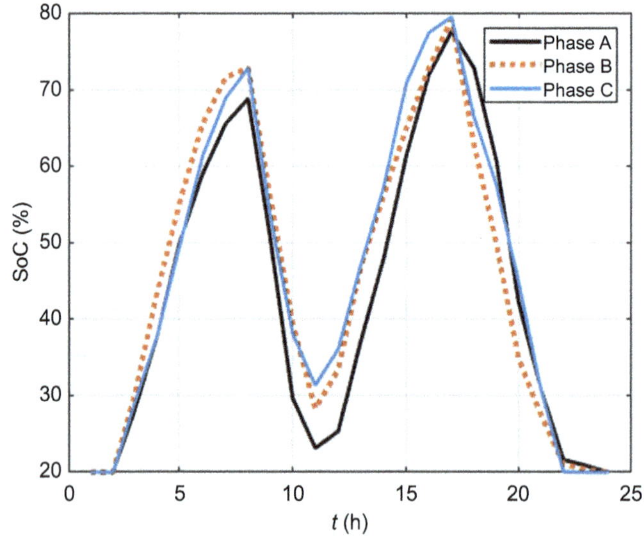


*Figure 3.9 Expected battery SoC during the day*

and (iii) with the BESS in the DN using the proposed OAPD algorithm. The results in Figure 3.10 clearly indicate that the proposed PSOS–CGSA-based approach for OAPD leads to a large reduction in costs for electricity compared to the base case and the approach for OAPD in [3].

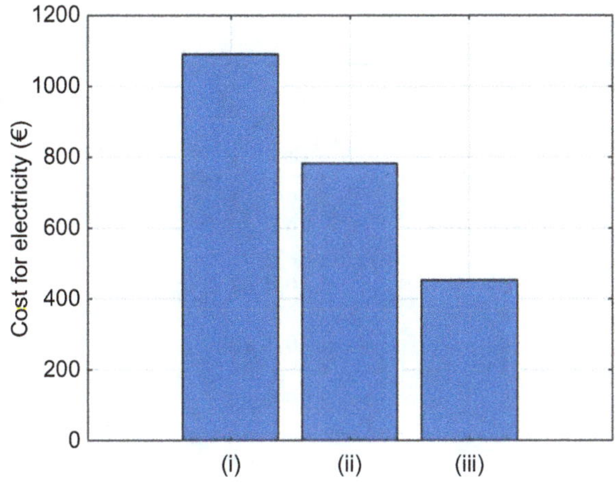


*Figure 3.10 Comparison of expected daily costs for electricity: (i) without BESS, (ii) with the BESS using approach in [3], and (iii) with the BESS using the proposed OAPD algorithm*

### *3.7.2 ORPD results*

Here, the optimal values of 21 control variables should be determined, as follows:

$$\mathbf{x} = \begin{bmatrix} Q^a_{634}, Q^b_{634}, Q^c_{634}, Q^b_{645}, Q^c_{645}, Q^b_{646}, Q^c_{646}, Q^a_{692}, Q^b_{692}, Q^c_{692}, \dots \\ Q^a_{675}, Q^b_{675}, Q^c_{675}, Q^c_{611}, Q^a_{652}, Q^a_{680}, Q^b_{680}, Q^c_{680}, V^a_{650}, V^b_{650}, V^c_{650} \end{bmatrix} \tag{3.30}$$

The simulation studies were performed for the following two test cases:

*Case 1*: Minimization of the active power losses (Fobj1);

*Case 2*: Minimization of the PVUR (Fobj2).

The voltages at the load buses must be within the range of [0.94–1.06] (p.u.). The voltage of the root bus ($V_0$) can be changed in the range of [0.9–1.1] (p.u.), while the limits of the reactive power of the converter are variable, according to (3.26).

The results are given in Tables 3.7 and 3.8. In the base case, the PV sources and BESS operate with unity power factor, and the root bus voltage is equal to 1 p.u. Tables 3.7 and 3.8 show the expected values of total active energy losses ($W_{loss}$) and maximum values of the phase voltage unbalance rate ($PVUR_{\max}$), respectively, for all ten reduced scenarios.

Applying the proposed procedure for ORPD achieves a reduction in expected energy losses of 22% compared to the base case, while the expected value of the maximum *PVUR* drops from 9.1% to 1.91%. There are significant energy savings, and it is ensured that the PVUR index level is below the maximum limit of 2% for all three-phase buses in the DN.

Based on the results shown in Tables 3.7 and 3.8, it can be seen that in the case of the minimization of active energy losses (*Fobj*1), the best scenario is 9, and in the case of the minimization of the voltage unbalance (*Fobj*2), the best scenario is scenario 1. Figures 3.11 and 3.12 show the optimal values of the control variables for these best scenarios.

*Table 3.7 ORPD results for Wloss*

| Scenario | Probability | $W_{loss}$ (kWh) | | |
|---|---|---|---|---|
| | | **Base** | **Case 1** | **Case 2** |
| 1 | 0.0420 | 969.62 | 725.76 | 849.90 |
| 2 | 0.0760 | 931.60 | 730.77 | 859.22 |
| 3 | 0.0565 | 878.93 | 681.53 | 813.92 |
| 4 | 0.0760 | 931.63 | 726.75 | 879.11 |
| 5 | 0.0580 | 973.22 | 761.71 | 882.67 |
| 6 | 0.1100 | 1016.80 | 805.72 | 951.66 |
| 7 | 0.2250 | 937.08 | 724.04 | 873.22 |
| 8 | 0.1265 | 990.97 | 776.68 | 919.96 |
| 9 | 0.1205 | 804.23 | **616.91** | 759.31 |
| 10 | 0.1095 | 977.61 | 772.15 | 905.42 |
| Expected | | 940.44 | **732.62** | 873.16 |

*Note:* The bold values indicate the best results.

*Table 3.8 ORPD results for $PVUR_{max}$*

| Scenario | Probability | $PVUR_{max}$ (%) | | |
|---|---|---|---|---|
| | | **Base** | **Case 1** | **Case 2** |
| 1 | 0.0420 | 12.54 | 7.70 | **1.14** |
| 2 | 0.0760 | 10.45 | 6.46 | 2.93 |
| 3 | 0.0565 | 8.42 | 7.43 | 2.05 |
| 4 | 0.0760 | 8.97 | 7.32 | 1.51 |
| 5 | 0.0580 | 11.91 | 6.46 | 2.75 |
| 6 | 0.1100 | 10.01 | 7.85 | 2.67 |
| 7 | 0.2250 | 8.34 | 6.46 | 1.71 |
| 8 | 0.1265 | 8.06 | 6.83 | 1.16 |
| 9 | 0.1205 | 6.46 | 7.58 | 1.46 |
| 10 | 0.1095 | 10.55 | 6.43 | 2.29 |
| Expected | | 9.10 | 6.96 | **1.91** |

*Note*: The bold values indicate the best results.

The voltage profiles at bus 671 for the best scenario in Case 1 and the best scenario in Case 2 are shown in Figures 3.13 and 3.14, respectively.

In Case 1, ORPD reduces the active energy losses compared to the base case, while the voltages are within the permissible limits, as shown in Figure 3.13. The phase voltage magnitudes are much closer to each other in Case 2 (Figure 3.14) than in Case 1 (Figure 3.13). Clearly, this is as a consequence of the *PVUR* minimization in Case 2.

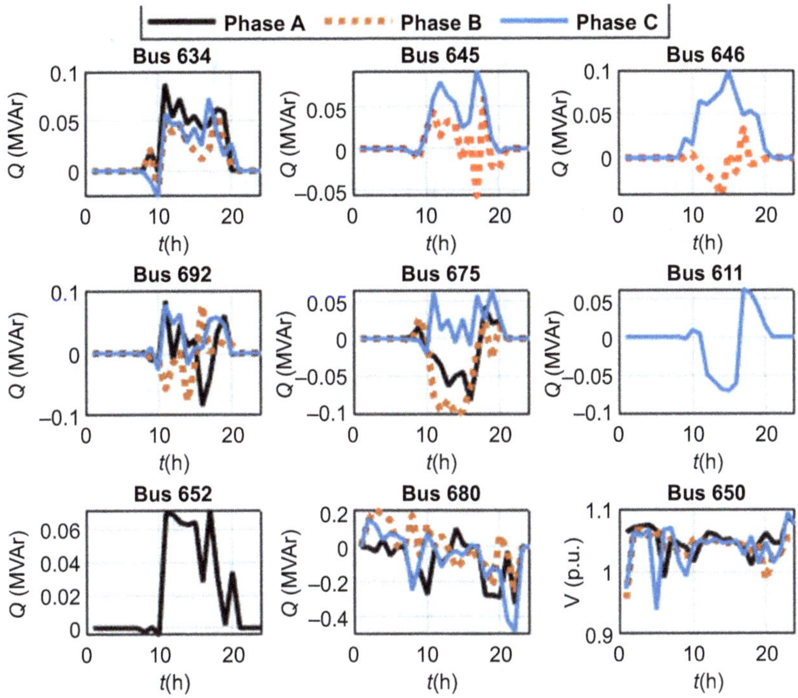


*Figure 3.11 Optimal values of control variables for the best scenario (9) in Case 1*

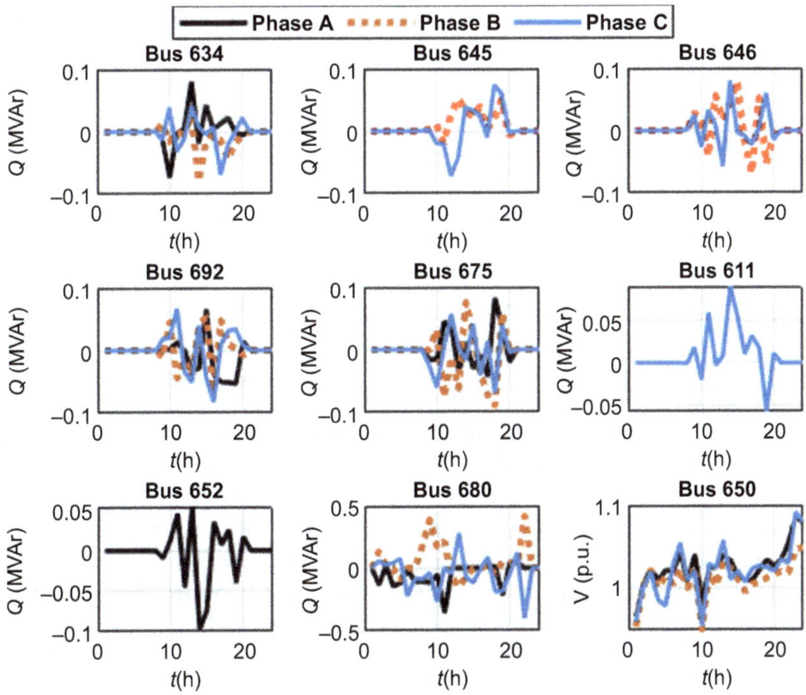


*Figure 3.12 Optimal values of control variables for the best scenario (1) in Case 2*

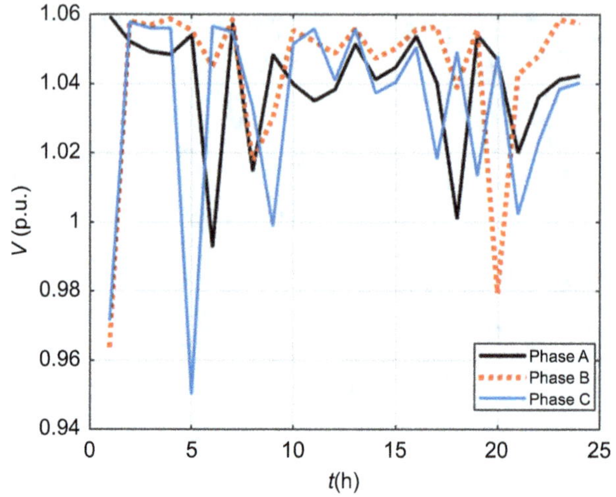


*Figure 3.13 Voltage profile at bus 671 in Case 1*

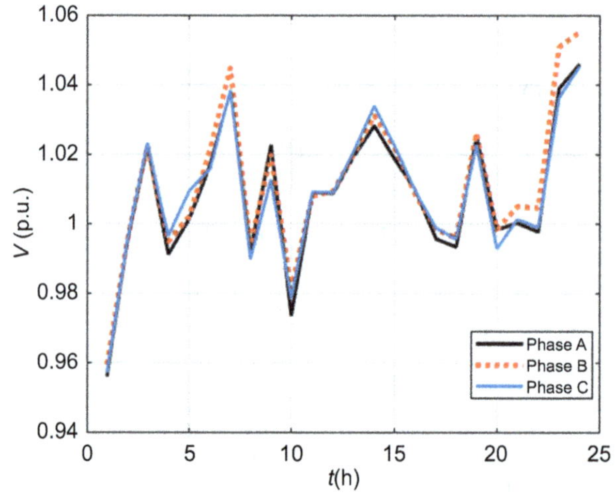


*Figure 3.14 Voltage profile at bus 671 in Case 2*

### *3.7.3 Evaluation of PSOS–CGSA*

The efficiency of the proposed solution algorithm (PSOS–CGSA) was confirmed by a comparison with well-established metaheuristic algorithms: GA, PSO, and the GSA; their recently improved versions, PSOS and CGSA, phasor PSO (PPSO), hybrid PSO and GSA (PSO–GSA), and hybrid PPSO and GSA (PPSO–GSA); and two recently proposed algorithms, namely the arithmetic optimization algorithm and the tunicate swarm algorithm. A brief description of all mentioned algorithms is given in the first part of the book (*Metaheuristic Optimization in Power*

*Engineering: Vol 1: Algorithms and Power Dispatch – using MATLAB-based software*, Chapters 1–5).

Algorithm parameters were adopted based on their default values. The population size and maximum iteration number were set as $N = 50$ and $it_{\max} = 200$ for all case studies. All algorithms were employed to solve the problem of OAPD under the same terms and input data.

The daily costs for electricity obtained for all ten reduced scenarios are shown in Figure 3.15. Furthermore, the expected daily costs for electricity obtained using different solution methods are shown in Figure 3.16. It is clear that the proposed

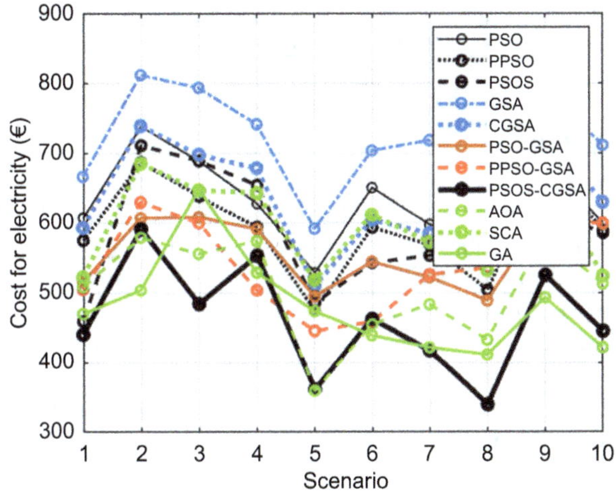


*Figure 3.15 Comparison of solution methods applied for OAPD*

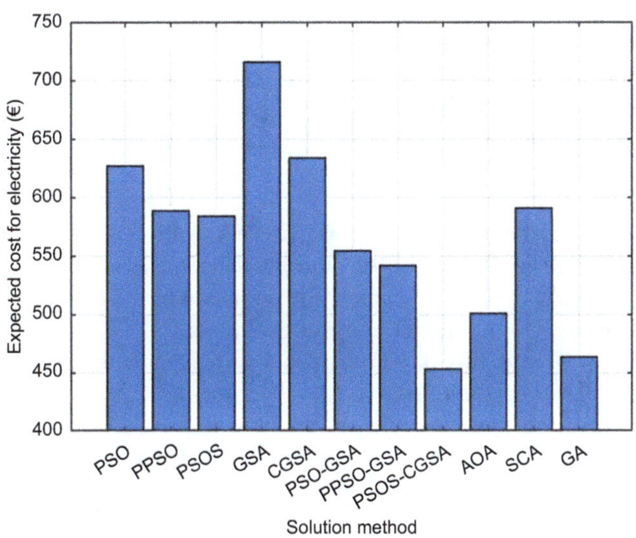


*Figure 3.16 Comparison of expected daily electricity costs obtained by different methods*

hybrid PSOS–CGSA method finds better solutions for the OAPD problem, i.e. provides lower costs in each of the ten scenarios and the least expected daily cost for electricity compared to other algorithms.

## 3.8 MATLAB® programs *oapdm* and *orpdm*

The presented two-stage multi-period OPF with OAPD-ORPD day-ahead approach is implemented in MATLAB programs *oapdm* and *orpdm*. The simulation results shown in Section 3.7 were obtained using these programs.

This program package is available as a supplementary file to this chapter on the book's home page in the IET Digital Library.

## 3.9 Conclusions

In this chapter, a two-stage multi-period OPF approach with day-ahead optimal active-reactive power dispatch in unbalanced DNs with high penetration of single-phase solar PV systems and BESS was proposed. The simulation results conducted on a modified IEEE 13-bus test system indicate the conclusions that can be summarized as follows:

- The proposed OAPD ensures minimum costs for electricity from the source grid for all scenarios generated by the MCS method and the simultaneous backward reduction algorithm.
- Single-phase PV inverters and BESSs can serve as efficient reactive power management resources.
- The proposed ORPD approach enables the reduction in active power losses in the DN and the reduction in the voltage unbalance on the three-phase load buses below the permitted limit.
- The proposed PSOS–CGSA solution technique provides better solutions for optimal power dispatch problems compared to other well-established metaheuristic algorithms, such as GA, PSO, GSA, and modified and hybrid versions of PSO and GSA.

The proposed methodology performed very well for the assumptions and limitations applied in this study. In future research, two of the most important assumptions should be relaxed, namely the inclusion of the operating costs of PV production in the OAPD model and the consideration of the ORPD model in terms of other power quality parameters, such as harmonic distortion. This can be achieved by extending the objective functions for the OAPD model and implementing an algorithm for harmonic power flows in the ORPD model, taking into account the nonlinear loads and the corresponding constraints.

## References

[1] Luo X., Wang J., Donner M., and Clarke J. 'Overview of current development in electrical energy storage technologies and application potential in power system operation'. *Applied Energy*. 2015;**137**:511–36.

[2] Ahmadi M., Adewuyi O. B., Sayed M., Danish S. M. S., Mandal P., Yona A., and Senjyu T. 'Optimum coordination of centralized and distributed renewable power generation incorporating battery storage system into the electric distribution network'. *Electrical Power and Energy Systems*. 2021;**125**:106–458.

[3] Radosavljević J., Milovanović M., Arsić N., Jovanović A., Perović B., and Vukašinović J. 'Optimal power dispatch in distribution networks with PV generation and battery storage'. In *Proceedings of the IX International Conference on Electrical, Electronic and Computing Engineering IcETRAN 2022*, Novi Pazar, Serbia, 6–9 June 2022.

[4] Gabash A. 'Active-reactive optimal power flow in distribution networks with embedded generation and battery storage'. *IEEE Transactions on Power Systems*. 2012;**27**:2026–2035.

[5] Mehrjerdi H., and Hemmati R. 'Modeling and optimal scheduling of battery energy storage systems in electric power distribution networks'. *Journal of Cleaner Production*. 2019;**234**:810–21.

[6] Adewuyi O.B., Shigenobu R., Ooya K., and Senjyu T. 'Static voltage stability improvement with battery energy storage considering optimal control of active and reactive power injection'. *Electric Power System Research*. 2019;**172**:303–12.

[7] Li X., Ma R., Gan W., and Yan S. 'Optimal dispatch for battery energy storage station in distribution network considering voltage distribution improvement and peak load shifting'. *Journal of Modern Power Systems and Clean Energy*. 2022;**10**:131–39.

[8] Milošević D., and Đurišić Ž. 'Technique for stability enhancement of microgrids during unsymmetrical disturbances using battery connected by single-phase converters'. *IET Renewevable Power Generation*. 2020; **14**:1529–40.

[9] Watson J.D., Watson N.R., and Lestas I. 'Optimized dispatch of energy storage systems in unbalanced distribution networks'. *IEEE Transactions on Sustainable Energy*. 2018;**9**:639–50.

[10] Das K., Grapperon A.L.T.P., Sorensen P.E., and Hansen A.D. 'Optimal battery operation for revenue maximization of wind-storage hybrid power plant'. *Electric Power System Research*. 2020;**188**:106–631.

[11] Prudhviraj D., Kiran P.B.S., and Pindoriya N.M. 'Stochastic energy management of microgrid with nodal price'. *Journal of Modern Power Systems and Clean Energy*. 2020;**8**:102–10.

[12] Giuntoli M., Subasic M., and Schmitt S. 'Control of distribution grids with storage using nested Benders' decomposition'. *Electric Power System Research*. 2021;**190**:106663.

[13] Khan H.A., Zuhaib M., and Rihan M. 'Voltage fluctuation mitigation with coordinated OLTC and energy storage control in high PV penetrating distribution network'. *Electric Power System Research*. 2022;**208**:107924.

[14] Radosavljević J. 'Voltage regulation in LV distribution networks with PV generation and battery storage'. *Journal of Electrical Engineering*. 2021; **72**:356–65.

[15] Udawalpola R., Masuta T., Yoshioka T., Takahashi K., and Ohtake H. 'Reduction of power imbalances using battery energy storage system in a bulk power system with extremely large photovoltaics interactions'. *Energies*. 2021;**14**:522.

[16] Radosavljević, J., Ktena A., Gajić M., Milovanović M., and Živić J. 'Dynamic optimal power dispatch in unbalanced distribution networks with single-phase solar PV units and BESS'. *Energies*. 2023;**16**:4356.

[17] Gröwe-Kuska N., Heitsch H., and Römisch W. 'Scenario reduction and scenario tree construction for power management problems'. In *Proceedings of the 2003 IEEE Bologna PowerTech—Conference Proceedings*, Bologna, Italy, 23–26 June 2003; Volume 3, pp. 152–58.

[18] LNesp. Scenred. GitHub. Available online: https://github.com/supsi-dacd-isaac/scenred (Accessed on 22 February 2023).

[19] Maheshwari A., Sood Y.R., and Jaiswal S. 'Flow direction algorithm-based optimal power flow analisys in the presence of stochastic renewable energy sources'. *Electric Power System Research*. 2023;**216**:109087.

[20] Biswas P.P., Suganthan P.N., and Amaratunga G.A.J. 'Optimal power flow solutions incorporating stochastic wind and solar power'. *Energy Conversion and Management*. 2017;**148**:1194–207.

[21] Ali A., Mahmoud K., Raisz D., and Lehtonen M. 'Probabilistic approach for hosting high PV penetration in distribution systems via optical oversized inverter with watt-var functions'. *IEEE Systems Journal*. 2021;**15**:684–93.

[22] Xavier L.S., Amorim W.C.S., Cupertino A.F., Mendes V.F., Boaventura W. C., and Pereira H.A. 'Power converters for battery energy storage systems connected to medium voltage systems: a comprehensive review'. *BMC Energy*.2019;**1**:7.

[23] Girigoudar K., and Roald L.A. 'On the impact of different voltage unbalance metrics in distribution system optimization'. *Electric Power System Research*. 2019;**189**:106656.

[24] IEEE Standard 141–1993; *Recommended Practice for Electric Power Distribution for Industrial Plants*. IEEE: New York, NY, USA, 1994; pp. 1–768.

[25] Khushalani S., Solanki K.M., and Shulz N.N. 'Development of three-phase unbalanced power flow using PV and PQ models for distributed generation and study of the impact of DG models'. *IEEE Transactions on Power Systems*. 2007;**22**:1019–25.

[26] Ullah Z., Elkadeem M.R., Wang S., and Radosavljević J. 'A novel PSOS–CGSA method for state estimation in unbalanced DG-integrated distribution systems'. *IEEE Access*. 2020;**8**: 113219–29.

[27] Tian D., Zhao X., and Shi Z. 'Chaotic particle swarm optimization with sigmoid-based acceleration coefficients for numerical function optimization'. *Swarm Evolutionary Computation*. 2019;**51**:100573.

[28] Mirjalili S., and Gandomi A.H. 'Chaotic gravitational constants for the gravitational search algorithm'. *Applied Soft Computing*. 2017;**53**:407–19.

[29] Radosavljević J., Arsić N., and Štatkić S. 'Dynamic economic dispatch considering WT and PV generation using hybrid PSOS–CGSA Algorithm'. In *Proceedings of the 20th International Symposium INFOTEH-JAHORINA (INFOTEH)*, East Sarajevo, Bosnia and Herzegovina, 17–19 March 2021; pp. 1–6.

[30] IEEE PES Test Feeder. Available online: https://cmte.ieee.org/pes-testfeeders/resources/ (Accessed on 18 February 2023).

# Chapter 4
# Optimal Volt/Var control in distribution networks

## 4.1 Introduction

One of the most important parameters of the power quality is the voltage magnitude. At each point of the distribution network (DN), the voltage must be within the permissible limits. Electrical devices are designed to have optimum operating characteristics at a certain, usually nominal voltage value. Deviation of the voltage beyond the allowable limits can cause irregular operation, reduction of efficiency, and, in extreme cases, the malfunction of electrical equipment. The voltage limits are standardized. The allowable voltage deviations are defined in percentages of the nominal voltage value. Depending on law and local practice, voltage deviations of $\pm 5\%$ or $\pm 10\%$ may be considered acceptable for the medium-voltage (MV) and low-voltage (LV) DNs. In order to maintain voltage within allowable limits under different load levels, various types of devices are used for voltage control. In conventional DNs, the voltage control is performed by changing the transmission ratio of the transformers and by compensating for the reactive power. Accordingly, the basic resources for voltage control in the DNs are

- Under-load tap-changing transformers (ULTCTs), usually this is the substation transformers (HV/MV) which connect the MV DN with a transmission (sub-transmission) supply network.
- Off-voltage tap-changing transformers (OVTCTs), usually this is distribution transformers (MV/LL) which connect the main MV DN with local LV DNs.
- Voltage regulators (VR), which are essentially transformers with tap-changers to adjust the voltage along the feeder, so as to compensate for the voltage drop over distance.
- Reactive power compensation devices: capacitors with unchangeable capacities (reactive power) and static VAR compensators that have the capability to change reactive power.

The ULTCT can be controlled locally by an automatic voltage regulator or remotely by a Supervisory control and data acquisition system [1,2]. The tap ratio control scheme for ULTCT is needed to achieve an appropriate change of voltage on the regulated bus according to load change. In addition to this basic requirement, when defining the control scheme for ULTCT, different factors should be considered, such as unbalanced load diversity on multiple feeders [3], the geometrical dimensions of the DN [4], the tap changing operating times [5], etc.

In addition to the regulating transformers, the control of the voltage in the DNs is accomplished to a large extent by regulating the reactive power flows, i.e. compensating the reactive power by using shunt capacitor banks and/or static var compensators (SVCs). In this case, the aim of Volt/Var control is to find an optimal combination of dispatch for capacitor banks, SVC reactive power output, and ULTCT setting at the substation to improve the voltage profile in the whole DN [6–11].

Distributed generation (DG) can have a significant influence on the power flow and voltage conditions in a DN. Adequate voltage control requires the consideration of the DG units' influence on the voltage profile. The impact of the DG units on the DN voltage depends on their size, location, and type [12–14].

The ability of a DG to participate in the voltage control process in a DN is highly dependent on the connection type and operation mode. There are three possible typical cases [15]:

1. The DG is located near a large load, and it is designated to supply this load, with constant active and reactive power output. In this case, the impact of the DG is brought down to decreasing in the total power load at the connection point. The bus at which the DG is connected can be represented as a PQ bus, whereas the DG can be modeled as a negative power load.
2. The DG injects active power with an approximately constant power factor. The voltage at the DG connection point increases together with the increase of the active power of the DG. In this case too, the bus at which the DG is connected can be represented as a PQ node, whereas the DG can be modeled as a negative power load.
3. The DG injects power with a specific voltage at the connection point. This DG controls the reactive power injection at the connection point, therefore controlling the voltage as well. The bus at which such a DG is connected, enabling constant active power and voltage magnitude, must be treated as a PV bus. Their participation in the voltage control of the DN can be significant.

For DNs with the supply ULTCT and integrated DG, which can control reactive power output, the optimal Volt/Var control implies the control scheme for coordination of the ULTCT actions with the reactive exchanges between the DG units and network [16,17].

Volt/Var control in DNs, which include the ULTCT, shunt capacitor banks, SVCs, and DG units, can be treated as an optimization problem with constraints. Solving this problem implies determining the control variables' values in order to achieve the set optimization criteria. Depending on the available regulation resources, the control variables can be the tap-changer settings of ULTCTs, OVTCTs, and VRs, the voltage magnitudes at PV buses, and the reactive powers of capacitors and SVCs. Dependent variables, which define an operation state of the DN, are the voltages at load buses. The objectives of optimal Volt/Var control can be different minimizing voltage deviation [18,19], minimizing total power loss

[20–28], minimizing operating costs [29–33], minimizing total number of switching operations [34,35], minimizing total harmonic distortion [36–38], and improving voltage stability [39].

Mathematically, the optimal Volt/Var control can be described as a non-linear large-scale static optimization problem with constraints. Metaheuristic optimization methods such as genetic algorithm (GA) [7,9,20,22,28,35–38], particle swarm optimization (PSO) [11,31–33], gravitational search algorithm (GSA) [26], and evolutionary programming [27] have proven to be very successful in solving the Volt/Var problem considering different objective functions.

This chapter deals with two methodologies for optimal voltage control. The first methodology related to determining the optimal transformer tap settings in radial (rural) DNs is based on the space and time decomposition of the voltage control problem. Thereafter, the Volt/Var control is formulated as a non-linear complex optimization problem with constraints and solved by using metaheuristic optimization methods.

## 4.2 Decomposition of the voltage control problem

The DNs often have a very large number of nodes and branches (several hundred), so there is also a large number of control variables to be considered. In addition, the specifics of DNs are the radial configuration and high *r/x* ratio. These specifics must be taken into account to create an efficient optimization model for optimal voltage control.

Taking into account the radial structure of the DN and the operation mode of transformers, a space and time decomposition of the voltage control problem can be achieved [1,2]. The voltage control activity performed by HV/MV (or MV/MV) ULTCTs influences voltages in the whole area "behind" (MV network, MV/LV transformers, and consequently their LV networks). At the same time, one ULTCT does not influence voltages in areas corresponding to other ULCTCs. Similarly, the voltage control performed by one MV/LV OVTCT influences voltages in corresponding small areas of the DN – the individual LV networks only; i.e. the control activities on the MV/LV transformer do not have any influence on voltages in LV networks, which belongs other MV/LV transformers. This disposition and influence of voltage control devices enable the space decomposition of the voltage control problem in radial DNs. Conversely, control action speeds of both transformer types differ to a great extent. Control actions of ULTCTs are "very fast," with a rate in minutes (hours). Changes in tap positions of OVTCTs are "very slow." They are usually performed a few times per year. Such speeds of transformer control actions enable the time decomposition of the voltage control problem [1].

### *4.2.1 Seasonal control of voltage*

Figure 4.1 shows a typical radial rural DN. It consists of a MV bus – the feed point of the network, OVTCT MV/MV supply transformer, an MV network, and several

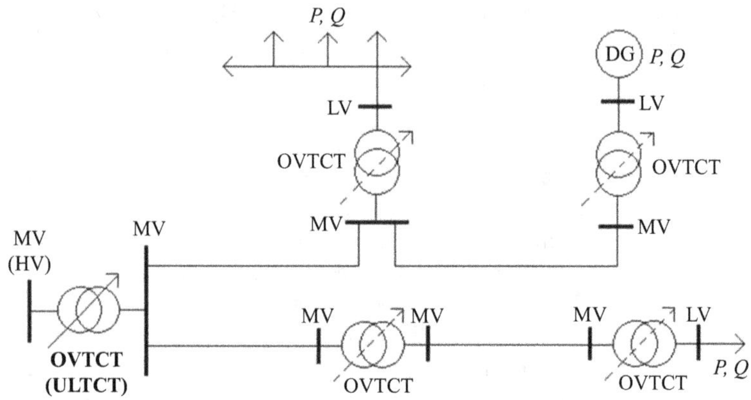


*Figure 4.1 Radial distribution network*

OVTCT MV/LV transformers and their belonging LV networks or LV feeders. Consumers and DGs are modeled as PQ buses.

Based on the space decomposition of the voltage control problem in radial DNs, the optimal seasonal control of voltage can be carried out in two stages. The first stage consists of determining the optimal tap-changer positions of the MV/LV transformers, while the second one implies determining the optimal tap-changer position of the power supply MV/MV transformer [40].

#### 4.2.1.1 The first stage of the optimal seasonal control of voltage

As already mentioned, due to the radial configuration of the DN, voltage control on the MV/LV transformers influences the voltage loads (as well as the DGs, which are modeled as "negative loads") connected to the transformer, but it does not have any influence on the voltage loads and DGs connected to the other transformers. This enables an independent determination of the optimal tap-changer positions for each MV/LV distribution transformer. For the given power of the LV load, the voltage at the MV bus of the MV/LV transformer is mainly constant in regard to the tap-changer position of the transformer. Accordingly, the optimal tap-changer position can be determined as the function of the load power (i.e. the DG) on the LV side of the transformer, the voltage on the MV bus of the transformer, and the desired voltage value on the LV side of the transformer. The issue of choosing the desired voltage value can be regarded as the optimization criterion. However, the optimal operation of the system is implicitly determined by the nominal voltage values. The aim of the seasonal control of voltage is to keep load voltages as approximate to nominal values as possible.

Figure 4.2 shows the part of the DN consisting of the MV/LV transformer, line, and load (i.e. the DG). The parameters and variables in this figure are given in (p.u.), where $\underline{Z}_T = R_T + jX_T$ is the transformer impedance,

*Figure 4.2 Segment of distribution network*

$\underline{Z}_\ell = R_\ell + jX_\ell$ is the line impedance, $\underline{V}_1$ is the voltage at MV bus, $\underline{V}_2$ is the voltage at the line end, $\underline{S} = P + jQ$ is the complex injected power, $T = 1 + n \cdot \Delta T$ is the transmission ratio of a transformer, $n$ is the tap-changer position, and $\Delta t$ is the step between tap-changer positions (p.u.).

The initial values of the bus voltages can be obtained by the power flow calculation for a given power of loads and DG units, and the transmission ratio of all transformers set to nominal value ($T = 1$). Assuming that the voltage magnitude $V_1$ is approximately constant, the relation between the voltages $V_1$ and $V_2$ can be written as follows:

$$\left(\frac{V_1}{T}\right)^2 = \left(V_2 + \frac{P \cdot R + Q \cdot X}{V_2}\right)^2 + \left(\frac{P \cdot X - Q \cdot R}{V_2}\right)^2 \tag{4.1}$$

where $R = R_T + R_\ell$ and $X = X_T + X_\ell$.

The transmission ratio $T$ is calculated from (4.1), for the specified voltage value $V_2$:

$$T = \frac{V_1}{\sqrt{\left(V_2 + \frac{P \cdot R + Q \cdot X}{V_2}\right)^2 + \left(\frac{P \cdot X - Q \cdot R}{V_2}\right)^2}} \tag{4.2}$$

It can now be calculated the optimal tap-changer position of the transformer:

$$n = \frac{T - 1}{\Delta t} \tag{4.3}$$

The real tap-changer position ($n_r$) is determined as the closest integer value from the range of the defined discrete values, based on the obtained value $n$ according to (4.3). At most MV/LV distribution transformers this range is $n_r = (-2, -1, 0, 1, 2)$. If the obtained value for $n$ is greater than the maximum value from the defined range, the maximum value from the range i.e. $n_r = n_{\max}$ is adopted as valid. In case the obtained value is smaller than the minimum value from the defined range, the minimum value $n_r = n_{\min}$ is adopted as valid for the tap-changer position. Since the real (adopted) position of the tap-changer $n_r$ is different from the calculated value $n$, the real value of the voltage $V_2$ is obtained by square root equation (4.1) at $V_2$ or by the power flow calculation in the whole DN, whereas the value of the transmission ratio is $T = 1 + \Delta t \cdot n_r$.

Since the power of load and the DGs power output are not constant during the day, month, or year, it is necessary to determine their representative values over the given period. One of the possible approaches is the medium power model. Depending on the shape, annual power diagrams of loads and DGs, they can be divided into seasons, usually two with the same or different durations. The seasonal power represents the mean value of the power segments that account for a season. In a general case, different power diagrams can be divided into different seasons. However, considering the fact that the seasonal voltage control implies simultaneous coordination of the tap-changer positions of the supply transformer and all MV/LV transformers, it is necessary that the seasons are the same (coinciding) for the whole system. It is possible to meet this demand only in case the following assumptions can be applied:

1. All loads have approximately the same normalized annual power diagrams;
2. DGs connected in the DN have similar annual power diagrams.

In order to apply the described procedure for determining the optimal tap-changer position of the MV/LV OVTCT, it is necessary to transform the LV network into the equivalent line model, as shown in Figure 4.2. The principle of equal voltage sags and the principle of equal power losses in the real (LV network) and equivalent (line model) systems are used for determining the equivalent line model impedance and equivalent injection power at the end of the equivalent line model. Figure 4.3 shows the radial LV line with $N$ nodes and sections as well as the equivalent line model. The impedance of section $i$, the injection power, and the voltage at node $i$ are marked with $\underline{Z}_i$, $\underline{S}_i$, and $\underline{V}_i$, respectively.

The equivalent line model impedance $\underline{Z}_{e1}$ is determined according to the principle of equal voltage sags:

$$\underline{Z}_{e1} = \frac{\underline{Z}_1 \cdot \frac{\underline{S}_1^*}{\underline{V}_1^*} + (\underline{Z}_1 + \underline{Z}_2) \cdot \frac{\underline{S}_2^*}{\underline{V}_2^*} + \ldots + (\underline{Z}_1 + \underline{Z}_2 + \ldots + \underline{Z}_N) \cdot \frac{\underline{S}_N^*}{\underline{V}_N^*}}{\sum_{i=1}^{i=N} \frac{\underline{S}_i^*}{\underline{V}_i^*}} \tag{4.4}$$

The voltage at the end of the equivalent line model is

$$\underline{V}_{e1} = \underline{V}_0 - \underline{Z}_{e1} \cdot \sum_{i=1}^{i=N} \frac{\underline{S}_i^*}{\underline{V}_i^*} \tag{4.5}$$

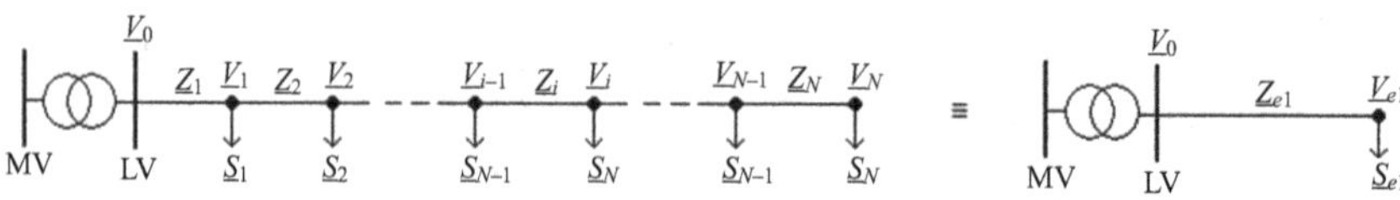


*Figure 4.3 Equivalent of a radial LV line*

Whereas the injection power at the end of the equivalent line model is

$$\underline{S}_{e1} = \underline{V}_{e1} \cdot \sum_{i=1}^{i=N} \frac{\underline{S}_i}{\underline{V}_i} \tag{4.6}$$

The node voltages in the per-unit system are close to 1, and therefore (4.4) and (4.6) can be simplified:

$$\underline{Z}_{e1} = \frac{\underline{Z}_1 \cdot \underline{S}_1 + (\underline{Z}_1 + \underline{Z}_2) \cdot \underline{S}_2 + \ldots + (\underline{Z}_1 + \underline{Z}_2 + \ldots + \underline{Z}_N) \cdot \underline{S}_N}{\sum_{i=1}^{i=N} \underline{S}_i} \tag{4.7}$$

$$\underline{S}_{e1} = \sum_{i=1}^{i=N} \underline{S}_i \tag{4.8}$$

The equivalent line model of $M$ radial LV lines, connected to the LV buses of the MV/LV transformers (Figure 4.4), can be determined based on the equal power loss principle:

$$\underline{Z}_e = \frac{\underline{Z}_{e1} \cdot \left|\frac{\underline{S}^*_{e1}}{\underline{V}^*_{e1}}\right|^2 + \underline{Z}_{e2} \cdot \left|\frac{\underline{S}^*_{e2}}{\underline{V}^*_{e2}}\right|^2 + \ldots + \underline{Z}_{eM} \cdot \left|\frac{\underline{S}^*_{eM}}{\underline{V}^*_{eM}}\right|^2}{\left|\sum_{j=1}^{j=M} \frac{\underline{S}^*_{ej}}{\underline{V}^*_{ej}}\right|^2} \tag{4.9}$$

The previous equation can be simplified:

$$\underline{Z}_e = \frac{\underline{Z}_{e1} \cdot |\underline{S}_{e1}|^2 + \underline{Z}_{e2} \cdot |\underline{S}_{e2}|^2 + \ldots + \underline{Z}_{eM} \cdot |\underline{S}_M|^2}{\left|\sum_{j=1}^{j=M} \underline{S}_{ej}\right|^2} \tag{4.10}$$

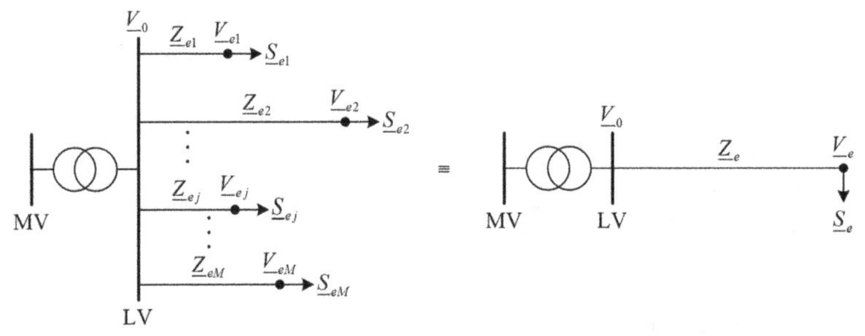


*Figure 4.4 Equivalent of several radial LV lines connected to the same bus*

Therefore, the injection power at the end of the line model is now:

$$\underline{S}_e = \sum_{j=1}^{j=M} \underline{S}_{ej} \tag{4.11}$$

where $\underline{Z}_i$ is the impedance of the $i$th section of the radial LV line, $i = 1, 2, ..., N$; $\underline{S}_i$ is the injection power at the node $i$ of the radial LV line; $\underline{V}_i$ is the voltage at the node $i$ of the radial LV line; $\underline{Z}_{ej}$ is the equivalent model impedance of the $j$th radial LV line, $j = 1, 2, ..., M$; $\underline{S}_{ej}$ is the equivalent model injection power of the $j$th radial LV line; $\underline{V}_{ej}$ is the equivalent model voltage of the $j$th radial LV line; $\underline{Z}_e$ is the equivalent model impedance of the $M$ radial LV lines (LV network); and $\underline{S}_e$ is the equivalent model injection power of the $M$ radial LV lines.

#### 4.2.1.2 The second stage of the optimal seasonal control of voltage

Determining the tap-changer position of the OVTCT MV/MV supply transformer located at the root of the MV network represents the second stage of the procedure for solving the problem of the optimal seasonal control of voltage. Actually, this is an optimization problem with the objective of minimizing the voltage deviation at load buses in the downstream network. The objective function can be expressed as follows:

$$ObjFun = VD = \sum_{i=1}^{NL} \left| V_i - V_i^{ref} \right| \tag{4.12}$$

where $V_i$ is the voltage at the bus $i$ of the DN, $V_i^{ref}$ is the reference (nominal) voltage value, and $NL$ is the number of load buses in the DN.

Carrying out this optimization task is brought down to the power flow calculation for all possible tap-changer positions of the supply transformer. During each power flow calculation, the tap-changer positions of the MV/LV transformers are determined for the current tap-changer position of the supply transformer according to the procedure described in Section 4.2.1.1. At the same time, the value of the objective function (4.12) is calculated. The minimum value of the objective function implicates the optimal solution, i.e. the optimal seasonal control plan of the distribution transformers in the given DN. The adopted optimization criterion determines the control plan. The flowchart of the proposed procedure is given in Figure 4.5. The procedure should be performed:

1. On a regular basis, once a year, based on the updated load diagrams from the current year.
2. Extraordinarily, after significant changes in the DN that have occurred, such as adding a new DG unit, network reconstruction by building a new line, transformer, etc.

#### 4.2.1.3 Optimal voltage control with ULTCTs

The optimal voltage control in the operation planning mode is performed for DNs with OVTCTs only, considering the mean power values over a given period.

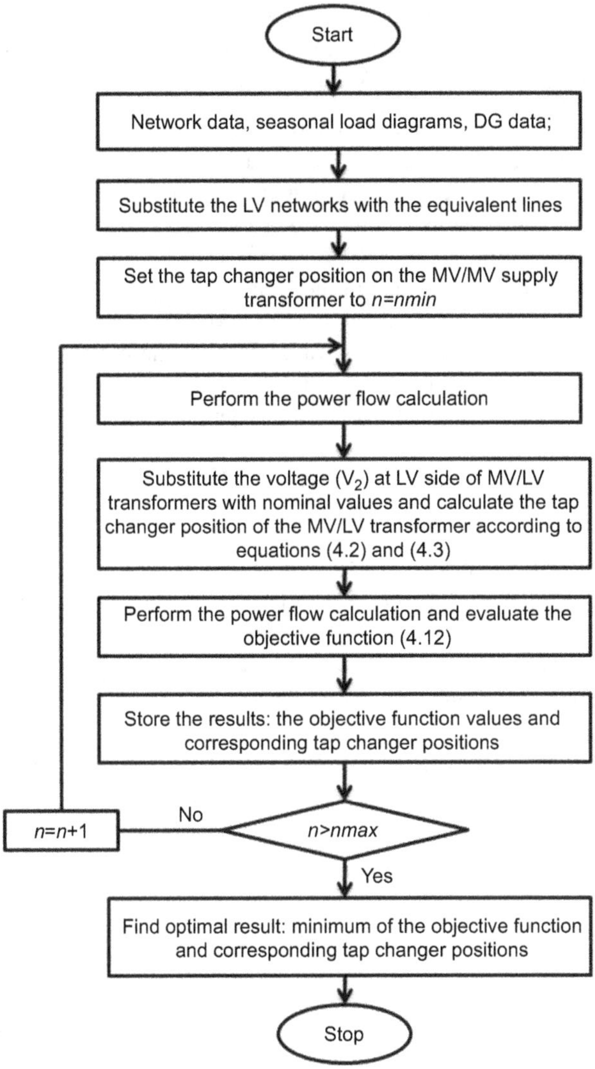


*Figure 4.5 Flow chart of optimal seasonal voltage control*

However, for DNs with ULTCT (Figure 4.1), the voltage control implies determining the optimal tap-changer settings of the ULTCT in accordance with the actual power of loads. Optimal real-time voltage control with ULTCT is based on the space and the time decomposition of the voltage control problem. The space decomposition enables a solution of the distribution voltage control problem for the MV network of each supply ULTCT (substation) separately. The time decomposition enables to solve of the problem in the operation planning mode (determining the optimal seasonal voltage control plan for OVTCTs in considered DN) and the real-time mode (determining the tap-changer position on the ULTCT). The objective function is defined by (4.12). The procedure for optimal voltage control in the real time by using the ULTCT can be described as follows:

*Step 1*: Define the DN configuration, lines data, transformers data, loads data, and DG units data. Use the per-unit system;
*Step 2*: Perform the algorithm for optimal seasonal control of voltage (Figure 4.5) and determine the optimal tap-changer positions for OVTCTs;
*Step 3*: Set the voltage value at the root bus (HV bus of ULTCT) and update the actual power of loads in the DN;
*Step 4*: Set the initial tap-changer position of ULTCT on $n = n_{min}$;
*Step 5*: Run the power flow program based on the backward/forward method and evaluate the objective function (4.12);
*Step 6*: Store the results: the objective function values and corresponding tap-changer positions of ULTCT;
*Step 7*: Set the tap-changer position of ULTCT on $n = n + 1$;
*Step 8*: Repeat Steps 5–8 until $n<n_{max}$; Stop.

#### 4.2.1.4 Simulation results

The procedure for optimal seasonal control of voltage is applied on a real DN located in the east of Serbia in the area of Stara Planina [41]. Figure 4.6 shows the DN containing 11 DGs. Those DGs are micro-hydropower plants (MHPs). The MV network is supplied through an OVTCT (35±2·2.5%)/10 kV/kV. The MV network supplies eight MV/LV OVTCTs (10±2·2.5%)/0.4 kV/kV. The data regarding the lines, transformers, and loads are given in Table 4.1. Figure 4.7 shows the annual power diagram of the MHPs.

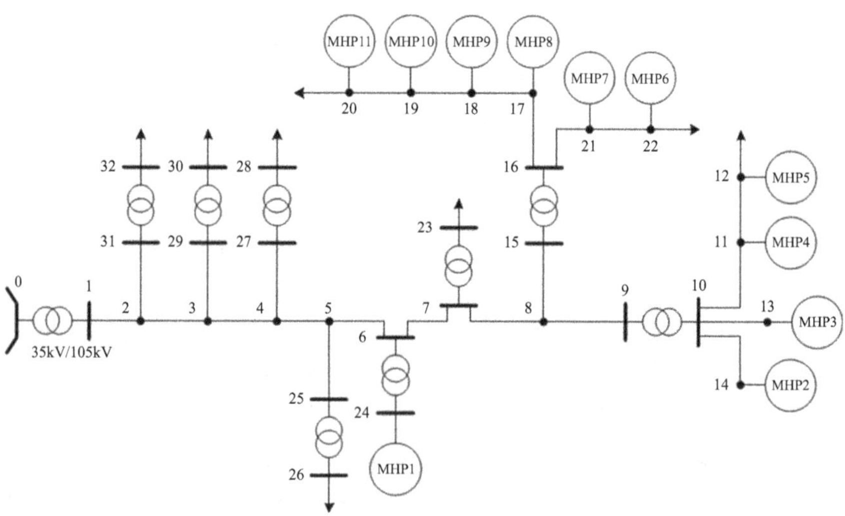


*Figure 4.6 Distribution network Stara Planina*

*Table 4.1 Data on the distribution network Stara Planina; Sbase = 1 MVA*

| Send. node | Rec. node | $R$ (p.u.) | $X$ (p.u.) | Load at Rec. node | |
|---|---|---|---|---|---|
| | | | | $P_L$ (p.u.) | $Q_L$ (p.u.) |
| 0 | 1 | 0.0021 | 0.0149 | | |
| 1 | 2 | 0.0083 | 0.0068 | 0 | 0 |
| 2 | 3 | 0.0066 | 0.0054 | 0 | 0 |
| 3 | 4 | 0.0072 | 0.0060 | 0 | 0 |
| 4 | 5 | 0.0027 | 0.0022 | 0 | 0 |
| 5 | 6 | 0.0145 | 0.0119 | 0 | 0 |
| 6 | 7 | 0.0010 | 0.0009 | 0 | 0 |
| 7 | 8 | 0.0376 | 0.0077 | 0 | 0 |
| 8 | 9 | 0.0517 | 0.0106 | 0 | 0 |
| 9 | 10 | 0.2000 | 0.3464 | 0 | 0 |
| 10 | 11 | 10.5262 | 3.2725 | 0 | 0 |
| 11 | 12 | 5.2631 | 1.6362 | 0.0074 | 0.0036 |
| 10 | 13 | 2.6316 | 0.8181 | 0 | 0 |
| 10 | 14 | 0.2256 | 0.0701 | 0 | 0 |
| 8 | 15 | 0.0451 | 0.0093 | 0 | 0 |
| 15 | 16 | 0.2000 | 0.3464 | 0 | 0 |
| 16 | 17 | 3.1579 | 0.9818 | 0 | 0 |
| 17 | 18 | 0.5263 | 0.1636 | 0 | 0 |
| 18 | 19 | 1.5789 | 0.4909 | 0 | 0 |
| 19 | 20 | 0.8271 | 0.2571 | 0.0014 | 0.0007 |
| 16 | 21 | 1.8045 | 0.5610 | 0 | 0 |
| 21 | 22 | 2.3308 | 0.7246 | 0.0008 | 0.0004 |
| 7 | 23 | 0.2000 | 0.3464 | 0.0028 | 0.0014 |
| 6 | 24 | 0.2000 | 0.3464 | 0 | 0 |
| 5 | 25 | 0.0199 | 0.0117 | 0 | 0 |
| 25 | 26 | 0.2000 | 0.3464 | 0.0096 | 0.0041 |
| 4 | 27 | 0.0024 | 0.00075 | 0 | 0 |
| 27 | 28 | 0.2000 | 0.3464 | 0.0126 | 0.0054 |
| 3 | 29 | 0.0042 | 0.0013 | 0 | 0 |
| 29 | 30 | 0.2000 | 0.3464 | 0.0178 | 0.0076 |
| 2 | 31 | 0.0120 | 0.0037 | 0 | 0 |
| 31 | 32 | 0.2000 | 0.3464 | 0.0099 | 0.0042 |

Two typical periods can be identified on the power diagrams of the MHPs: the period of the higher power (January–June) and the period of the lower power (July–December). The mean values of the MHP powers have been taken into consideration during the two seasons. It is assumed that the loads are constant; due to the fact the rural area has a slight change in load power during the year. The voltages $V_2$ (4.2), $V_{ref}$ (4.12), and root node voltage are specified to 1 (p.u.).

The impedances and injection powers of the equivalent line models of the LV networks, which belong to transformers 9–10 and 15–16, are shown in Table 4.2. The optimal seasonal voltage control plan is given in Table 4.3. Figure 4.8 shows the voltage profiles for the period of higher MHP powers (January–June). The

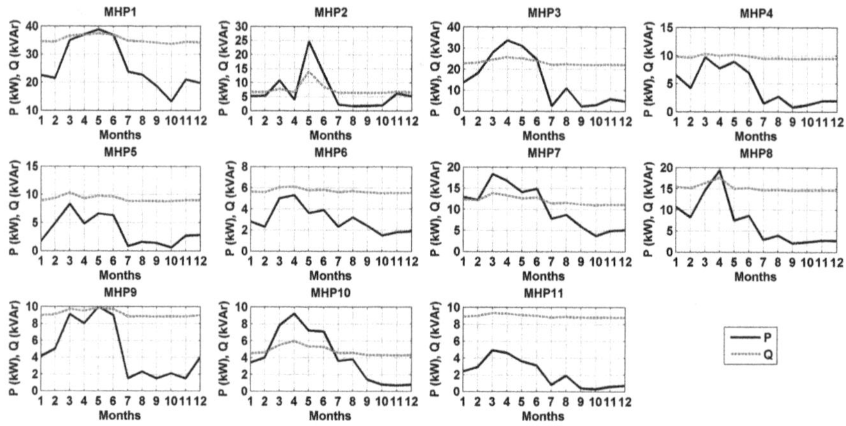


*Figure 4.7 Annual power diagrams of MHPs*

*Table 4.2 Equivalent model impedances and injection powers of the LV networks*

| LV network | | January–June | | July–December | |
|---|---|---|---|---|---|
| **TF** | **Nodes** | $\underline{Z}_e$**(p.u.)** | $\underline{S}_e$ **(p.u.)** | $\underline{Z}_e$ **(p.u.)** | $\underline{S}_e$ **(p.u.)** |
| 9–10 | 11,12,13,14 | 1.8899 +*j*0.4162 | –0.0408 + *j*0.0243 | 5.6586 +*j*1.2793 | –0.004 + *j*0.0193 |
| 15–16 | 17,18,19,20,21,22 | 1.9149 + *j*0.5245 | –0.0456 + *j*0.0283 | 2.0187 +*j*0.5035 | –0.0136 + *j*0.0231 |

*Table 4.3 Optimal seasonal voltage control plan*

| | Season: January–June | | |
|---|---|---|---|
| Transformer | 9–10 | 15–16 | Other transformers |
| Tap changer position $n_r$ | 2 | 2 | 0 |
| | Season: July–December | | |
| Transformer | 9–10 | Other transformers | |
| Tap changer position $n_r$ | –2 | 0 | |

upper and lower voltage limits in the MV and LV networks are specified according to the engineering standards at 1.05 (p.u.) and 0.95 (p.u.), respectively. Figure 4.8 clearly shows that the correction of the voltage ratio at transformers 9–10 and 15–16 in accordance with the optimal seasonal control of voltage enables reducing the voltages at nodes 13, and 17–20 within the allowable limits (0.95 p.u.–1.05 p.u.). This is the maximum improvement of the voltage profile, which implicates a limited domain of the seasonal voltage control.

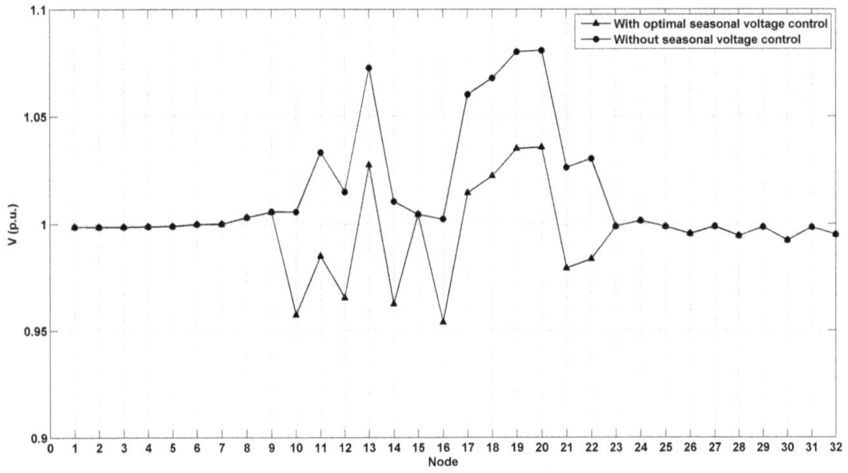


*Figure 4.8 Voltage profiles in the cases without voltage control (the tap-changer positions of all transformers are n = 0) and with optimal seasonal control of voltage*

## 4.3 Optimal Volt/Var control using metaheuristic optimization

Optimum Volt/Var control implies the coordination of control variables of DNs in real time with the aim of minimizing voltage deviation and power loss. This is possible for automated DNs with remote control and coordination of the system components. According to the available regulation devices, the control variables can be discrete (tap positions and number of capacitors) or continuous (voltage magnitudes and reactive power outputs). Dependent variables, which define an operation state of the DN, are the voltages at load buses. The objective function in this optimization problem can take different forms such as deviation of the voltage from the reference value, the costs of consumers due to voltage deviation from the nominal value, the voltage regulation costs (DG, capacitors, power losses, reactive power), power losses in the network, etc. Some objective functions can be in collision, which means that satisfying one objective can have the opposite effect on the other objective function. For example, minimizing voltage deviations can be at the expense of increasing power losses and vice versa. A compromise can be the choice of a multi-objective function that includes several individual objective functions that are weighted by appropriate coefficients in a way that depends on the goal set and the degree of importance of individual members of that function. The solution of the optimum voltage regulation must satisfy the operating constraints regarding the voltage limits at buses and the maximum loadability of the branches in the DN.

### 4.3.1 Problem formulation

Optimal values of control variables imply values of dependent variables that meet the optimization criteria. This definition of the optimal Volt/Var control can be considered as one of the forms of the optimal power flow problem in DNs. For a given network configuration and load level, each combination of control variable values defines a certain stationary state of the DN.

Generally, the optimal Volt/Var control problem can be formulated as follows:

$$\min F(\mathbf{x}, \mathbf{y}) \tag{4.13}$$

Subject to:

$$g(\mathbf{x}, \mathbf{y}) = 0 \tag{4.14}$$

$$h(\mathbf{x}, \mathbf{y}) \leq 0 \tag{4.15}$$

$$\mathbf{x} \in \mathbf{X} \tag{4.16}$$

where $F$ is the objective function; the objective function $F$ can take different forms. The main objective in the optimal Volt/Var control is the voltage profile improvement that is the minimization of the load bus $VDs$ (4.12). However, starting from the premise that the problem of voltage control is solved when the voltage values in all nodes of the network are within the permissible limits of min–max, the problem of optimal Volt/Var control can be viewed in a wider context. For example, minimizing active power losses or simultaneous minimization of voltage deviation and power losses can be adopted as objectives in the optimal Volt/Var control.

$\mathbf{x}$ is the vector of control variables, which generally consists of the following elements

- The voltage magnitude of the root node $V_0$; actually this is the specified voltage on the secondary side of the supply ULTCT located at the root of the network. Based on $V_0$ and known (constant) primary voltage, the tap-changer position of ULTCT is determined.
- The voltage magnitude of DG units modeled as PV buses, $V_{DG}$
- The tap settings of regulating distribution transformers, $T$, and
- The reactive power of shunt VAR compensators, $Q_C$.

Hence, $\mathbf{x}$ can be expressed as

$$\mathbf{x} = [V_0, V_{G1} \ldots V_{GNG}, T_1 \ldots T_{NT}, Q_{C1} \ldots Q_{CNC}] \tag{4.17}$$

where $NG$ is the number of DGs modeled as PV bus, $NT$ is the number of regulating distribution transformers, and $NC$ is the number of VAR compensators in DN.

$\mathbf{y}$ is the vector of dependent variables consisting of load bus voltages $V_L$, DGs reactive power outputs $Q_G$, and branch loadings $S_l$. Accordingly, vector $\mathbf{y}$ can be expressed as

$$\mathbf{y} = [V_{L1} \ldots V_{LNL}, Q_{G1} \ldots Q_{GNG}, S_{l1} \ldots S_{lNB}] \tag{4.18}$$

where $NL$ and $NB$ are number of load buses and network branches, respectively.

The equality constraints (4.14) are the typical non-linear power flow equations.

$$\underline{J}_i = \underline{I}_{Li} + \underline{I}_{Ci} - \underline{I}_{DGi} + \underline{Y}_i^o \cdot \underline{V}_i + \sum_{\substack{\ell \in \alpha_{\ell i} \\ \ell \neq i}} \underline{J}_\ell \quad i = NB, NB-1, \ldots, 1; \tag{4.19}$$

$$\underline{V}_i = \underline{V}_m - \underline{Z}_i \underline{J}_i \tag{4.20}$$

where $\underline{J}_i$ is the current in branch $i$ between nodes $m$ and $i$ (see Figure 1.3 in Chapter 1), $\underline{I}_{Li}$ is the complex current of load at bus $i$, $\underline{I}_{Ci}$ is the complex current of a Var compensator at bus $i$, $\underline{I}_{DGi}$ is the current injections at node $i$ corresponding to DG, $\underline{J}_\ell$ is the complex current in branch $\ell$ emanating from bus $i$, $\underline{Y}_i^o$ is the shunt admittance at bus $i$, $\alpha_{\ell i}$ is the set of buses connected with bus $i$, $\underline{V}_i$ is the complex voltage at bus $i$, $\underline{V}_m$ is the complex voltage at bus $m$, and $\underline{Z}_i$ is the series impedance of branch $i$.

Inequality constraints (4.15) are the functional operating constraints, such as load bus voltage magnitude limits, DG reactive power output limits, and branch flow limits

$$V_{Li}^{\min} \le V_{Li} \le V_{Li}^{\max} \quad i = 1, \ldots, NL \tag{4.21}$$

$$Q_{Gi}^{\min} \le Q_{Gi} \le Q_{Gi}^{\max} \quad i = 1, \ldots, NG \tag{4.22}$$

$$S_{li} \le S_{li}^{\max} \quad i = 1, \ldots, NB \tag{4.23}$$

The constraints related to voltage at load buses (4.21) and power flow through network branches (4.23) are also known as the security constraints.

Constraints (4.16) define the feasibility region of the problem control variables, such as root bus voltage limits, DG bus voltage magnitude limits, transformer tap setting limits, and shunt VAR compensation limits. Actually, those define the space of possible solutions for the Volt/Var control problem.

$$V_0^{\min} \le V_0 \le V_0^{\max} \tag{4.24}$$

$$V_{Gi}^{\min} \le V_{Gi} \le V_{Gi}^{\max} \quad i = 1, \ldots, NG \tag{4.25}$$

$$T_i^{\min} \le T_i \le T_i^{\max} \quad i = 1, \ldots, NT \tag{4.26}$$

$$Q_{Ci}^{\min} \le Q_{Ci} \le Q_{Ci}^{\max} \quad i = 1, \ldots, NC \tag{4.27}$$

It is worth mentioning that the control variables ($\mathbf{x}$) are self-constrained. The hard inequality constraints of the dependent variables consisting of load bus voltages $V_L$ are restricted by adding them as the quadratic penalty terms to the objective function. The new expanded objective function to be minimized becomes

$$F_p(\mathbf{x}, \mathbf{y}) = F(\mathbf{x}, \mathbf{y}) + \lambda_V \sum_{i=1}^{NL} \left(V_{Li} - V_{Li}^{\lim}\right)^2 \tag{4.28}$$

where $\lambda_V$ is the penalty factor. $V_{Li}^{\lim}$ is defined as follows:

$$V_{Li}^{\lim} = \begin{cases} V_{Li}^{\min}, & V_{Li} < V_{Li}^{\min} \\ V_{Li}^{\max}, & V_{Li} > V_{Li}^{\max} \\ V_{Li}, & V_{Li}^{\min} \leq V_{Li} \leq V_{Li}^{\max} \end{cases} \tag{4.29}$$

### 4.3.2 Solution method

In the population-based metaheuristic optimization algorithms, the population is defined by a set of individuals (agents) which represent potential solutions to the optimization problem. The number of agents ($N$) is named as the size of the population. In general, an agent can be represented as a vector whose elements are the values of the control variables of the optimization problem. The number of control variables ($n$) is the search space dimension of the optimization problem.

The essence of metaheuristic methods is iterative correction of the solution, i.e. generating a new population by applying algorithmic operators with stochastic search mechanisms on agents from the current population. The way in which are defined the algorithmic operators constitutes the essence of a particular metaheuristic optimization method. The efficiency and performance of metaheuristic optimization methods are dependent on the proper setting of the corresponding algorithmic parameters.

In recent years, various population-based metaheuristic optimization methods such as GA, PSO, GSA, etc., has been suggested for solving the different power system problems.

The main performances of metaheuristics are fast search of large solution spaces, ability to find global solutions and avoiding local optimum. Their main advantage compared to the classical (exact) optimization methods is that they are not limited to requirements for differentiability, non-convexity, and continuity of the objective function or types of control variables. Moreover, these methods can be used for practical optimization problems, taking into account various types of objective function and constraints.

The basic elements of metaheuristic optimization methods can be defined as follows:

*Agent*, **x**(t): It is a candidate solution represented by an $n$-dimensional vector, where $n$ is the number of control variables. At time (iteration) $t$, the $i$th agent $\mathbf{x}_i$(t) can be described as $\mathbf{x}_i(t) = [x_i^1(t), ..., x_i^d(t), ..., x_i^n(t)]$, where $x_i^d(t)$ is the position of the $i$th agent with respect to the $d$th dimension, i.e. the value of the $d$th control variable in the $i$th candidate solution at iteration $t$.

*Population*, **POP**(t): It is a set of $N$ agents at time (iteration) $t$, i.e. $\mathbf{POP}(t) = [\mathbf{x}_1(t), ..., \mathbf{x}_N(t)]^T$.

*Space of possible solutions*, **X**: It is a $n$-dimensional solution space that is defined by lower and upper limits of control variables.

*Fitness* is a direct metric of the performance of the individual population member (agent). The fitness of each agent of the population is calculated from the value of the function being optimized.

General structure of metaheuristic optimization methods can be represented as follows:

*Initialization*

1. Defining the objective function $F(\mathbf{x}_i)$ and the space of possible solutions $\mathbf{X}$.
2. Generate initial population of $N$ agents: $\mathbf{POP}(1) = [\mathbf{x}_1(1), ..., \mathbf{x}_i(1), ..., \mathbf{x}_N(1)]^T \subseteq \mathbf{X}$.
   Usually, the initial positions of each agent are randomly selected between minimum and maximum values of the control variables.
   Set the iteration counter: $t = 1$.

*Iterative procedure*

1. Calculate the fitness value $F(\mathbf{x}_i(t))$ for each agent $\mathbf{x}_i(t)$, $i = 1, ..., N$ in the current population $\mathbf{POP}(t)$.
2. Generate new population $\mathbf{POP}(t+1) = [\mathbf{x}_1(t+1), ..., \mathbf{x}_i(t+1), ..., \mathbf{x}_N(t+1)]^T \subseteq \mathbf{X}$ by applying the algorithmic operators on search agents from the current population $\mathbf{POP}(t)$.
3. Repeat the iterative procedure until the stop criteria is reached.
4. The optimal solution $\mathbf{x}^*$ is determined.

*End*

#### 4.3.2.1 Implementation for optimal Volt/Var control

A general procedure of applying metaheuristic population-based optimization methods to solve the optimal Volt/Var control can be described in the following steps:

*Step 1*: Define the DN configuration, lines data, transformers data, shunt VAR compensators data, loads data, and DG units data. Use the per-unit system.

*Step 2*: Specify the control variables $\mathbf{x}$ (4.17) and their lower and upper limits (4.24–4.27); Specify the dependent variables $\mathbf{y}$ (4.18), and their lower and upper limits (4.21–4.23); Specify the objective function to be optimized $F(\mathbf{x,y})$ (4.13), and the penalty factor ($\lambda_V$) to form the expanded objective function $F_p(\mathbf{x,y})$ (4.28).

*Step 3*: Initialize algorithm parameters, such as the population size ($N$), maximum iteration number ($t_{\max}$), and other parameters depending on the selected optimization method.

*Step 4*: Generate an initial random population of $N$ agents:

$$\mathbf{POP}(0) = \begin{bmatrix} \mathbf{x}_1(0) \\ \mathbf{x}_2(0) \\ \vdots \\ \mathbf{x}_i(0) \\ \vdots \\ \mathbf{x}_N(0) \end{bmatrix} = \begin{bmatrix} x_1^1(0), ..., x_1^d(0), ..., x_1^n(0) \\ x_2^1(0), ..., x_2^d(0), ..., x_2^n(0) \\ \vdots \\ x_i^1(0), ..., x_i^d(0), ..., x_i^n(0) \\ \vdots \\ x_N^1(0), ..., x_N^d(0), ..., x_N^n(0) \end{bmatrix} \tag{4.30}$$

The initial positions of each agent are randomly selected between the minimum and maximum values of the control variables. Therefore, the initial value of the $d$th control variable in the $i$th particle (candidate solution) is obtained as follows:

$$x_i^d(0) = x_{i,\min}^d + r \cdot \left(x_{i,\max}^d - x_{i,\min}^d\right) \tag{4.31}$$

where $r$ is a random number in [0,1] and $x_{i,\min}^d$ and $x_{i,\max}^d$ are the lower and upper limits of the $d$th control variable. According to (4.17), the $i$th agent can be written as follows:

$$\mathbf{x}_i = \left[V_0^1, V_{G1i}^{\cdots}, \ldots, V_{GNGi}, T_{T1i}, \ldots, T_{NTi}^d, Q_{C1i}^{d+1}, \ldots, Q_{CNCi}^n\right] \tag{4.32}$$

where $N$ is the number of particles (population size) and $n$ is the number of control variables.

*Step 5:* Run the power flow program based on the backward/forward sweep method for each agent $x_i(t)$ from the current population POP($t$) and calculate the corresponding values of the objective function.

*Step 6:* Update the position of each agent by applying algorithmic operators of the selected optimization method.

*Step 7:* Repeat Steps 5 and 6 until the stop criterion is reached, that is the maximum number of iterations $t_{\max}$.

*Step 8*: Return best solution obtained in the last iteration; stop.

## 4.3.3 Simulation results

The proposed approach was implemented on the IEEE 31-bus test system shown in Figure 4.9. The data of the network including nominal powers of loads are given in

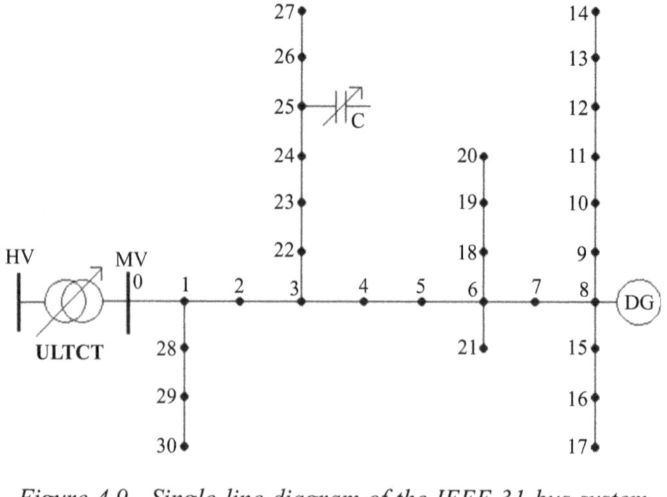


*Figure 4.9 Single-line diagram of the IEEE 31-bus system*

*Table 4.4 Data for IEEE 31-bus test system*

| Branch | $R$ (p.u.) | $X$ (p.u.) | $P_p$ (p.u.) | $Q_P$ (p.u.) |
|---|---|---|---|---|
| 0–1 | 0.000963 | 0.003219 | – | – |
| 1–2 | 0.000414 | 0.000022 | 0.522 | 0.174 |
| 2–3 | 0.000659 | 0.000651 | – | – |
| 3–4 | 0.002221 | 0.001931 | 0.936 | 0.312 |
| 4–5 | 0.001045 | 0.000908 | – | – |
| 5–6 | 0.003143 | 0.001770 | – | – |
| 6–7 | 0.002553 | 0.001438 | – | – |
| 7–8 | 0.002553 | 0.001438 | – | – |
| 8–9 | 0.002506 | 0.001412 | 0.189 | 0.063 |
| 9–10 | 0.002506 | 0.001412 | – | – |
| 10–11 | 0.007506 | 0.004229 | 0.336 | 0.112 |
| 11–12 | 0.003506 | 0.001975 | 0.657 | 0.219 |
| 12–13 | 0.001429 | 0.000805 | 0.783 | 0.261 |
| 13–14 | 0.002909 | 0.001639 | 0.729 | 0.243 |
| 8–15 | 0.000898 | 0.000781 | 0.477 | 0.159 |
| 15–16 | 0.001377 | 0.000775 | 0.549 | 0.183 |
| 16–17 | 0.002467 | 0.001390 | 0.477 | 0.159 |
| 6–18 | 0.000915 | 0.000795 | 0.432 | 0.144 |
| 18–19 | 0.003005 | 0.002612 | 0.672 | 0.224 |
| 19–20 | 0.002909 | 0.001639 | 0.495 | 0.165 |
| 6–21 | 0.001143 | 0.000994 | 0.207 | 0.069 |
| 3–22 | 0.001066 | 0.001054 | 0.522 | 0.174 |
| 22–23 | 0.000649 | 0.000641 | 1.917 | 0.639 |
| 23–24 | 0.001083 | 0.000941 | – | – |
| 24–25 | 0.002760 | 0.002399 | 1.116 | 0.372 |
| 25–26 | 0.002009 | 0.001746 | 0.549 | 0.183 |
| 26–27 | 0.002857 | 0.001609 | 0.792 | 0.264 |
| 1–28 | 0.000881 | 0.000047 | 0.882 | 0.294 |
| 28–29 | 0.003091 | 0.001741 | 0.882 | 0.294 |
| 29–30 | 0.002106 | 0.001187 | 0.882 | 0.294 |

*Note:* The parameter values in (p.u.) were calculated for a base power of 1 MVA and a base voltage of 23 kV.

Table 4.4 and can also be found in the supplementary material (MATLAB® files) for this chapter. The network is fed through the ULTCT that maintains the voltage at the bus 0 on the specified value of $V_0$. The DG connected at node 8 has a constant active power output of 5 p.u. and it operates in PV mode with reactive power in the range of −4 (p.u) to 4 (p.u.). The reactive power compensator connected to node 25 has the capability to generate reactive power in the range of 0–1.35 (p.u.). The limits of voltage $V_0$ at the secondary bus of ULTCT were set as follows: $0.97\,(\text{p.u}) \le V_0 \le 1.05\,(\text{p.u})$.

The task of optimal Volt/Var control in this case study is to determine the optimal voltage value of $V_0$, i.e. tap-changer position of the ULTCT, the optimal voltage at DG bus, i.e. the reactive power injection of the DG, and the optimum reactive power of the compensator. For solving this problem, it is assumed that all consumers in the DN follow a daily load diagram shown in Figure 4.10.

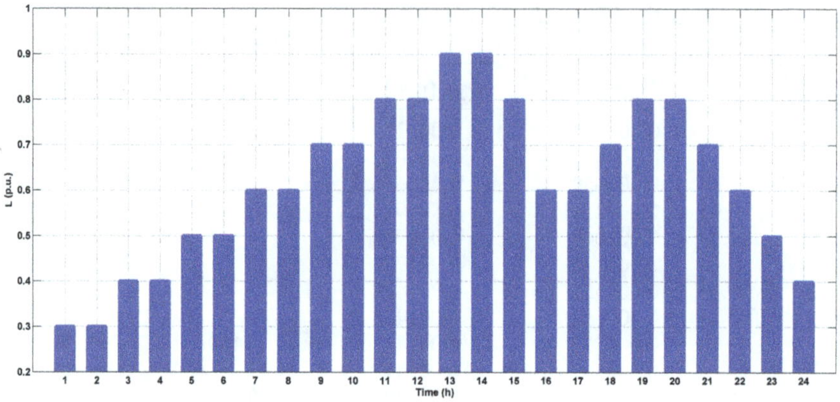

Figure 4.10 Daily load diagram

The optimal Volt/Var control problem was solved for three different objectives as follows:

*Case 1*: minimization of voltage deviation at load buses

$$\min_{\mathbf{x}} F(\mathbf{x},\mathbf{y}) = \min_{\mathbf{x}} VD(\mathbf{x},\mathbf{y}) = \min_{\mathbf{x}} \sum_{i=1}^{NL} \left| V_i - V_i^{ref} \right| \tag{4.33}$$

*Case 2*: minimization of total power loss in the network

$$\min_{\mathbf{x}} F(\mathbf{x},\mathbf{y}) = \min_{\mathbf{x}} Ploss(\mathbf{x},\mathbf{y}) = \min_{\mathbf{x}} \sum_{i=1}^{NB} R_i \cdot J_i^2 \tag{4.34}$$

*Case 3*: simultaneous minimization of the voltage deviation and power loss

$$\min_{\mathbf{x}} F(\mathbf{x},\mathbf{y}) = \min_{\mathbf{x}} \{ w_V VD(\mathbf{x},\mathbf{y}) + w_P Ploss(\mathbf{x},\mathbf{y}) \} \tag{4.35}$$

where $w_V$ and $w_P$ are the weighting factors.

The results shown were obtained by using the *opfdm* program package described in Section 2.7. This software package is also given as a supplementary file to this chapter. Figure 4.11 shows an example program for Case 1 (minimization of voltage deviation at load buses), with load level of L = 0.9 p.u., and PSO as the solution method.

Complete results are given in Figures 4.12–4.15 in the form of daily diagrams of optimal values of control variables.

When scenario 1 is applied, there are high values of the reactive power of the compensator ($Q_C$) and a wide range of reactive power changes of DG ($Q_{DG}$), which are in function of maintaining the voltages at buses 25 and 8 on specified values to minimizing the voltage deviation in the rest of network. Voltage control is directly related to the compensation of reactive power. In Case 2, high values for $V_0$ and

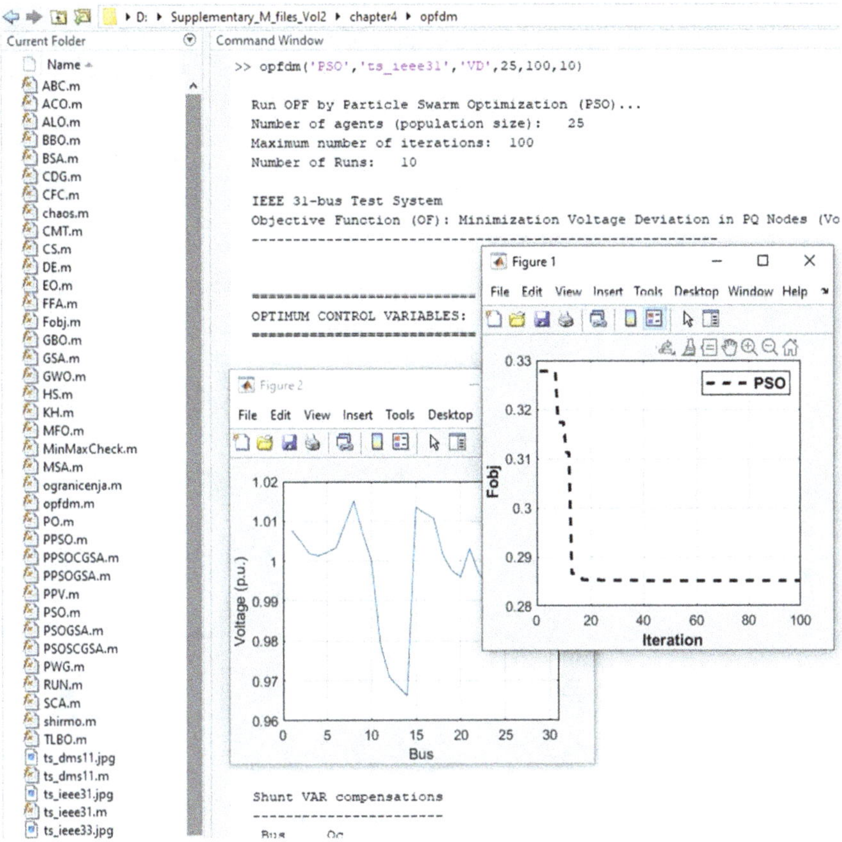


Figure 4.11 Optimal volt/var using the opfdm program

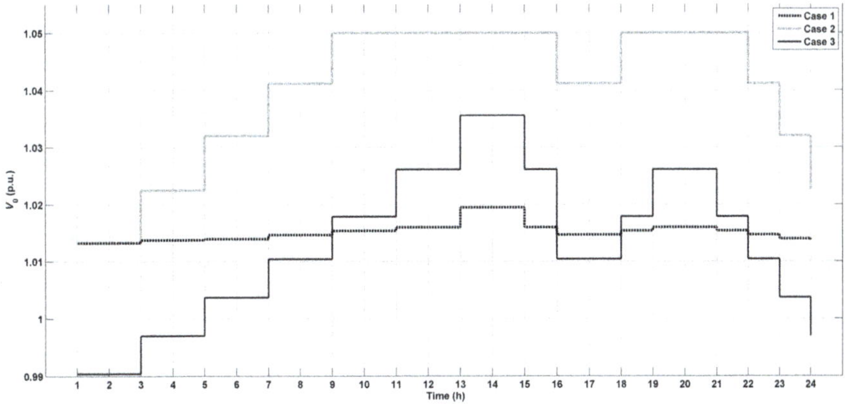


Figure 4.12 Optimal voltage at the root bus, $V_0$

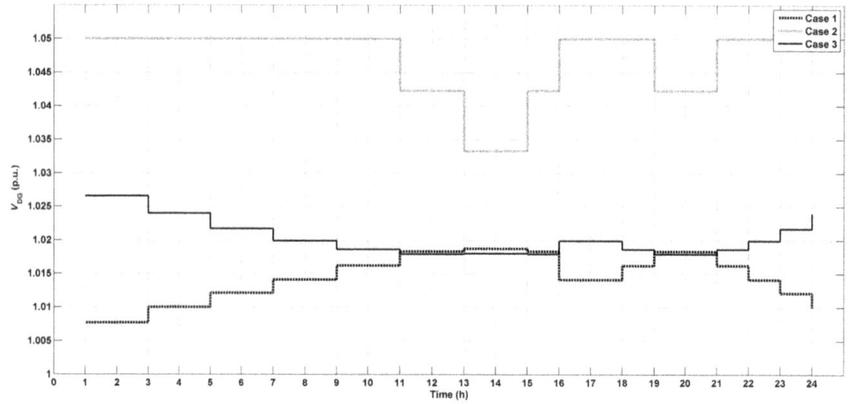


*Figure 4.13 Optimal voltage at the DG bus, $V_{DG}$*

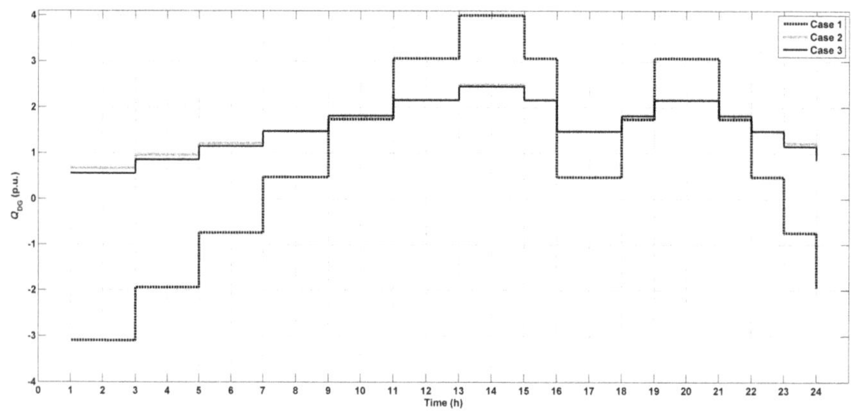


*Figure 4.14 Reactive power output of the DG, $Q_{DG}$*

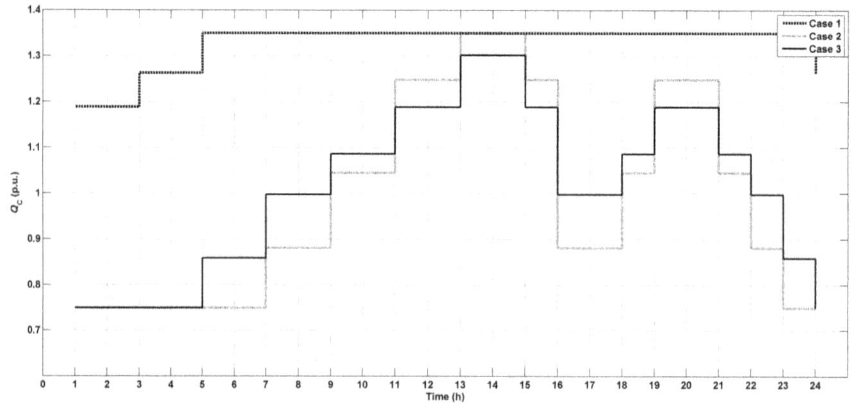


*Figure 4.15 Optimal reactive power of the compensator at bus 25, $Q_C$*

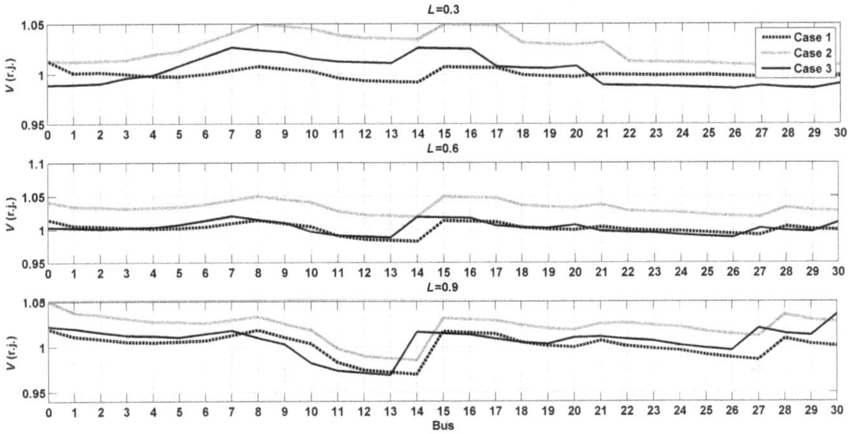


*Figure 4.16 Voltage profile for different optimization cases and load levels*

$V_{DG}$ are obtained throughout the load range. This is because the loads were modeled with constant active and reactive power (PQ model). Since power losses are proportional to the square of the branch current, for constant load power minimizing power losses is achieved by increasing the bus voltage.

Figure 4.16 shows the voltage profiles of the network for three load levels ($L = 0.3$, $L = 0.6$, and $L = 0.9$) for all three cases. As can be seen from Figure 4.16, Case 3 can be considered as a compromise solution, as it also reduces power losses and voltage deviations at the same time.

## 4.4 Conclusion

This chapter deals with two methodologies for optimal voltage control: (i) the methodology for optimal seasonal voltage control based on the space and time decomposition of the voltage control problem, and (ii) the methodology for optimal Volt/Var control in DNs based on metaheuristic optimization methods.

The main characteristics of the procedure for determining the optimal seasonal voltage control plan in rural DNs with DGs are

- Spatial decomposition principle of the voltage control problem;
- Bringing down the LV network to the equivalent line with the aim of applying the optimal setting of the transfer ratio of MV/LV distribution transformer;
- Additional improvement of the network voltage profile by correcting the voltage at PV buses.

The case studies have shown the practical application of the described procedure in the DNs with a larger number of DGs modeled as PQ and PV buses.

The application of the procedure on a real 33-node DN with DGs modeled as PQ nodes has shown the following:

- The outstanding variation of the generated power during the year and a significant space dispersion of the MHPs caused the voltage variation out of the standard limits.
- The optimal model of the seasonal voltage control at transformers enables the control of the voltage values within the standard limits, except for some outlying nodes.

Simulation results presented in the second part of this chapter show that the metaheuristic optimization methods are very suitable for solving the problem of optimal Volt/Var control in active DNs.

## References

[1] Strezoski V.S., Katic N.A., and Janjic D.S. 'Voltage control integrated in distribution management system'. *Electric Power Systems Research*, 2001; **60**:85–97.

[2] Švenda G., Simendić Z., and Strezocki V. 'Advanced voltage control integrated in DMS'. *Electrical Power and Energy Systems*. 2012;**43**:333–43.

[3] Choi J.H., and Kim J.C. 'The online voltage control of ULTC transformer for distribution voltage regulation'. *Electrical Power and Energy Systems*. 2001; **23**:91–98.

[4] Saied M.M. 'The global voltage regulation: a suggested measure for the supply quality in distribution networks'. *Electrical Power and Energy Systems*. 2001;**23**:427–34.

[5] Choi J.H., and Moon S.I. 'The dead band control of LTC transformer at distribution substation'. *IEEE Transactions on Power Systems*. 2009;**24**(1): 319–26.

[6] Liu Y., Zhang P., and Qiu X. 'Optimal Volt/Var control in distribution systems.' *Electrical Power and Energy Systems*. 2002;**24**:271–76.

[7] Hu. Z., Wang X., Chen H., and Taylor G.A. 'Volt/Var control in distribution systems using a time-unterval based approach'. *IEE Proceedings – Generation, Transmission and Distribution*. 2003;**155**(5): 548–54.

[8] Kim G.W., and Lee K.Y. 'Coordination control of ULTC transformer and STATCOM based on an artificial neural network'. *IEEE Transactions on Power Systems*. 2005;**20**(2):580–86.

[9] Park J.Y., Nam S.R., and Park J.K. 'Control of a ULTC considering the dispatch schedule of capacitors in a distribution system'. *IEEE Transactions on Power Systems*. 2007;**22**(2):755–61.

[10] Liang R.H., Chen Y.K., and Chen Y.T. 'Volt/Var control in a distribution system by a fuzzy optimization approach'. *Electrical Power and Energy Systems*. 2011;**33**:278–87.

[11] Mokgonyana L., Zhang J., Zhang L., and Xia X. 'Coordinated two-stage volt/var management in distribution networks'. *Electric Power System Research*. 2016;**141**:157–64.

[12] Kojovic L. 'Impact of DG on voltage regulation'. *Proceedings of IEEE Power Engineering Society Summer Meeting*; Chicago, USA, July 2002. IEEE; 2005. pp. 97–102.

[13] Kashem M.A., and Ledwich G.L. 'Multiple distributed generators for distribution feeder voltage support'. *IEEE Transactions on Energy Conversion*. 2005;**20**(3):676–84.

[14] Mollen M.H.J., and Sannino A. 'Voltage control with inverter-based distributed generation'. *IEEE Transactions on Power Delivery*. 2005;**20**(1): 519–20.

[15] Cheng C.S., and Shirmohammadi D. 'A three-phase power flow method for real-time distribution system analysis'. *IEEE Transactions on Power Systems*. 1995;**10**: 671–79.

[16] Kim T.E., and Kim J.E. 'Voltage regulation coordination of distributed generation system in distribution system'. *Proceedings of Power Engineering Society Summer Meeting*; Vancouver, Canada, July 2001. IEEE; 2002. pp. 480–84.

[17] Robbins B.A., and Dominguez-Garcia A.D. 'Optimal reactive power dispatch for voltage regulation in unbalanced distribution systems'. *IEEE Transactions on Power Systems*. 2016;**31**(4): 2903–13.

[18] Vaiwan F.A., Sannino A., and Daalder J. 'Voltage control with on-load tap changers in medium voltage feeders in presence of distributed generation'. *Electric Power System Research*. 2007;**77**:1314–22.

[19] Castro J.R., Saad M., Lefebvre S., Asber D., and Lenoir L. 'Optimal voltage control in distribution network in the presence of DGs'. *Electrical Power and Energy Systems*. 2016;**78**:239–47.

[20] Niknam T., Ranjbar A.M., and Shirani A.R. 'Impact of distributed generation on Volt/Var control in distribution networks'. *Proceedings of IEEE Bologna PowerTech Conference*; Bologna, Italy, June 2003. IEEE; 2003. pp. 1–7.

[21] Vaiwan F.A., and Karlsson D. 'Combined local and remote voltage and reactive power control in the presence of induction machine distributed generation'. *IEEE Transactions on Power Systems*. 2007;**22**(4):2003–12.

[22] Hong Y.Y., and Luo Y.F. 'Optimal VAR control considering wind farms using probabilistic load-flow and Gray-based genetic algorithm'. *IEEE Transactions on Power Delivery*. 2009;**24**(3):1441–49.

[23] Liu M.B., Canizares C.A., and Huang W. 'Reactive power and voltage control in distribution systems with limited switching operations'. *IEEE Transactions on Power Systems*. 2009;**24**(2):889–99.

[24] Mohapatra A., Bijwe P.R., and Panigrahi B.K. 'An efficient hybrid approach for Volt/Var control in distribution systems'. *IEEE Transactions on Power Delivery*. 2014;**29**(4):1780–88.

[25] Kolenc M., Papič I., and Blažič B. 'Coordinated reactive power control to achieve minimal operating costs'. *Electrical Power and Energy Systems*. 2014;**63**:1000–7.

[26] Wang S., Wang N., Wang Q., Yang Y., and Huang G. 'Reactive power optimization in distribution network by using improved gravitational search algorithm'. *Proceedings of 5th International Conference on Electric Utility Deregulation and Restructuring and Power Technologies*; Changsha, China, Nov 2015. IEEE; 2015. pp. 178–82.

[27] Madureira A.G., and Lopes J.A.P. 'Coordinated voltage support in distribution networks with distributed generation and microgrids'. *IET Renewable Power Generation*. 2009;**3**(4):439–54.

[28] Senjyu T., Miyazato Y., Yona A., Urasaki N., and Funabashi T. 'Optimal distribution voltage control and coordination with distributed generation'. *IEEE Transactions on Power Delivery*. 2008;**23**(2):1236–42.

[29] Villacci D., Bontempi G., and Vaccaro A. 'An adaptive local learning-based methodology for voltage regulation in distribution networks with dispersed generation'. *IEEE Transactions on Power Systems*. 2006;**21**(3): 1131–40.

[30] Galdi V., Vaccaro A., and Villacci D. 'Voltage regulation in MV networks with dispersed generations by a neural-based multiobjective methodology'. *Electric Power System Research*. 2008;**78**:785–93.

[31] Niknam T., Firouzi B.B., and Ostadi A. 'A new fuzzy adaptive particle swarm optimization for daily Volt/Var control in distribution networks considering distributed generators'. *Applied Energy*. 2010;**87**:1919–28.

[32] Cheng S., and Chen M.Y. 'Multi-objective reactive power optimization strategy for distribution system with penetration of distributed generation'. *Electrical Power and Energy Systems*. 2014;**62**:221–28.

[33] Chen S., Hu W., Su C., Zhang X., and Chen Z. 'Optimal reactive power and voltage control in distribution networks with distributed generators by fuzzy adaptive hybrid particle swarm optimization method'. *IET Generation, Transmission & Distribution*. 2015;**9**(11)1096–1103.

[34] Daratha N., Das B., and Sharma J. 'Robust voltage regulation in unbalanced radial distribution system under uncertainty of distributed generation and loads'. *Electrical Power and Energy Systems*. 2015;**73**:516–27.

[35] Zhang L., Tang W., Liang J., Cong P., and Cai Y. 'Coordinated day-ahead reactive power dispatch in distribution network based on real power forecast errors'. *IEEE Transactions on Power Systems*. 2016;**31**(3):2472–80.

[36] Ulinuha A., Masoum M.A.S., and Islam S. 'Hybrid genetic-fuzzy algorithm for volt/var/total harmonic distortion control of distribution systems with high penetration of non-linear loads'. *IET Generation, Transmission & Distribution*. 2011;**5**(4)425–439.

[37] Jashfar S., and Esmaeili S. 'Volt/var/THD control in distribution networks considering reactive power capability of solar energy conversion'. *Electrical Power and Energy Systems*. 2014;**60**:221–33.

[38] Jashfar S., Hosseini-Biyouki M.M., and Esmaeili S. 'A stochastic programming to Volt/VAR/total harmonic distortion control in distribution networks including wind turbines'. *Electric Power Components and Systems*. 2015; **43**(7):733–46.

[39] Seeling B.H.T., de Souza L.J., and Prada R.B. 'Voltage control in heavily loaded networks'. *International Transactions on Electrical Energy Systems*. 2015;**25**:3275–87.

[40] Radosavljević J., Jevtić M., and Klimenta D. 'Optimal seasonal voltage control in rural distribution networks with distributed generators'. *Journal of Electrical Engineering*. 2010;**61**(6):321–31.

[41] Radosavljević J., Jevtić M., and Klimenta D. 'The analysis of a rural distribution network with distributed generation in catchment area of Stara Planina'. *Facta Universitatis, Series: Electronics and Energetics*. 2009; **22**(1):71–89.

# Chapter 5
# Voltage regulation in LV DNs with PV generation and BESS

## 5.1 Introduction

Modern distribution networks (DNs) are evolving into smart grids that integrate distributed generation (DG), nonlinear loads, information and communication technologies, and power electronics technology. A large share of renewable energy-based DG sources can cause deterioration of the power quality. Improvement of the power quality in various aspects becomes one of the most important tasks of electricity distribution companies to increase competitiveness in the liberalized electricity market.

The most important parameter of power quality is the voltage magnitude. The lower and upper voltage magnitude limits in DNs are defined by appropriate technical standards, such as the European voltage disturbances standard E 50160 [1]. Accordingly, the voltage limit for low-voltage (LV) DNs in the EU is 400 V $\pm$ 10%, whereas in the United Kingdom, the limit is 400 V + 10%,−6%.

One of the key actions of a distribution system operator (DSO) is voltage regulation, which consists of a set of control actions to ensure that the voltage magnitudes at all buses of the DN are within the permissible limits.

As well known, the integration of DG units such as inverter-based photovoltaic (PV) systems into the LV DNs changes the nature of these networks from passive (unidirectional power flow) to active DNs with bidirectional power flow. Depending on the size and location of the PV units, they could positively impact the network and consumers such as improving supply reliability, improving voltage profile, and minimizing power losses [2]. However, if PV generation significantly exceeds local load demand (e.g. during a few hours at midday), it can lead to unacceptable voltage rise at load buses. The stochastic nature of PV generation as well as load demand may cause significant voltage fluctuations during the day. In addition, voltage rise exceeding the G83/1 limits may cause an unplanned switch-off of the PV system from the network [3]. Therefore, the voltage regulation of LV DNs with integrated PV systems is a very important issue that cannot be solved in a traditional way, i.e. using seasonal voltage control for passive DNs (using the MV/LV off-load tap-changing transformers and shunt capacitor banks) [4].

In recent years, the authors have proposed the application of various equipment based on power electronic converters for voltage regulation in LV networks, as well as appropriate algorithms for their control and coordination. Four different solutions based on using power electronic converters for voltage regulation in LV networks are presented in [3]. The performances of these power electronic solutions, namely power electronic substation (PES), online tap changer (OLTC), active power filter (APF), and mid-feeder compensator (MFC), were evaluated and compared to a representative LV test network. The authors in [5] proposed a compact power electronic converter that can compensate for an under-voltage or over-voltage condition on an LV feeder. It is considered a power electronic solution to the problem of mid-feeder regulation in LV networks. Power electronic compensator (PEC) in output voltage control mode was considered in [6] for voltage regulation in LV networks with high penetration of PV generation and electric vehicle charging stations. The authors proposed the placement of the PEC between the feeder and each customer (point-of-load compensation, PoLC) and compared this solution with the already proposed solution where the PEC is placed somewhere on the feeder (MFC).

Many authors in their research have focused on the development of efficient methods for optimal coordination of regulation resources in LV DNs. Some of these methods just aim to maintain the voltage within the permissible limits. Another group of approaches are methods for optimal volt/var regulation aiming to minimize voltage deviation and/or power loss in the networks. Different control schemes proposed for voltage regulation in DNs can be classified into four main groups: local control, centralized control, distributed control, and decentralized control. References [7–10] give a comprehensive overview of these approaches.

Conti *et al.* [11] proposed a local control strategy based on PV generation curtailment to prevent overvoltage at the point of common coupling. The active power produced by the PV generators is adjusted according to the local bus voltage. In [12], a distributed control scheme was proposed to solve overvoltage problems in LV DNs due to high PV generation. The control scheme is based on adjusting the active and reactive power output of PV inverters to prioritize the use of reactive power and minimize the active power curtailment. Battery energy storage systems (BESSs) have been proposed in [13,14] to solve the voltage deviation problems in LV DNs with high penetration of rooftop PVs. The authors considered the distributed control based on a consensus algorithm for bus voltage regulation, whereas a local control scheme was employed to regulate the state of charge (SoC) of each BESS within the desired SoC range. A similar approach was proposed in [15], where a coordinated control strategy is a combination of the local droop-based control method and a distributed control scheme based on consensus algorithms, each of which has specific objectives in regulating the charge/discharge of BESS and ensuring that bus voltages remain within specified limits. Also, a distributed control strategy based on two consensus algorithms is proposed in [16] to achieve cooperative control of plug-in electric vehicles and the active power curtailment of PV generators for voltage regulation in LV DNs.

The authors in [17] consider voltage regulation in LV DNs with PV units using OLTC-fitted transformers. For this purpose, they propose a practical remote voltage estimation method for the end nodes of LV feeders without remote monitoring. Kostolas *et al.* [18] proposed a control framework for optimal coordination of OLTC distribution transformers, battery energy storage systems, and distributed energy resources, including rooftop PV installations, to assure admissible voltage magnitudes in LV DNs. The methodology comes down to a multi-objective multi-period optimal power flow solved by the interior-point algorithm. A centralized approach for volt/var control in DN with PV generation is presented in [19]. This approach uses particle swarm optimization (PSO) for optimal coordination of step voltage regulators and distribution of static synchronous compensators with objectives to minimize the active power loss while the voltages remain within the allowable limits. In [20], optimal reactive power management and active power curtailment of PV inverters have been defined as a multi-objective optimal power flow problem with objectives to simultaneously improve voltage magnitude and minimize power loss and solved using PSO.

Garcia *et al.* [21] analyzed a combined centralized/decentralized voltage regulation method consisting of coordination of control actions performed by the DSO with local control at the PV inverter level. Using this strategy, the authors determine reactive power references for PV inverters integrated into an LV DN. To mitigate the voltage rise issues in LV networks with PV generation, the authors in [22] proposed a coordinated volt/var control scheme between the LV distribution transformer and PV inverters to optimize the PV generation level, considering two standard concepts for reactive power control, $\cos\varphi(P)$ and $Q(U)$. A multi-objective genetic algorithm solved this optimization problem. The problem of coordinating the operation of PV inverters, BESS, and tap-changers has been considered in [23] as a multi-objective OPF problem. The optimization problem is then solved by applying the epsilon-constrained technique to simultaneously improve the voltage profile, minimize power loss, and reduce peak load power.

As noted above, the traditional voltage regulation in LV networks using an off-load tap changing transformer is very limited. A potential solution for LV DNs is a hybrid distribution transformer (HDT). This is a concept that represents a combination of a classic distribution transformer and a fractionally rated power electronic converter [24]. The HDT can provide additional control capabilities such as dynamic voltage regulation and reactive power compensation. Therefore, the HDT can be used to continually adjust the voltage magnitude on an LV feeder. On the other hand, it has been shown [13–16] that in LV networks with a large share of PV sources, the impermissible voltage deviation can be solved by using distributed BESS to balance power injection at the buses where PV sources are connected. It seems that the combination of these two approaches can give the best results in voltage regulation of LV networks with a large share of PV generators.

This chapter discusses the voltage regulation problem in LV networks equipped with the HDT and with high penetration of rooftop PV sources as proposed in [25]. A two-stage algorithm for voltage regulation is presented. In the first stage, a local (distributed) voltage control is performed by minimizing the injection power of the PV-BESS-load system at their common connection point. In this way, the power flow between the nodes of the DN is minimized, and thus, the voltage drops are minimized, which implies less deviation of bus voltages from nominal values, i.e. tends to keep the bus voltages within permissible limits. In the second stage, which can be activated if the first stage fails to bring the voltages within permissible limits, optimal coordination is performed between the HDT and the local voltage control. In fact, the second stage is an optimal voltage regulation problem. The aim is to minimize the voltage deviations at load buses by optimal setting the voltage support of the HDT. To solve this optimization problem, a PSO algorithm has been proposed.

## 5.2 Local voltage control

Well-known characteristics of LV DNs are radial topology, high *R/X* ratio, and negligible shunt admittance. Accordingly, a part of LV DN can be presented as in Figure 5.1. A PV source, a BS, and a local AC load are connected at bus *i*.

The voltage drop $\Delta V_i$ along the line between two adjusted buses (*j* and *i*) can be expressed as follows:

$$\Delta \underline{V}_i = \frac{R_i \cdot P_i + X_i \cdot Q_i}{\underline{V}_i^*} + j\frac{X_i \cdot P_i - R_i \cdot Q_i}{\underline{V}_i^*} \tag{5.1}$$

As noted above, LV DNs have a high *R/X* ratio. In addition, the power factor in LV networks is high, typically 0.95, which means that active power flow ($P_i$) dominates over reactive power flow ($Q_i$). Accordingly, the imaginary part of (5.1) can be neglected. Assuming the DG inverter operates with a unity power factor ($Q_{DG} = 0$), the voltage drop for the LV network can be expressed

$$\Delta V_i = \frac{R_i \cdot (P_{Li} - P_{DGi} + \sum_{k \in \alpha_{ki}} P_k) + X_i \cdot (Q_{Li} + \sum_{k \in \alpha_{ki}} Q_k)}{V_i} \tag{5.2}$$

where $P_{Li}$ and $Q_{Li}$ are the active and reactive power of load at the *i*th bus, $P_{DGi}$ is the active power of DG consisting of PV and BS at the *i*th bus, and $P_k$ and $Q_k$ are the active and reactive power in branch *k* emanating from bus *i*.

According to (5.2), the voltage drop can be controlled by controlling the injection/absorption of active power at the *i*th bus:

$$P_{inj/abs} = P_{Li} - P_{DGi} = P_{Li} - (P_{PVi} \pm P_{BESSi}) \tag{5.3}$$

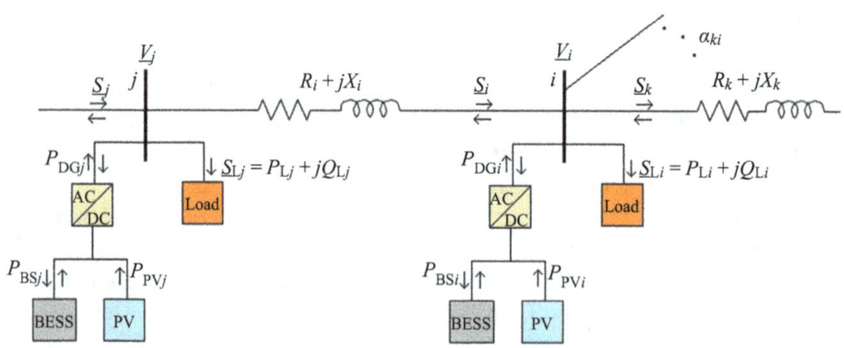


*Figure 5.1 A part of LV distribution network with PV and BS*

The active power of DG, $P_{DGi}$, depends on PV generation and the control strategy of BESS charging/discharging. It is clear that the appropriate control of the $P_{BESS}$ can enable the minimization of active power injection at the *i*th bus, and consequently, less voltage drop $\Delta V_i$, which means less voltage deviation at the *i*th bus in relation to the nominal value.

The aim of the local control strategy is to reduce the power imbalance between PV generation ($P_{PVi}$) and load demand ($P_{Li}$) to overcome the impermissible voltage deviation. Whereas the BESSs function is to reduce the bus power injections during the high PV generation and reduce the power absorption from the network during the night [13]. The goal is to the energy that is produced locally to be consumed locally, that is to minimizing energy exchange between the customer (an entity consisting of PV-BESS and Load) and the DN.

The proposed strategy of local voltage regulation for the customer on the *i*th bus is shown in the form of a functional scheme in Figure 5.2. The solid lines show the power flow, whereas the dotted lines denote the communication among equipment. The PV panel is connected to the DC link through the boost converter controlled by the MPPT algorithm to achieve the maximum power generation under different conditions of solar irradiance and ambient temperature. In this work, a residential Vanadium redox battery (VRB) model is adopted for energy storage purposes within the BS. The parameters for the VRB are shown in Table 5.1 [14]. The BESS is connected to the DC link through the bidirectional converter to enable the charging/discharging process. The active power is transferred between the DC link, the AC load, and the DN through the DC/AC inverter. The inverter can enable a reverse process, i.e. transferring the active power from the DN to the BS.

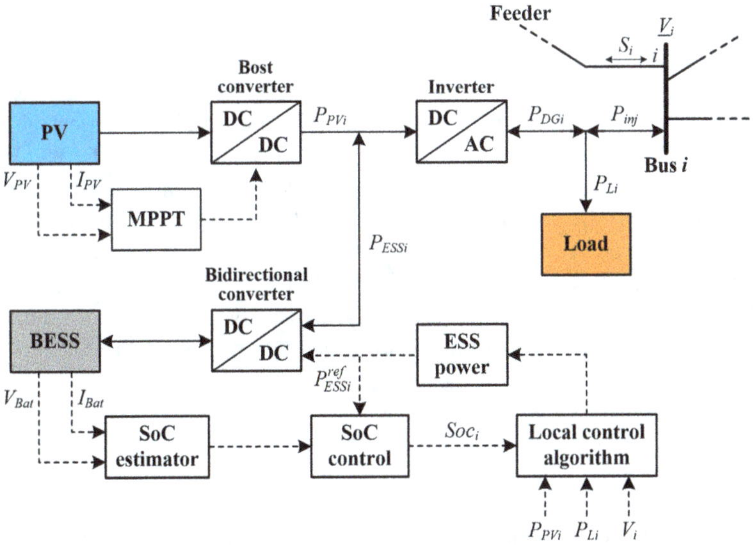


*Figure 5.2 Schematic diagram of the proposed strategy for local voltage control*

*Table 5.1 Parameters of VRB*

| | |
|---|---|
| Configuration | 38 series cell stack |
| Rated power, $P^{rated}$ | 5 kW |
| Rated capacity, $C_B$ | 20 kWh |
| Operation voltage | 42–57 V |
| SoC region ($SoC^{min}$, $SoC^{max}$) | 15%, 85% |

High PV generation in periods with low power of load can cause overvoltage at the *i*th bus due to the high value of power injection ($P_{inj}$). Active power injection from PV can be limited by storing the extra power in the BESS. In contrast, in periods with low (or zero) PV generation and high load, when the voltage at the *i*th bus can drop below the lower limit, the stored energy in BESS can be used for the local load and thus reduce the voltage drop in the network.

The aim of the local voltage regulation algorithm is to control the process of charging/discharging of the BESS under various operating conditions, such as the periods of peak PV generation coinciding with low load periods and vice versa. This is a way to ensure that the power injection into the *i*th bus is minimal, and thus, the voltage deviation at the *i*th bus is minimal.

As explained in [14], the *SoC* of the BESS is calculated as follows:

$$SoC_t = SoC_{t-\Delta t} + \Delta SoC \tag{5.4}$$

$$\Delta SoC = \frac{P_{BESS} \cdot \Delta t}{C_B} = \frac{V_{bat} \cdot I_{bat} \cdot \Delta t}{C_B} \tag{5.5}$$

where $C_B$ is the total energy capacity of the VRB, $\Delta t$ is the time interval, and $V_{bat}$ and $I_{bat}$ are the output voltage and current of the VRB, respectively.

Unlike the distributed control based on the consensus algorithm [13–16], there is no need for communication between neighboring customers. Local voltage regulation is performed at each individual customer based on local information on the current power of the PV source, consumption, voltage, and condition of the BESS. The proposed local voltage regulation algorithm provides a determination of the reference power of the BS ($P^{ref}_{BESS}$) at the time *t* based on the estimated *SoC* of the BESS, PV generation ($P_{PV}$), power of load ($P_L$), and voltage magnitude ($V$) at the time ($t - \Delta t$), evaluated through the communication system. As noted above, $\Delta t$ is the time interval adopted for the time-series power flow analysis (usually $\Delta t$ =1, 5, or 10 min).

The efficient operation of the BESS means that its power and *SoC* are within defined limits:

$$-P^{rated}_{BESS} \leq P_{BESS,t} \leq P^{rated}_{BESS} \tag{5.6}$$

$$SoC^{\min} \leq SoC_t \leq SoC^{\max} \tag{5.7}$$

To take into account the value of the voltage $V_{i,t}$ in determining the $P^{ref}_{BESS,t}$, a control signal $e_t$ is introduced. The purpose of this control signal is that with a larger voltage deviation, the charging/discharging rate of the battery will be higher in relation to the period when the voltage is very close to the nominal value. Therefore, the power of the BESS should be maximum when the voltage is outside the allowable limits. The control signal $e_t$ is defined as follows:

$$e_t = \begin{cases} 0 & if \quad |1 - V_{i,t}| < V_{th\ \min} \\ 1 & if \quad |1 - V_{i,t}| > V_{th\ \max} \\ 1 - k_v(V_{th\ \max} - |1 - V_{i,t}|) & if \quad V_{th\ \min} \leq |1 - V_{i,t}| \leq V_{th\ \max} \end{cases} \tag{5.8}$$

where $V_{th\ \min}$ and $V_{th\ \max}$ are the minimum and maximum thresholds. In this work, $V_{th\ \min}$ is adopted as 0.01 p.u., whereas $V_{th\ \max}$ can be considered as max permissible voltage drop (we adopted $V_{th\ \max} = 0.05$ p.u.).

The current state of the $i$th customer ($P_{PVi}$, $P_{Li}$, $SoC_i$, and $V_i$) is evaluated by the local communication system. Considering the current state and the defined power limits and $SoC$ limits, the required output power of the BESS is determined by the control algorithm, which can be explained by the flow chart in Figure 5.3.

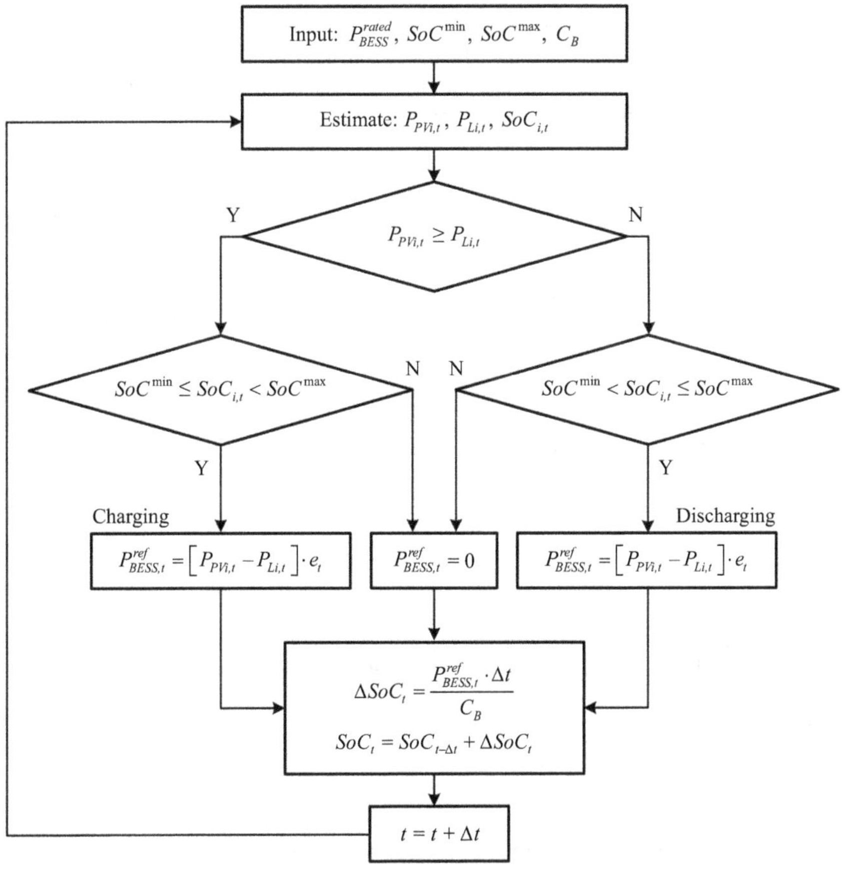


*Figure 5.3 Flow chart of the local voltage control*

## 5.3 Optimal voltage regulation

### 5.3.1 Hybrid distribution transformer

The development of power electronics has enabled the development of various solutions to mitigate voltage problems in LV DNs with high penetration of stochastically DG such as rooftop PV systems. One such solution is the application of the concept of a hybrid distribution transformer (HDT) proposed in [24].

An HDT, as its name suggests, consists of a classic distribution transformer with built-in partial voltage electronic converter on the secondary side, as illustrated in Figure 5.4. The partial voltage electronic converter allows the HDT to have continuous voltage regulation in a defined range.

High efficiency at lower cost and complexity are the main advantages of the proposed HDT concept compared to power electronic transformers [3,24].

### 5.3.2 Problem formulation

In this work, optimal voltage regulation implies optimal coordination between the HDT and the local voltage control using PSO, as illustrated in Figure 5.4.

The objective function of optimal voltage regulation is the minimization of voltage deviations at load buses via optimal settings of the control variable – voltage support of the HDT, under various technical constraints. Generally, the optimal voltage regulation problem can be expressed by the following equations:

$$\min F(\mathbf{x}, \mathbf{y}) \text{ subject to : } \begin{cases} g(\mathbf{x}, \mathbf{y}) = 0 \\ h(\mathbf{x}, \mathbf{y}) \leq 0 \\ \mathbf{x} \in \mathbf{X} \end{cases} \tag{5.9}$$

where $F$ is the objective function to be minimized and $\mathbf{x}$ and $\mathbf{y}$ are vectors of control and dependent variables, respectively.

For DN under study, the vector of control variables ($\mathbf{x}$) consisting of root bus voltage $V_0$, that is voltage support of the HTD. Therefore, the vector of control variables can be expressed as

$$\mathbf{x} = [V_0] \tag{5.10}$$

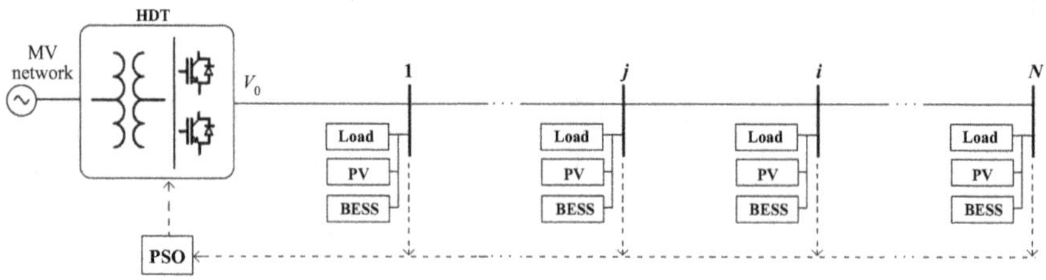


*Figure 5.4 Optimal voltage regulation in LV network*

The vector of dependent variables (**y**) consisting of the active power $P_{gr}$ that is delivered through the root bus of the DN; load bus voltage, $V_L$; power of the BESS, $P_{BESS}$; $SoC$ of the BESS; branch flow $S_l$. Therefore, **y** can be expressed as

$$\mathbf{y} = \left[P_{gr}, V_1 \ldots V_N, P_{BESS1}, \ldots, P_{BESS_N}, SoC_1, \ldots, SoC_N, S_{l1} \ldots S_{lN}\right] \quad (5.11)$$

The objective function can take different forms. However, in this study, the minimization of voltage deviation at load buses is considered objective function

$$F = \sum_{i=1}^{N} |1 - V_i| \quad (5.12)$$

where $V_i$ is the voltage magnitude at load bus $i$ and $N$ is the total number of buses in the DN.

The equality constraints in (5.9) represent power flow equations. The inequality constraints are the functional operating constraints:

$$V_i^{\min} \leq V_i \leq V_i^{\max} \quad i = 1, \ldots, N \quad (5.13)$$

$$-P_{BESSi}^{rated} \leq P_{BESSi} \leq P_{BESSi}^{rated} \quad i = 1, \ldots, N \quad (5.14)$$

$$SoC_i^{\min} \leq SoC_i \leq SoC_i^{\max} \quad i = 1, \ldots, N \quad (5.15)$$

$$S_i \leq S_i^{\max} \quad i = 1, \ldots, N \quad (5.16)$$

Constraint $\mathbf{x} \in \mathbf{X}$ defines the feasibility region of the problem control variable, i.e. root bus voltage limits:

$$V_0^{\min} \leq V_0 \leq V_0^{\max} \quad (5.17)$$

### *5.3.3 Solution method*

In this work, a well-known metaheuristic optimization method – particle swarm optimization (PSO) – is proposed to solve the problem of optimal voltage regulation. Detailed descriptions of the basic PSO algorithm and its improved, adaptive, and hybrid versions are given in the first part of the book (*Metaheuristic Optimization in Power Engineering: Vol 1: Algorithms and Power Dispatch – using MATLAB®-based software*, Chapter 3).

Figure 5.5 shows a flow chart of the proposed algorithm for optimal voltage regulation using PSO.

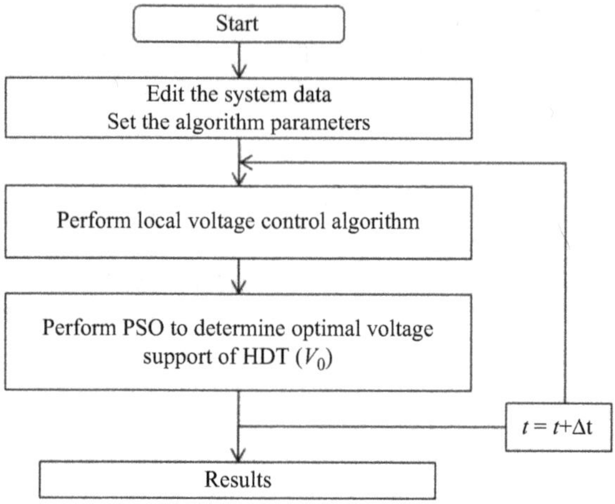


*Figure 5.5 Flow chart of the proposed algorithm*

## 5.4 Simulation results

Two test cases have been considered to evaluate the proposed voltage regulation approach. These test cases are as follows:

*Test case A*: Local voltage control only;
*Test case B*: Optimal voltage regulation (coordination of HDT and local voltage control).

The parameters of the local voltage control algorithm and parameters of PSO used for optimal voltage regulation are given in Tables 5.2 and 5.3, respectively.

To assess the effectiveness of the proposed approach for voltage regulation in large-scale realistic DNs, the IEEE European LV test feeder is considered, which is a typical topology for European DNs. It is a radial LV DN at the voltage level of 416 (phase-to-phase) connected to the MV network through a transformer MV/LV (11/0.4 kV/kV) at the substation. The test feeder has 906 buses. In total, 55 lateral buses are connected customers. Complete data on the IEEE European LV test feeder are given in [26]. In addition to network parameters, load profiles with a 1-minute time resolution over 24 h are provided for each of the 55 loads, i.e. for time-series simulation.

In this work, the original IEEE European LV test feeder is modified as follows:

1. The test feeder is considered symmetrical/balanced.
2. All loads are changed from single-phase to three-phase with a power factor of 0.95.
3. Load shapes are transformed from a 1-minute time resolution to a 5-minute time resolution (using 5-minute mean values), giving 288 load data points for 24 hours.

4. The rooftop PV system with the capacity of 4 kWp and the BESS based on VRB with data, given in Table 5.1, are integrated into each of 55 customers.
5. The output power profile shown in Figure 5.6 is considered identical for all PV systems.

*Table 5.2 Parameters of local voltage control*

| | |
|---|---|
| $SoC^{min}$ | 20% |
| $SoC^{max}$ | 80% |
| $V_{thsh\ \mathrm{min}}$ | 0.01 |
| $V_{thsh\ \mathrm{max}}$ | 0.06 |
| $k_v$ | 10 |

*Table 5.3 Parameters of PSO*

| | |
|---|---|
| $C_1$ | 2 |
| $C_2$ | 2 |
| $w_{\mathrm{min}}$ | 0.4 |
| $w_{\mathrm{max}}$ | 0.9 |
| $N$ | 10 |
| $t_{\mathrm{max}}$ | 20 |

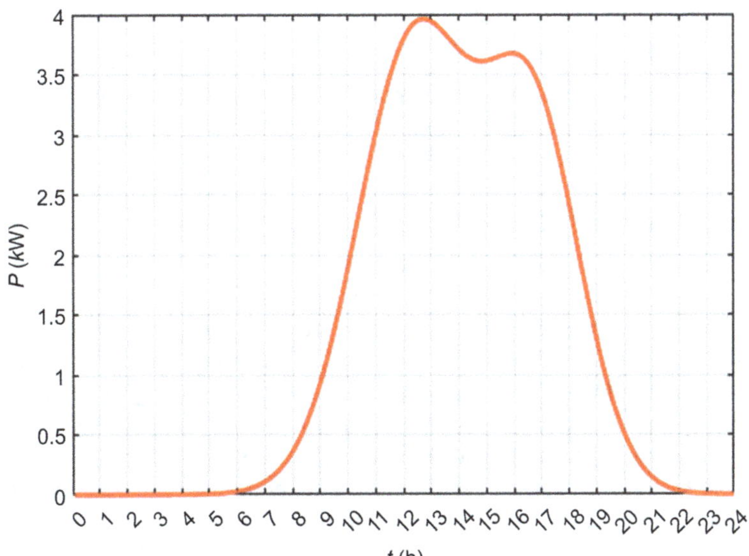


*Figure 5.6 PV generation profile*

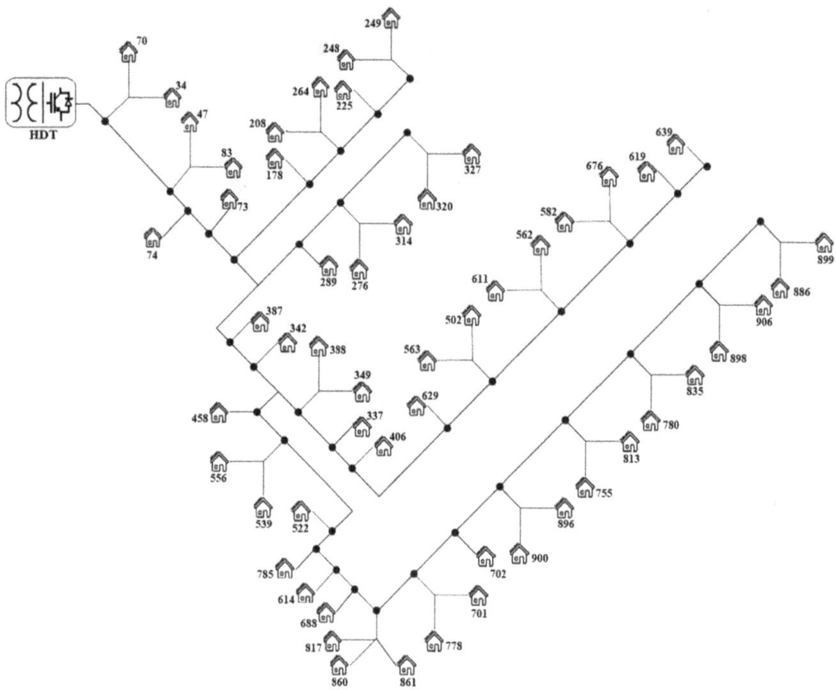


*Figure 5.7 Single-line diagram of the IEEE European LV test feeder [16]*

6. The MV/LV transformer at substation is consiederd as the HDT.
7. The maximum voltage deviation in the DN is considered as $\pm 5\%$.

The single-line diagram of the modified IEEE European LV test feeder is shown in Figure 5.7. The three typical load profiles shown in Figure 5.8 were distributed to the customers at random.

## *5.4.1 Test case A: local voltage control*

The proposed local voltage control algorithm is tested for an initial SoC value of 50% for each BESS and a constant voltage magnitude at the HDT of 1 p.u. Figures 5.9–5.11 show the voltage profiles at buses 562, 906, and 249, which belong to different laterals of the LV network. The voltages at buses 562 and 906 exceed the upper limit of 1.05 p.u. in the period with high PV generation (from 12 to 16 h). During the period of increased load in the morning and especially in the evening (from 6 pm to midnight), the voltage values at the considered buses are below the lower limit value of 0.95 p.u. By applying the proposed algorithm for local voltage control, these voltages are improved during the entire period of 24 h. However, during the peak load period, these voltages are still less than the allowable value of 0.95 p.u. In the period with high solar irradiation, the PV generator supplies the local load, and the excess power is used to charge the BESS. Then,

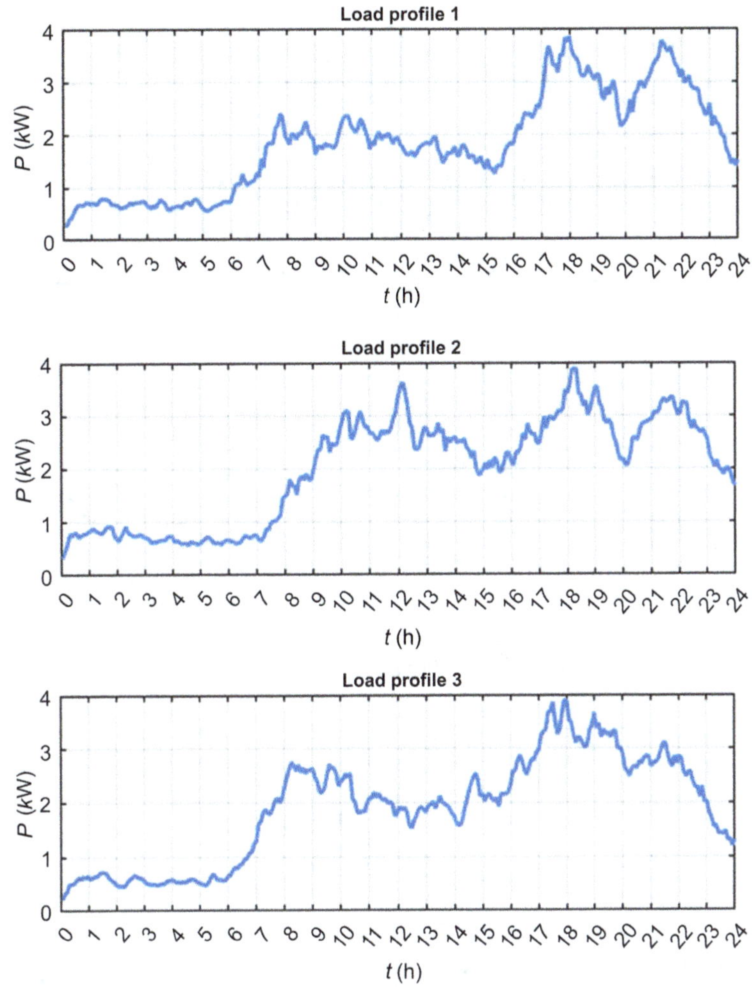


*Figure 5.8 Daily load profiles*

during the peak load period in the evening, the accumulated energy of the BESS is used for the supply of the local load, enabling some improvement of the voltage, but the available battery power is insufficient to cover the entire local load and provide complete voltage recovery. Figures 5.12 and 5.13 show the corresponding SoC profiles and the power outputs of the BESSs at buses 562, 906, and 249, resulting in the local voltage control. These figures clearly show that the battery charging and discharging modes, as well as their power outputs, are determined by the ratio of local loads and PV generation in accordance with the proposed local voltage control algorithm in Figure 5.3.

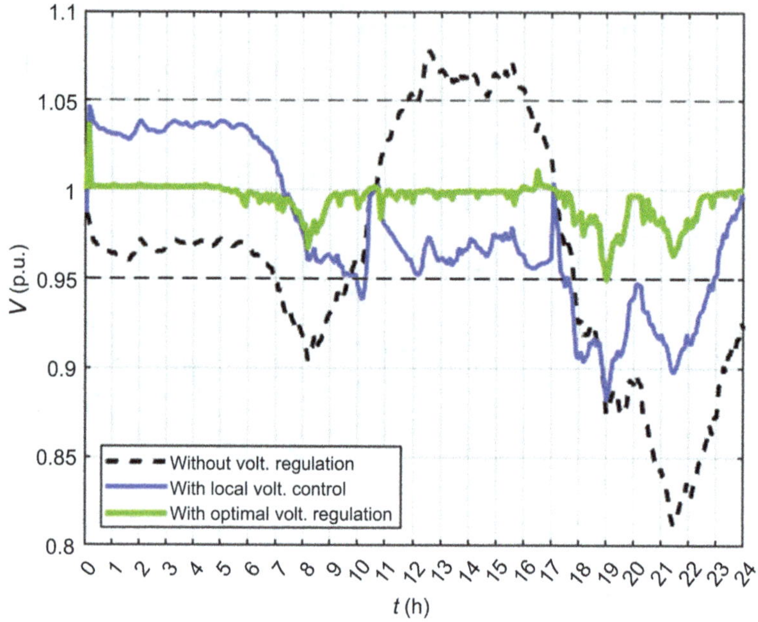


*Figure 5.9 Voltage profiles at bus 562*

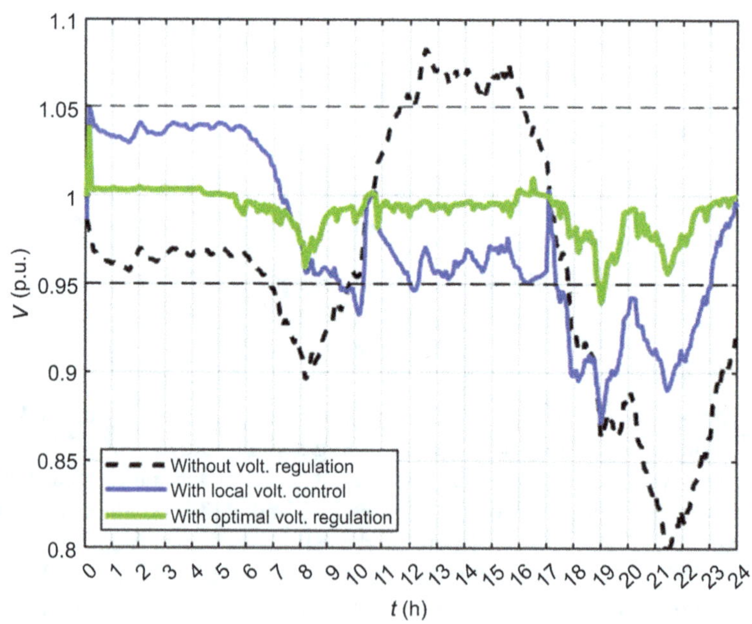


*Figure 5.10 Voltage profiles at bus 906*

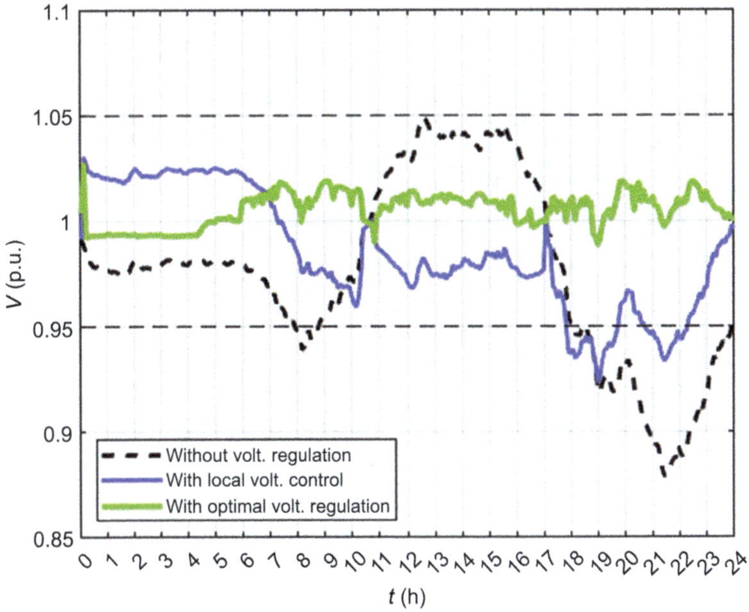


*Figure 5.11 Voltage profiles at bus 249*

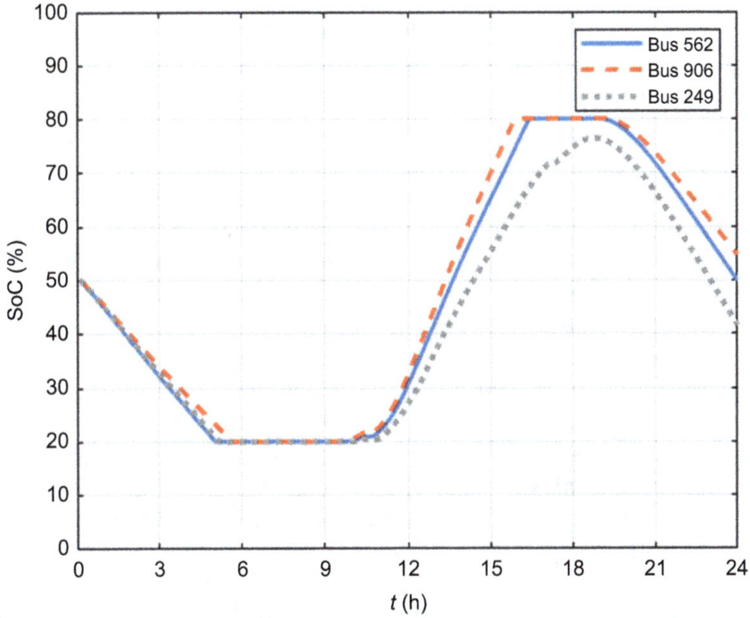


*Figure 5.12 SoC of BESS at different buses*

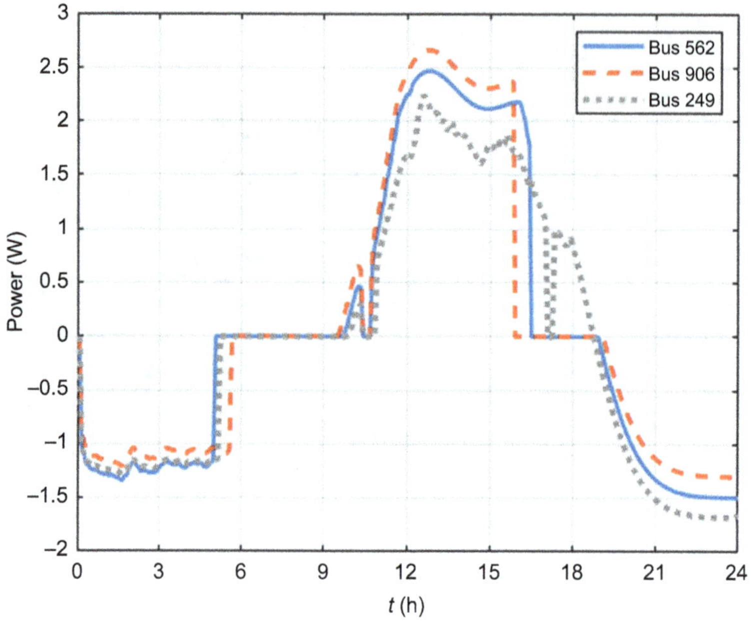


*Figure 5.13 Power output of BESS at different buses*

## 5.4.2 Test case B: optimal voltage regulation

It is supposed that the HDT may continuously control terminal voltage ($V_0$) in the range of 0.95–1.05 p.u. Figure 5.14 presents the optimal voltage support of the HDT for the IEEE European test feeder obtained by the proposed optimal voltage regulation algorithm.

Figures 5.9–5.11 compare the voltages at buses 562, 906, and 249 for the basic case-without regulation, Case A – local voltage control, and Case B – optimal voltage regulation. As can be seen, optimal voltage regulation maintains the voltage magnitude very close to the nominal value (1 p.u.). A very clear effect of optimal voltage regulation (Case B) on voltages in the whole DN can be seen in Figures 5.15 and 5.16, presenting the total voltage deviation at load buses and the statistical indicators of load bus voltages during the day, respectively. The box plots presented in Figure 5.16 cover all the load bus voltages over the day, giving a notion about statistical indicators, i.e. minimum, maximum, mean values, and the deviation from the mean values in different cases. The tops and bottoms of each "box" are the 25th and 75th percentiles of the samples, respectively. The line in the middle of each box is the sample median of all load bus voltages at time $t$. It is clear that the optimal coordination of the HDT and local voltage control enables minimum voltage deviation at load buses.

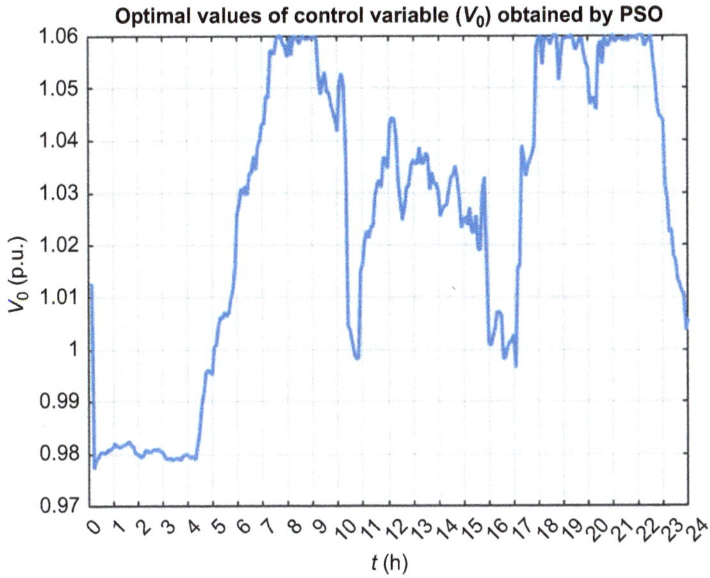


*Figure 5.14 Optimal values of the control variable ($V_0$)*

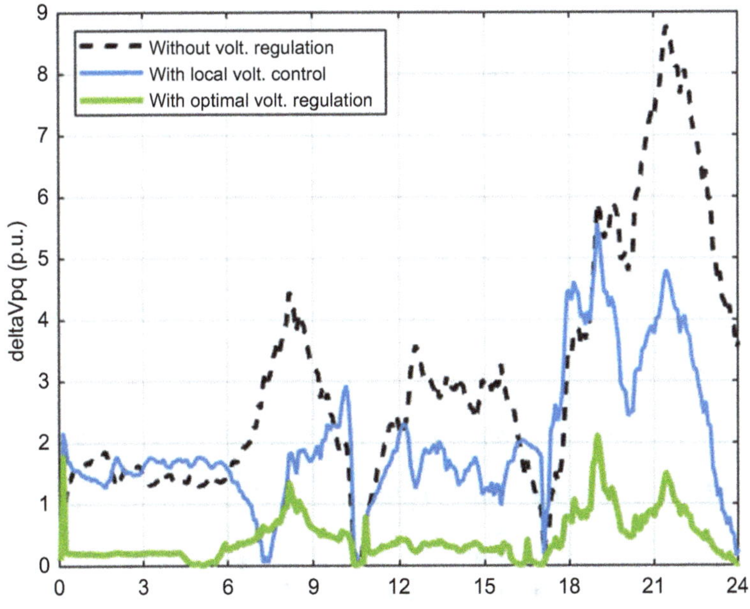


*Figure 5.15 Total voltage deviation in the distribution network*

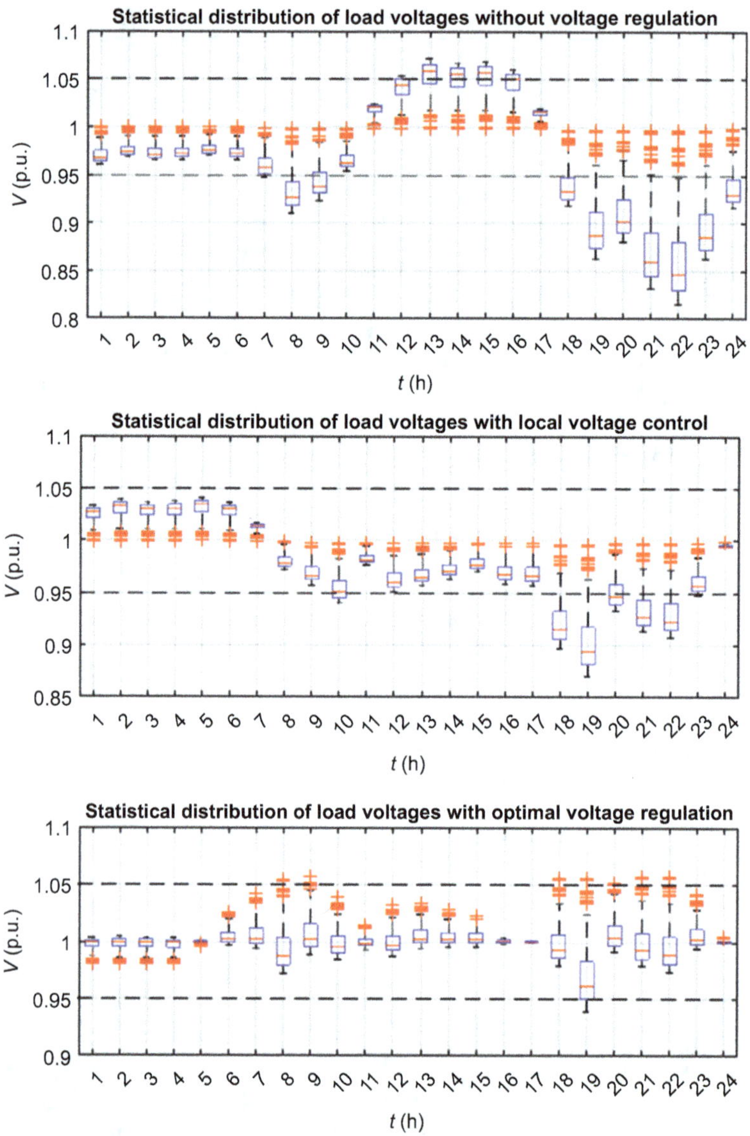


*Figure 5.16 Statistical indicators for load bus voltages*

## 5.5 MATLAB® program *vrlvdnm*

The proposed two-stage algorithm for voltage regulation in LV DNs is coded in MATLAB software and implemented on the IEEE European LV test feeder. All results presented above were obtained using the *vrlvdnm* program package, which is available as supplementary material of this book on the book's homepage at the IET digital library.

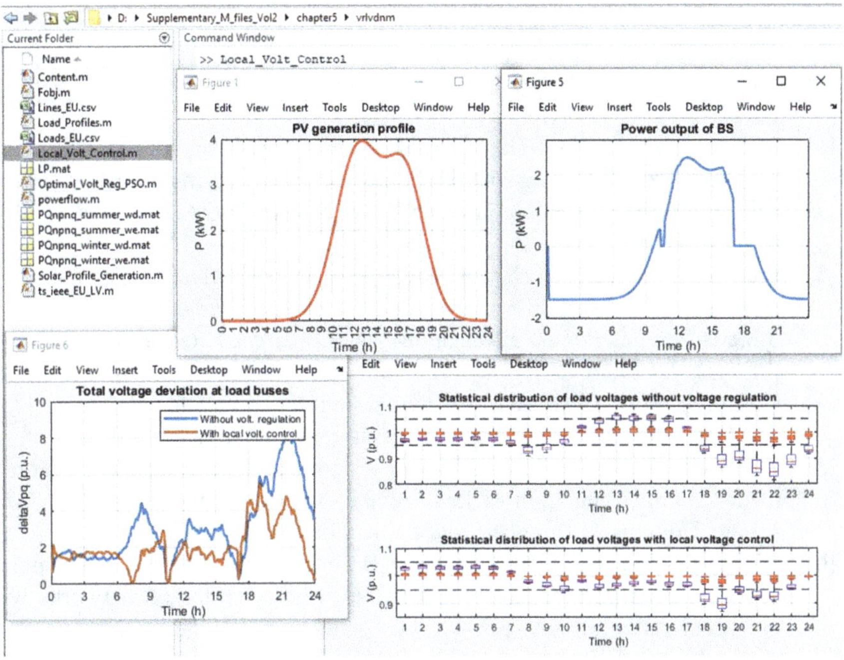


*Figure 5.17 Local voltage control using vrlvdnm program*

An example of using this software package in the case of local voltage control is shown in Figure 5.17.

## 5.6 Conclusion

In this chapter, a two-stage voltage regulation approach for LV DNs with rooftop PV generation and BESS has been considered. The proposed algorithms have been tested on the IEEE European LV test feeder. The simulation results lead to conclusions that can be summarized as follows:

- Local voltage control reduces the power imbalance between PV generation and load demand by managing the output power of the BESS. In this way, the energy exchange between the customer and the DN is minimized, enabling improving voltage profile of the network.
- Unlike the distributed control based on a consensus algorithm, the proposed local voltage control does not require communication between neighboring customers.
- Optimal voltage regulation based on optimal coordination of the HDT voltage support and local voltage control using PSO provides minimum voltage deviation at all load buses in whole daytime.
- The proposed approach can be easily implemented for any practical LV DN with PV sources.

## References

[1] Markiewicz H., and Klajn A. *'Voltage Disturbances Standard EN 50160'*, 2014.

[2] Radosavljević J., Arsić N., Milovanović M., and Ktena A. 'Optimal placement and sizing of renewable distributed generation using hybrid metaheuristic algorithm'. *Journal of Modern Power Systems and Clean Energy*. 2020; **8**(3):499–510.

[3] Frost T., Mitcheson P. D., and Green T. C. 'Power electronic voltage regulation in LV distribution networks'. *Proc. 2015 IEEE 6th International Symposium on Power Electronics for Distributed Generation Systems* (PEDG), 22–25, 2015.

[4] Radosavljević J. *Metaheuristic Optimization in Power Engineering*. London: The Institution of Engineering and Technology (IET), 2018.

[5] Silversides K., Green T., and Merlin M. 'A high density converter for mid feeder voltage regulation of low voltage distribution feeders'. *Proc. IEEE Energy Conversion Congress and Exposition* (ECCE), 2014.

[6] Akhtar Z., Opatovsky M., Chaudhuri B., and Hui S. 'Comparison of point-of-load versus mid-feeder compensation in LV distribution networks with high penetration of solar photovoltaic generation and electric vehicle charging stations'. *IET Smart Grid*. 2019;**2**(2):283–92.

[7] Antoniadou-Plytaria K. E., Kouveliotis-Lysikatos I. N., Georgilakis P. S., and Hatziargyriou N. D. 'Distributed and decentralized voltage control of smart distribution networks: models, methods, and future research'. *IEEE Transactions on Smart Grid*. 2017;**8**(6):2999–3008.

[8] Lan Z., Long Y., and Rao Y. 'Review of voltage control in low voltage distribution networks with high penetration of photovoltaics'. *Proc. of the 2nd International Conference on Information Technologies and Electrical Engineering (ICITEE)*, 2019.

[9] Dong J., Xue Y., Olama M., Kuruganti T., Nutaro J., and Winstead C. 'Distribution voltage control: current status and future trends'. *Proc. 2018 9th IEEE International Symposium on Power Electronics for Distributed Generation Systems* (PEDG), 2018.

[10] Bletterie B., Kadam S., Bolgaryn R., and Zegers A. 'Voltage control with PV inverters in low voltage networks – In depth analysis of different concepts and parameterization criteria'. *IEEE Transactions on Power Systems*. 2017; **32**(1):177–85.

[11] Conti S., Greco A., Messina N., and Raiti S. 'Local voltage regulation in LV distribution networks with PV distributed generation'. *Proc. International Symposium on Power Electronics, Electrical Drives, Automation and Motion* (SPEEDAM), 2006.

[12] Olivier F., Aristidou P., Ernst D., and Van Cutsem T. 'Active management of low-voltage networks for mitigation overvoltages due to photovoltaic units'. *IEEE Transactions on Smart Grids*. 2016;**7**(2):926–36.

[13] Wang Y., Wang B. F., and So P. L. 'A voltage regulation method using distributed energy storage systems in LV distribution networks'. *Proc. 2016 IEEE International Energy Conference* (ENERGYCON), 2016.

[14] Wang Y., Tan K. T., Peng X. Y., and So P. L. 'Coordinated control of distributed energy storage systems for voltage regulation in distribution networks'. *IEEE Transactions on Power Delivery*. 2016;**31**(3):1132–41.

[15] Zeraati M., Golshan M. E. H., and Guerrero J. M. 'Distributed control of battery energy storage systems for voltage regulation in distribution networks with high PV penetration'. *IEEE Transactions on Smart Grid*. 2018;**9**(4):3582–93.

[16] Zeraati M., Golshan M. E. H., and Guerrero J. M. 'A consensus-based cooperative control of PEV battery and PV active power curtailment for voltage regulation in distribution networks'. *IEEE Transactions on Smart Grid*. 2019;**10**(1):670–80.

[17] Procopiou A. T., and Ochoa L. F. 'Voltage control in PV-rich LV networks without remote monitoring'. *IEEE Transactions on Power Systems*. 2016; **32**(2):1224–36.

[18] Kostalos K., Miranda I., Dominguez-Garcia J. L., Leite H., Silva N., and Hatziargyriou N. 'Exploiting OLTC and BESS operation coordinated with active network management in LV networks'. *Sustainability*. 2020;**12**(8):1–25.

[19] Zafar R., and Ravishankar J. 'Coordinated control of step voltage regulator and D-STATCOM in the presence of distributed photovoltaic systems'. *Proc. 2016 IEEE International Conference on Power System Technology* (POWERCON), 2016.

[20] Su X., Masoum M. A. S., and Wolfs P. 'PSO based multi-objective optimization of unbalanced LV distribution networks by PV inverter control'. *Proc. 2014 China International Conference on Electricity Distribution* (CICED 2014), 2014.

[21] Garcia A.M., Mastromauro R.A., and Liserre M. 'A combined centralized/decentralized voltage regulation method for PV inverters in LV distribution networks'. *Proc. IEEE PES General Meeting Conference & Exposition*, 2014.

[22] Juamperez M., Yang G., and Kjaer S. B. 'Voltage regulation in LV grids by coordinated volt-var control strategies'. *Journal of Modern Power Systems and Clean Energy*. 2014;**2**:319–328.

[23] Hashemipour N., Aghaei J., Lotfi M., Niknam T., Askarpour M., Shafie-Khah M., and Catalao J.P.S. 'Multi-objective optimization method for coordinating battery storage systems, photovoltaic inverters and tap changers'. *IET Renewable PowerGeneration*. 2020;**14**(3):475–483.

[24] Bala S., Das D., Aeloiza E., Maitra A., and Rajagopalan S. 'Hybrid distribution transformer: concept development and field demonstration'. *Proc. 2012 IEEE Energy Conversion Congress and Exposition* (ECCE), 2012.

[25] Radosavljević J. 'Voltage regulation in LV distribution networks with PV generation and battery storage'. *Journal of Electrical Engineering*. 2021; **72**(6):356–65.

[26] 'European Low Voltage Test Feeder', https://site.ieee.org/pes-testfeeders/resources/ (Accessed in 23 September 2021).

# Chapter 6

# Optimal energy and operation management of microgrids

## 6.1 Introduction

In the last two decades, distributed generation (DG) of energy has been steadily increasing. The integration of DG units with renewable energy sources such as photovoltaic (PV) systems, and small wind turbines (WTs) is dependent on geographical and meteorological conditions. Moreover, their output powers change over time and are difficult to predict. Conversely, the DG units with nonrenewable energy, such as fuel cells (FCs), gas-fired microturbines (MTs), diesel generators (DEs), and battery energy storage systems (BESS), can be linked to any point in the distribution network and produce predictable output power. Depending on the DG unit technology, the primary energy of DG units may be injected into the distribution network via either a synchronous or asynchronous electric machine that is directly connected to the grid, a combination of an electric machine and a power electronic interface, or only a power electronic interface [1].

In general, DG units, if properly planned and controlled, can improve power quality, minimize power loss, enhance reliability, and lower emissions. A good way to realize the maximum potential of DG is coordinated operation and control together with storage devices and controllable loads. This brings about the concept of microgrids (MGs) [2–8].

Accordingly, an MG can be defined as a local distribution network comprising various DG units, controllable loads, and BESS, which can operate either interconnected or isolated from the main distribution grid as a controlled entity. The MGs represent the core of smart distribution networks within modern power systems.

In order to achieve the full benefits from the operation of MGs, such as better economics and reduction dependence on the main grid, it is important that the integration of the DG units into MG, and their relation with the main distribution network upstream will contribute to optimize the general operation of the system [2]. In order to achieve this goal, it is necessary to use the appropriate control system architecture. The usual control system architecture of the MG comprises three hierarchical control levels. As shown in Figure 6.1, the different control levels are [2–6]

1. Local controllers (LC),
2. Microgrid system central controller (MGCC), and
3. Distribution management system (DMS).

The LC takes advantage of the power electronic interface of the DG units and can be enhanced with various degrees of intelligence. LCs are responsible for controlling components of an MG and have to be adapted to each type of component such as DG units (PV, WT, FC, MT, etc.), BESS, and controllable loads. It uses local information to control the voltage and frequency in case isolated operating mode of MG. When the MG is connected to the main distribution grid LCs follow the demands from the MGCC and perform local optimization of the DG power production.

The MGCC acts as an interface between the MG and the rest of the power system. Its functions can range from monitoring the actual active and reactive power of DG units to full responsibility for the maximization of the MG's value and the optimization of its operation by sending control signal settings to the DG units, BESS and controllable loads.

Conventional approaches to DMSs need to be enhanced with new features related to the operation MGs connected to the main distribution grid. MGs connected on the feeders of DMS should ideally look like concentrated loads. In a network where more than one MG exists, MGCCs communicate with DMS to ensure a harmony, reliable, and safe operation mode of the MGs.

In the grid-connected mode of operation, the decision on local generation is based on the maximization of the MG value, according to the availability of the primary energy sources and the energy prices. Depending on the decision-making scheme, the MGs control systems can be categorized as centralized and decentralized. Centralized control tends to maximize local production according to market prices. This requires two-way communication between the MGCC and each LC. Conversely, in decentralized control, the LCs act smartly and communicate with each other to increase the revenue and the performance of the MG [4]. Therefore, the main optimization problem for the MGs connected to the main grid is energy and operation management (EOM). Network limits, namely the capacity of the supply transformer interconnecting the MG with the main grid, or MG network congestion, must be observed [2].

Many researchers have focused on formulating the EOM problem and developing efficient solution methodologies for optimal EOM. There are two general approaches for EOM of MG. Those are deterministic and probabilistic approaches. The deterministic EOM of the MGs assumes that the PV and the WT output powers, as well as load power and market prices, are equal to their forecast values. When some of the input variables are uncertain, the EOM problem becomes probabilistic. Because of the probabilistic nature of wind speed and solar irradiance, the power output of the WT and the PV units is also random. Moreover, because of unexpected disturbance, forecast error, or load variation, it is very unlikely that the forecast of the load demand will be exactly correct. Likewise, in the open access market, there are always errors in forecast values for market prices [9].

Recently, numerous methods have been proposed by researchers to solve both deterministic and probabilistic EOM of MGs. An overview of metaheuristic techniques proposed for solving the EOM problem is given in the first part of the book (*Metaheuristic Optimization in Power Engineering: Vol 1: Algorithms and Power Dispatch – Using MATLAB®-Based software*, Chapter 6, Section 6.6).

The rest of this chapter describes the application of population-based meta-heuristic algorithms to solve deterministic EOM of MGs. Different MG operation scenarios including DG units, loads, and energy storage systems were considered.

## 6.2 Problem formulation of EOM

A typical low-voltage (LV) grid-connected MG comprising different types of DG sources, storage devices, and consumers are shown in Figure 6.1. The MG is supplied from the utility (MV distribution grid) through an MV/LV transformer. The MGCC is located downstream of the transformer, and the LCs are installed close to the DGs. The MGCC optimizes the MG operation based on the open market prices, bids received by the DG sources, and forecast loads and sends signals to the LCs of DG sources to be committed and, if applicable, to determine the level of their production. The optimization procedure implemented in MGCC depends on the market policy adopted in the MG operation. For example, two market policies are proposed in [3,10].

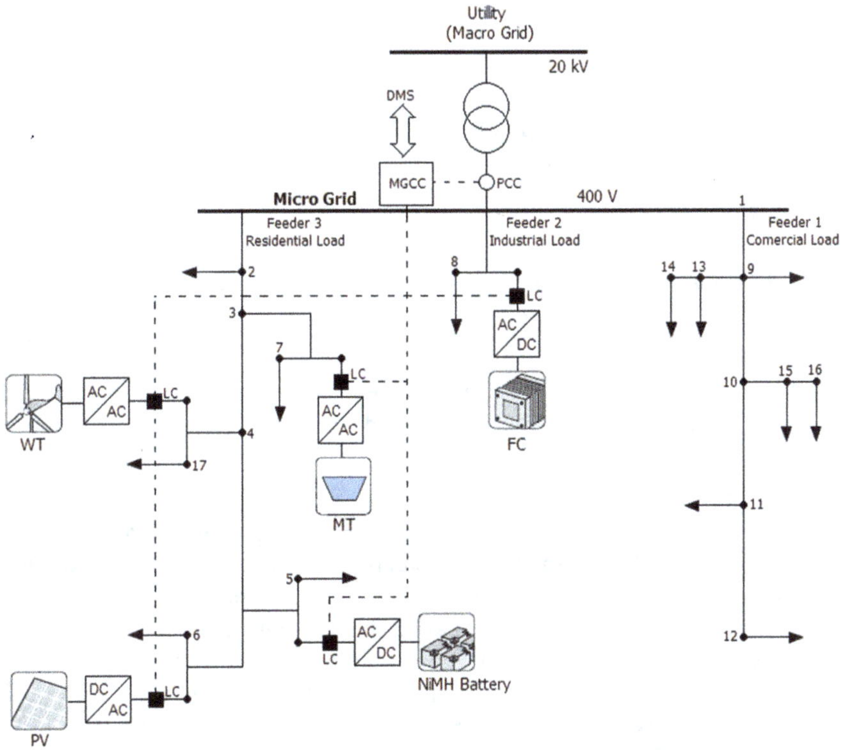


*Figure 6.1 Typical low-voltage microgrid*

In the first policy, the MGCC aims to serve the total demand of the MG, using its local production, as much as possible, without exporting power to the utility. From the consumers' point of view, the MGCC minimizes the operational cost of the MG while taking into account open market prices, demand, and DG bids.

In the second policy, the MG participates in the open market, buying and selling active power to the grid, probably via an aggregator or similar energy service provider. According to this policy, the MGCC tries to maximize the value of the MG, i.e. maximize the corresponding revenues of the aggregator, by exchanging power with the grid.

In this work, the EOM problem is defined according to the first market policy. Therefore, the EOM problem solution in a typical MG aims to minimize the total operating cost of the MG through optimal adjustment of the DG's power generation while satisfying various system operating constraints. The total operating cost of the MG includes the DG bids and the market prices of power exchange between the MG and the utility. The mathematical model of such a problem can be expressed as follows.

### 6.2.1 Objective function

$$\begin{aligned} \min_{\mathbf{X}} F &= \min_{\mathbf{X}} \sum_{t=1}^{NT} Cost(\mathbf{x}_t, y_t) \\ &= \min_{\mathbf{X}} \left\{ \sum_{t=1}^{NT} \sum_{i=1}^{NG} B_{Gi}\left(P_{Gi,t}\right) + \sum_{t=1}^{NT} \sum_{i=1}^{NB} B_{Bi}\left(P_{Bi,t}\right) + MP_t \cdot P_{Grid,t} \right\} \end{aligned} \tag{6.1}$$

where $\mathbf{x}_t$ is the vector of control variables, including the active power of the DG units within the MG and can be expressed as

$$\mathbf{x}_t = \left[P_{G1,t}, P_{G2,t}, \ldots, P_{GNG,t}\right] \tag{6.2}$$

$y_t$ is the dependent variable consisting of the charging/discharging power of the BESS and active power which is bought (sold) from (to) the utility

$$y_t = \left[P_{B1,t}, \ldots, P_{BNB,t}, P_{Grid,t}\right] \tag{6.3}$$

where $NT$ is the total number of time intervals (hours) in the considered period, $NG$ is the total number of DG, $NB$ is the total number of BESS, $P_{Gi,t}$ is the active power outputs of the $i$th DG at time $t$, $P_{Bi,t}$ is the charging/discharging power of the $i$th BESS at time $t$, $B_{Gi}\left(P_{Gi,t}\right)$ is the bid of the $i$th DG unit as a function of its active power at time $t$, $B_{Bi}\left(P_{Bi,t}\right)$ is the bid of the $i$th BESS as a function of its power at time $t$, $P_{Grid,t}$ is the active power which is bought (sold) from (to) the utility at time $t$, and $MP_t$ is the market price of power exchange between the MG and the utility at time $t$.

The cost function (6.1) is generally a nonlinear function because the DG bids are typically either piecewise linear or smooth quadratic functions of $P_{Gi}$ [11].

### 6.2.2 Constraints

#### 6.2.2.1 Power balance

The total output power of DGs, BESS, and the utility must cover the total load demand in the MG for each time interval $t$. The real power loss in the MG is neglected. Accordingly, the power balance constraint can be represented as follows:

$$P_{Grid,t} = \sum_{D=1}^{ND} P_{L_D,t} - \sum_{i=1}^{NG} P_{Gi,t} + \sum_{i=1}^{NB} P_{Bi,t} \tag{6.4}$$

where $P_{L_D}$ is the amount of the $D$th load and $ND$ is the total number of loads in the MG.

#### 6.2.2.2 Real power generation capacity

For stable operation, the real power output of each unit in the MG, including the utility, is restricted by minimum and maximum power limits as follows:

$$P_{Gi,\,\min} \le P_{Gi,t} \le P_{Gi,\,\max} \tag{6.5}$$

$$P_{Grid,\,\min} \le P_{Grid,t} \le P_{Grid,\,\max} \tag{6.6}$$

where $P_{Gi,\,\min}$ and $P_{Grid,\,\min}$ are the minimum active powers of the $i$th DG and the utility and $P_{Gi,\,\max}$ and $P_{Grid,\,\max}$ are the maximum active powers of the $i$th DG and the utility.

#### 6.2.2.3 BESS limits

Since there are some limitations on the charge and discharge rate of BESS during each time interval, the following equations and constraints can be expressed for a typical battery [12]:

$$SoC_t = SoC_{t-\Delta t} + \left( \eta_{ch} P_{ch,t-\Delta t} - \frac{P_{dch,t-\Delta t}}{\eta_{dch}} \right) \frac{\Delta t}{C_B} \tag{6.7}$$

$$\begin{cases} SoC_{\min} \le SoC_t \le SoC_{\max} \\ P_{ch,t} \le P_{ch,\,\max};\, P_{dch,t} \le P_{dch,\,\max} \end{cases} \tag{6.8}$$

where $SoC_t$ and $SoC_{t\text{-}\Delta t}$ are the amounts of energy storage inside the battery (i.e. state of charge) at time $t$ and $t-\Delta t$, respectively; $P_{ch}$ and $P_{dch}$ are the charging and discharging powers during a definite period of time ($\Delta t$), respectively; $\eta_{ch}$ and $\eta_{dch}$ are the efficiency of the battery during charge and discharge process, respectively; $SoC_{\min}$ and $SoC_{\max}$ are the lower and upper limits on amount of energy storage inside the battery, respectively; $P_{ch,\max}$ and $P_{dch,\max}$ are the maximum charging and discharging powers of the battery, respectively; and $C_B$ is the total capacity of the battery.

#### 6.2.2.4 Calculation of the active power from (to) the utility

To enforce the active power balance constraint given in (6.4), the active power from (to) the utility (MV distribution grid) is considered as a dependent variable.

The obtained $P_{Grid,t}$ is checked as to whether it satisfies the constraint defined in (6.6) or not. Consequently, the variable $P^{\lim}_{Grid,t}$ is defined as

$$P^{\lim}_{Grid,t} = \begin{cases} P_{Grid,\max} & \text{if} \quad P_{Grid,t} > P_{Grid,\max} \\ P_{Grid,\min} & \text{if} \quad P_{Grid,t} < P_{Grid,\min} \\ P_{Grid,t} & \text{if} \quad P_{Grid,\min} \le P_{Grid,t} \le P_{Grid,\max} \end{cases} \tag{6.9}$$

It is worth mentioning that the control variables are self-constrained while the dependent variable, i.e. $P_{Grid,t}$, must be added to the objective function as a quadratic penalty term. In this term, a penalty factor multiplied by the square of the difference between the actual value and the limiting value of the dependent variable is added to the objective function, and any unfeasible solution obtained during the optimization process is ignored. The new expanded objective function to be minimized becomes

$$\min_{\mathbf{x}} F_p = \min_{\mathbf{x}} \sum_{t=1}^{NT} Cost(\mathbf{x}_t, y_t) + \lambda_p \left(P_{Grid,t} - P^{\lim}_{Grid,t}\right)^2 \tag{6.10}$$

where $\lambda_p$ is the penalty factor.

#### 6.2.2.5 BESS operating mode

BESS mode means the time schedule and dynamics of charging/discharging of the battery during the considered time period (e.g. 24 h). The BESS operating mode in each time segment of the observed period is determined in accordance with the adopted function of the BESS in the MG. This can be, for example, minimizing the cost of electricity from the source grid, reducing peak loads, maximizing the use of energy from RES to supply local loads in the MG, etc. The power of the BESS at time $t$ depends on the adopted battery mode operation. It seems that the minimization of the operational cost of the MG (6.1) is the most appropriate function of BESS in the MG.

This means that during periods of the day when the electricity price is low, the BESS is used to store the energy from RES and the source grid, working in the charging mode. During periods of the day with a high price of the energy from the source grid, the BESS is used as an additional power source, i.e. it operates in the discharge mode. It is assumed that the electricity price from the source grid changes during the day according to a predetermined diagram, i.e. the electricity price is known in each time segment (hour) of the considered time period.

It is clear that in the charging mode, BESS behaves as a load and that according to the adopted reference directions, the battery power is positive, and in the discharge mode, BESS acts as a generator in the MG, and the power $P_B$ is negative. In order to define the terms "high price" and "low price," it is assumed that the electricity price at a given time $t$ is high if it is higher than the mean price in the observed time period (day) and that the electricity price at the given time $t$ is small if it is less than the mean value of the electricity price in the considered period.

In order to define the charging/discharging power of the BESS, i.e. to define the value of $P_B(t)$, the principle adopted is that the greater the difference in the energy price at a given moment $t$, $MP(t)$, in relation to the mean electricity price during the day, it is also the higher power of the battery $P_B(t)$ in relation to the rated power of the battery $P_{Brated}$. Therefore, the BESS operating mode at the time $t$ can be defined as follows:

$$P_B(t) = \begin{cases} 0 & \text{if} \quad MP(t) \le MP_{mean} \\ & \text{and} \quad SoC(t) = SoC_{\max} \\ a_f(t) \cdot P_{Brated} & \text{if} \quad MP(t) \le MP_{mean} \\ & \text{and} \quad SoC_{\min} \le SoC(t) < SoC_{\max} \\ 0 & \text{if} \quad MP(t) > MP_{mean} \\ & \text{and} \quad SoC(t) = SoC_{\min} \\ -a_f(t) \cdot P_{Brated} & \text{if} \quad MP(t) > MP_{mean} \\ & \text{and} \quad SoC_{\min} < SoC(t) \le SoC_{\max} \end{cases} \tag{6.11}$$

where the coefficient $a_f$ shows the difference in the market electricity price at the time $t$ from the mean value during the observed day [13]

$$a_f(t) = \frac{|MP(t) - MP_{mean}|}{MP_{\max}} \tag{6.12}$$

$MP_{mean}$ and $MP_{\max}$ are the mean and maximum electricity prices during the day, respectively.

## 6.2.3 Distributed generation bids calculation

DG bids are considered quadratic, as presented in (6.13), according to the cost function of the units, if any, the feedback from the market prices, and the need for making some profit necessary for the annual depreciation of the installation cost [10]

$$B_{Gi} = a_i \cdot (P_{Gi})^2 + b_i \cdot P_{Gi} + c_i \tag{6.13}$$

### 6.2.3.1 Microturbine and FC

The MT and FC bids in €/h can be calculated as follows [14]:

$$B_G = C_{fuel} \frac{P_G}{\eta_G} + C_{inv} \tag{6.14}$$

where $P_G$ is the electrical output power (kW) of DG, e.g. MT and FC; $\eta_G$ is the DG electric efficiency; $C_{fuel}$ is the price of fuel (natural gas) to supply the DG (€/kWh); and $C_{inv}$ is the hourly payback amount for the investment cost of the DG (€/h).

$C_{inv}$ is a function of the annual investment cost for depreciation ($AC$) (€/kW-year), the annual production ($AP$) (kWh/kW), and the nominal power of the DG (kW):

$$C_{inv} = AC \frac{P_{Gnom}}{AP} \tag{6.15}$$

$$AC = \frac{i(1+i)^n}{(1+i)^n - 1} \cdot IC \tag{6.16}$$

where $i$ is the interest rate, $n$ is the depreciation period in years, and $IC$ is the installation cost of the DG.

The efficiency of the MT increases with the increase of the output power. The electric efficiency characteristic of the MT can be estimated as a quadratic function of its output power. The coefficients of this quadratic function can be obtained by fitting the electrical efficiency curve of the MT based on manufacturers' data.

The FC efficiency is a nonlinear function of power level and can be expressed as follows (for a proton-exchange membrane FC (PEM FC)):

$$\eta_{FC} = \frac{1}{2.964}\left[V_0 + \sqrt{V_0^2 - 4\left(V_0 V_{Pnom} - V_{Pnom}^2\right) \cdot \xi}\right] \tag{6.17}$$

where $V_0$ is the theoretical or thermodynamic FC potential (0.9 V), $V_{Pnom}$ is the selected cell potential at nominal power (0.45–0.75 V), $\xi = P_{FC}/P_{nom}$ is the power level, and $P_{nom}$ is the nominal power of the FC.

The coefficients of a quadratic function (6.13) can be obtained by curve fitting, constructed according to (6.14) for a power range of $P_G$, e.g. $P_{G\,\min} \le P_G \le P_{G\,\max}$.

#### 6.2.3.2 Diesel generator

The diesel fuel consumption data (l/h) at 25%, 50%, 70%, and 100% of the DE power rating (kW) are usually given by the manufacturer. Based on these data, the fuel consumption characteristic of the DE can be estimated as a quadratic function of its active power output:

$$Fuel_{DE} = a_f P_{DE}^2 + b_f P_{DE} + c_f \tag{6.18}$$

where $Fuel_{DE}$ is the DE fuel consumption (l/h), $P_{DE}$ is the DE power output (kW), and $a_f$, $b_f$, and $c_f$ are coefficients of the fuel consumption characteristic. Accordingly, the DE bids (€/h) can be expressed as

$$B_G = C_{fuel} Fuel_{DE} + C_{inv} \tag{6.19}$$

where $C_{fuel}$ is the diesel fuel price to supply the DE (€/l) and $C_{inv}$ is the hourly payback amount for the investment cost of the DE (€/h) defined by (6.15).

#### 6.2.3.3 WT and PV

The bid functions of WT and PV consider the annual investment cost for depreciation of equipment ($AC$) (€/kW) and the annual energy production per kW ($AP$) (kWh/kW):

$$B_G = \frac{AC}{AP} P_G \tag{6.20}$$

$$AC = \frac{i(1+i)^n}{(1+i)^n - 1} \cdot IC \tag{6.21}$$

where $i$ is the interest rate, $n$ is the depreciation period in years, and $IC$ is the installation cost of DGs, e.g. WT and PV.

Renewable source-based DGs, e.g. WT and PV cannot be regulated, and their output is determined by the availability of the primary source, i.e. wind or solar irradiance. To account for their production in the optimization functions, WT and PV forecasting is required.

To calculate the power output of a WT, two main factors must be known: the wind speed at a certain location and the power curve of the WT. According to [15,16], the power curve of a WT can be modeled by means of a function split into four different parts:

$$P_{WT} = \begin{cases} 0, & v \leq v_{ci} \\ \dfrac{v^2 - v_{ci}^2}{v_{nom}^2 - v_{ci}^2} \cdot P_{WT,nom}, & v_{ci} < v \leq v_{nom} \\ P_{nom}, & v_{nom} < v \leq v_{co} \\ 0, & v > v_{co} \end{cases} \tag{6.22}$$

where $P_{WT,nom}$, $v_{nom}$, $v_{ci}$, and $v_{co}$ are nominal power, nominal wind speed, cut-in wind speed, and cut-out wind speed of the WT, respectively; $P_{WT}$ is the power output of the WT; and $v$ is the wind speed.

The power output of the PV module is dependent on the solar irradiance and ambient temperature of the site, as well as the characteristics of the module itself. The following equation can be used to calculate the power output of the PV module, $P_{PV}$ [16,17]:

$$P_{PV} = P_{STC}\frac{Is}{1000}[1 + \gamma(T_c - 25)] \tag{6.23}$$

where $P_{STC}$ is the PV module maximum power under standard test condition (STC) (W), $Is$ is the solar irradiance on the PV module surface (W/m$^2$), $\gamma$ is the PV module temperature coefficient for power (°C$^{-1}$), and $T_c$ is the PV cell (module) temperature (°C). The PV module temperature can be calculated as a function of solar irradiance and ambient temperature based on the module's nominal operating cell temperature (NOCT). The NOCT model equation is [17]

$$T_c = T_a + \frac{Is}{800} \cdot (T_{NOCT} - 20) \tag{6.24}$$

where $T_a$ is the ambient temperature (°C) and $T_{NOCT}$ is the nominal operating cell temperature (°C) of the module.

Bids coming from the batteries can be defined by (6.20).

## 6.3 Solution method

This section describes the application of metaheuristic optimization methods for solving the EOM problem in MGs. The control variables of the EOM problem are given by (6.2). The vector of control variables (6.2), containing the active power of dispatchable DG units within the MG, represents the search agent in the metaheuristic optimization paradigm. In a system with $N$ agents, the position of the $i$th agent in the time segment $t$ is defined by

$$\mathbf{x}_i = \left[P_i^1, ..., P_i^d, ..., P_i^n\right] \quad i = 1, 2, ..., N \quad n = NG \tag{6.25}$$

The application of population-based metaheuristic algorithms (described in Volume 1 of the book: Chapters 1–5), for optimal EOM for MG can be summarized in the following steps:

**Input data**

*Step 1*: Define the MG system configuration, DG units data, storage units data, loads data, and control strategy. Choice of metaheuristic algorithm.

*Step 2*: Specify the control variables $\mathbf{x}$ (6.2) and their lower and upper limits (6.5);
Specify the dependent variable $y$ (6.3) and its lower and upper limit (6.6);
Specify the objective function to be optimized $F(\mathbf{x,y})$ (6.1), and the penalty factor ($\lambda_P$) to form the expanded objective function $F_p(\mathbf{x,y})$ (6.10);

**Initialization**

*Step 3*: Defining algorithm parameters, such as population size ($N$), maximum number of iterations ($t_{\max}$), and other parameters depending on the chosen solution technique.

*Step 4*: Generate an initial random population of $N$ agents:

$$\mathbf{POP}(0) = \begin{bmatrix} \mathbf{x}_1(0) \\ \mathbf{x}_2(0) \\ \vdots \\ \mathbf{x}_i(0) \\ \vdots \\ \mathbf{x}_N(0) \end{bmatrix} = \begin{bmatrix} x_1^1(0), ..., x_1^d(0), ..., x_1^n(0) \\ x_2^1(0), ..., x_2^d(0), ..., x_2^n(0) \\ \vdots \\ x_i^1(0), ..., x_i^d(0), ..., x_i^n(0) \\ \vdots \\ x_N^1(0), ..., x_N^d(0), ..., x_N^n(0) \end{bmatrix} \tag{6.26}$$

The initial positions of each agent are randomly selected between the minimum and maximum values of the control variables. Therefore, the initial value of the $d$th control variable in the $i$th agent (candidate solution) is obtained as follows:

$$x_i^d(0) = x_{i,\min}^d + r \cdot \left(x_{i,\max}^d - x_{i,\min}^d\right) \tag{6.27}$$

where $r$ is a random number in [0,1], $x_{i,\min}^d$ and $x_{i,\max}^d$ are the lower and upper limits of the $d$th control variable. Initialize the velocity of agents by a zero matrix of dimension $N \times n$, where $N$ is the number of agents (population size) and $n$ is the number of control variables ($n = NG$).

**Iterative procedure**

*Step 5*: Determine the BESS operating mode (6.11), calculate the active power from (to) the utility (6.4), and check the operating constraints for each search agent in the current population

*Step 6*: Calculate the fitness value for each agent using (6.1) and (6.10).

*Step 7*: Generate new population $\mathbf{POP}(t+1) = [\mathbf{x}_1(t+1), \ldots, \mathbf{x}_i(t+1), \ldots, \mathbf{x}_N(t+1)]^T \subseteq \mathbf{X}$ by applying the algorithmic operators of the selected metaheuristic algorithm on search agents from the current population $\mathbf{POP}(t)$

*Step 8*: Repeat Steps 5–7 until the stop criterion is reached. That is a predefined number of iterations, $t_{max}$.

*Step 9*: The best solution – optimal daily power dispatching in MG. Stop.

## 6.4 Simulation results

The MATLAB® program package *eommg* was developed to solve deterministic EOM and implemented in two different MGs. This program is supplementary material to this chapter.

### 6.4.1 Microgrid MG1

The MG1 is a typical LV grid-connected MG, as shown in Figure 6.1. The MG1 consists of different types of DG sources such as gas-fired MT, PEM-FC, WT, PV, and storage devices (BESS). It is assumed that all DG sources produce active power at the unity power factor. Furthermore, there is a power exchange link between the utility and the MG1 to trade energy during the day based on decisions of the MGCC. The load demand within the MG1 for a typical day comprises one primarily residential area, one industrial feeder serving a small workshop, and one feeder with light commercial consumers, which is equivalent to a total energy demand of 1695 kWh for that day. Table 6.1 offers the minimum and maximum production limits and the bid coefficients of the DG units in the MG1. The hourly forecast load demands, market energy prices, and WT and PV power outputs are given in Table 6.2. The system data are taken from [18]. Table 6.3 gives the BESS data adopted for these simulations.

The best results presented in Table 6.4 were obtained using the *eommg* program package. Within this program package, more than 20 different metaheuristic algorithms are offered as a solution method. In this particular case, the Runge–Kutta algorithm (RUN) was chosen as the solution method.

According to the BESS operating mode given by (6.11) and (6.12), corresponding daily diagrams of battery charging/discharging were obtained, as presented in Figures 6.2 and 6.3.

*Table 6.1 The power limits and coefficients of bid functions of the installed DG units*

| ID | Type | Min power (kW) | Max power (kW) | $a_i$ (€ct/kWh$^2$) | $b_i$ (€ct/kWh) | $c_i$ (€ct/h) |
|---|---|---|---|---|---|---|
| 1 | MT | 6 | 30 | 0 | 0.457 | 0 |
| 2 | FC | 3 | 30 | 0 | 0.294 | 0 |
| 3 | PV | 0 | 25 | 0 | 2.584 | 0 |
| 4 | WT | 0 | 15 | 0 | 1.073 | 0 |
| 5 | Battery | −30 | 30 | 0 | 0.38 | 0 |

*Table 6.2 The forecast values of the load demand, market price, and WT and PV power production*

| Time (h) | Load (kW) | MP (€ct/kWh) | WT (kW) | PV (kW) |
|---|---|---|---|---|
| 1 | 52 | 0.23 | 1.7850 | 0 |
| 2 | 50 | 0.19 | 1.7850 | 0 |
| 3 | 50 | 0.14 | 1.7850 | 0 |
| 4 | 51 | 0.12 | 1.7850 | 0 |
| 5 | 56 | 0.12 | 1.7850 | 0 |
| 6 | 63 | 0.20 | 0.9150 | 0 |
| 7 | 70 | 0.23 | 1.7850 | 0 |
| 8 | 75 | 0.38 | 1.3050 | 0.200 |
| 9 | 76 | 1.50 | 1.7850 | 3.750 |
| 10 | 80 | 4.00 | 3.0900 | 7.525 |
| 11 | 78 | 4.00 | 8.7750 | 10.450 |
| 12 | 74 | 4.00 | 10.410 | 11.950 |
| 13 | 72 | 1.50 | 3.9150 | 23.900 |
| 14 | 72 | 4.00 | 2.3700 | 21.050 |
| 15 | 76 | 2.00 | 1.7850 | 7.875 |
| 16 | 80 | 1.95 | 1.3050 | 4.225 |
| 17 | 85 | 0.60 | 1.7850 | 0.550 |
| 18 | 88 | 0.41 | 1.7850 | 0 |
| 19 | 90 | 0.35 | 1.3020 | 0 |
| 20 | 87 | 0.43 | 1.7850 | 0 |
| 21 | 78 | 1.17 | 1.3005 | 0 |
| 22 | 71 | 0.54 | 1.3005 | 0 |
| 23 | 65 | 0.30 | 0.9150 | 0 |
| 24 | 56 | 0.26 | 0.6150 | 0 |

*Table 6.3 Parameters of the installed BESS*

| $P_{Brated}$ (kW) | $C_B$ (kWh) | $SoC_{\min}$ (%) | $SoC_{\max}$ (%) | $SoC_{\text{initial}}$ (60%) | $\eta_c$ | $\eta_d$ |
|---|---|---|---|---|---|---|
| 30 | 60 | 20 | 100 | 60% | 0.95 | 0.92 |

*Table 6.4 Best solutions obtained for deterministic EOM problem using the RUN algorithm*

| Time (h) | Power (kW) | | | | | | | Cost (€ct/h) |
|---|---|---|---|---|---|---|---|---|
| | Load | MT | FC | PV | WT | BESS | Utility | |
| 1 | 52 | 6 | 3 | 0 | 1.7850 | 6.8578 | 48.0728 | 19.2020 |
| 2 | 50 | 6 | 3 | 0 | 1.7850 | 7.1428 | 46.3578 | 17.0616 |
| 3 | 50 | 6 | 3 | 0 | 1.7850 | 7.4991 | 46.7141 | 14.9289 |
| 4 | 51 | 6 | 3 | 0 | 1.7850 | 7.6416 | 47.8566 | 14.1859 |
| 5 | 56 | 6 | 3 | 0 | 1.7850 | 0 | 45.2150 | 10.9651 |

*(Continues)*

*Table 6.4 (Continued)*

| Time (h) | Power (kW) | | | | | | | Cost (€ct/h) |
|---|---|---|---|---|---|---|---|---|
| | Load | MT | FC | PV | WT | BESS | Utility | |
| 6 | 63 | 6 | 3 | 0 | 0.9150 | 0 | 53.0850 | 15.2228 |
| 7 | 70 | 6 | 3 | 0 | 1.7850 | 0 | 59.2150 | 19.1588 |
| 8 | 75 | 6 | 30 | 0.200 | 1.3050 | 0 | 37.4950 | 27.7272 |
| 9 | 76 | 30 | 30 | 3.750 | 1.7850 | −2.5068 | 7.9582 | 47.0252 |
| 10 | 80 | 30 | 30 | 7.525 | 3.0900 | −22.8872 | −13.5022 | −0.0216 |
| 11 | 78 | 30 | 30 | 10.450 | 8.7750 | −22.8872 | −24.1122 | −28.8034 |
| 12 | 74 | 30 | 30 | 11.950 | 10.410 | 0 | −8.3600 | 31.1387 |
| 13 | 72 | 30 | 30 | 23.900 | 3.9150 | 0 | −15.8150 | 64.7659 |
| 14 | 72 | 30 | 30 | 21.050 | 2.3700 | 0 | −11.4200 | 33.7862 |
| 15 | 76 | 30 | 30 | 7.875 | 1.7850 | 0 | 6.3400 | 57.4743 |
| 16 | 80 | 30 | 30 | 4.225 | 1.3050 | 0 | 14.4700 | 63.0642 |
| 17 | 85 | 30 | 30 | 0.550 | 1.7850 | 4.2216 | 26.8866 | 43.6026 |
| 18 | 88 | 6 | 30 | 0 | 1.7850 | 5.5753 | 55.7903 | 38.4700 |
| 19 | 90 | 6 | 30 | 0 | 1.3020 | 6.0028 | 58.7008 | 35.7854 |
| 20 | 87 | 6 | 30 | 0 | 1.7850 | 5.4328 | 54.6478 | 39.0403 |
| 21 | 78 | 30 | 30 | 0 | 1.3005 | 0.1603 | 16.8598 | 43.7123 |
| 22 | 71 | 30 | 30 | 0 | 1.3005 | 4.6491 | 14.3486 | 33.4403 |
| 23 | 65 | 6 | 30 | 0 | 0.9150 | 6.3591 | 34.4441 | 25.2935 |
| 24 | 56 | 6 | 3 | 0 | 0.6150 | 6.6441 | 53.0291 | 20.5962 |
| | | | | | | | Total Cost (€ct) | 686.8223 |

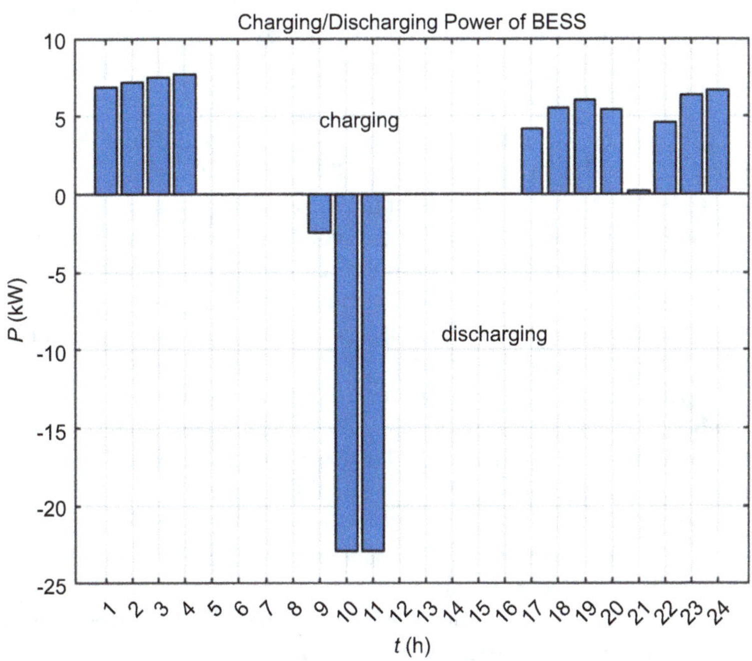


*Figure 6.2 The power variation of the BESS during the day*

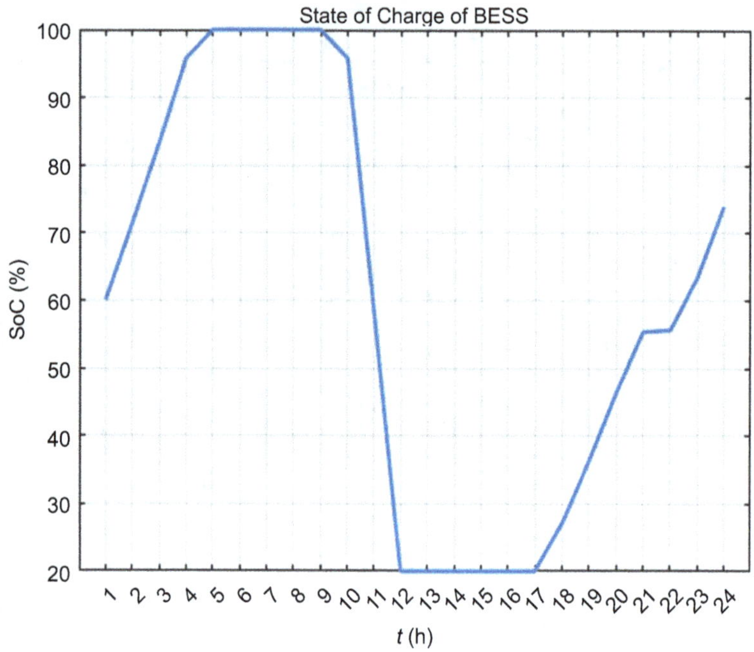


*Figure 6.3 The SoC variation of the BESS during the day*

Likewise, other DGs such as FC and MT adjust their generation set-points according to load levels during each hour of the day in an economical manner.

## 6.4.2 Microgrid MG2

The MG2 is considered in order to show how the bids of the different DG units can be determined. The MG2 is an MG consisting of five different DG units, i.e. MT, FC, DE, WT, PV, and electric loads. Table 6.5 summarizes the data assumed for the DG units. The data were used to calculate the DG bids coefficients presented in Table 6.6. The 24-hour load demand profile and the forecasted market energy prices are given in Table 6.7. Similarly, as in the MG1, it is assumed that all DG sources produce active power at the unity power factor. There is a power-exchange link between the utility and the MG2 in order to trade energy during the day based on decisions of the MGCC.

The *eommg* program package is implemented to find optimal solutions to the EOM problem for the MG2 under a scenario of operation as follows: It is supposed that the renewable energy sources (WT and PV) act at their available maximum power outputs during each hour of the day and the rests of DGs including MT, FC, and DE can operate just within their power limits while satisfying the needed constraints. All DGs with related characteristics produce electricity within the MG2, and additional demand or surplus of energy inside the grid is exchanged with the utility from the point of common coupling. The utility behaves as an unconstraint unit and exchanges energy with the MG2 without any limitation.

*Table 6.5 DG units data for MG2*

| | |
|---|---|
| MT | Tip: CGT301-302; $P_{nom}$ = 300 kW;<br>Efficiency characteristic [19]: $P_{MT}$ = [ 90 120 150 180 210 240 270 300] kW;<br>$\eta_{MT}$ = [33.6 35.4 37.5 39.0 39.7 40.2 40.5 40.0] %<br>$C_{fuel}$ = 1.5 €ct/kWh; $IC$ = 1200 €/kW; $i$ = 7%; $n$ = 10 year; $AP$ = 7884 kWh/kW; |
| FC | Tip: PEM-FC; $P_{nom}$ = 300 kW; $V_0$=0.9 V; $V_{Pnom}$ = 0.6 V;<br>$C_{fuel}$ = 1.5 €ct/kWh; $IC$ = 4000 €/kW; $i$ = 7%; $n$ = 10 year; $AP$ = 7884 kWh/kW; |
| DE | Tip: MDJW 410 T6; $P_{nom}$ = 369 kW;<br>Fuel consumption characteristic [20]: $P_{DE}$ = [92.5 184.5 276.75 369.0] kW;<br>$Fuel_{DE}$ = [29.5 51.9 74.8 100.9] L/h<br>$C_{fuel}$ = 0.6 €/L; $IC$ = 500 €/kW; $i$ = 7%; $n$ = 10 year; $AP$ = 4380 kWh/kW; |
| WT | Tip: Vestas V44/600 kW; $P_{nom}$ = 600 kW; $v_{ci}$ = 4 m/s; $v_{nom}$ = 16 m/s; $v_{co}$ = 20 m/s;<br>$IC$ = 1600 €/kW; $i$ = 7%; $n$ = 10 year; $AP$ = 3504 kWh/kW;<br>Figure 6.4 shows the wind speed data used to calculate the power generated by the WT |
| PV | Tip: SW 250 mono; $P_{nom}$ = 250 kW; $P_{STC}$ = 250 W; $\gamma$ = -0.0045 °C$^{-1}$; $T_{NOCT}$ = 46 °C;<br>$IC$ = 6000 €/kW; $i$ = 7%; $n$ = 20 year; $AP$ = 1300 kWh/kW; Figure 6.4 shows the ambient temperature and solar irradiance data used to calculate the output power generated by the PV modules |

*Table 6.6 Power limits and calculated coefficients of bid functions of the installed DG units in MG2*

| ID | Type | Min power (kW) | Max power (kW) | $a_i$ (€ct/kWh$^2$) | $b_i$ (€ct/kWh) | $c_i$ (€ct/h) |
|---|---|---|---|---|---|---|
| 1 | MT | 90 | 300 | 0.0018 | 2.63 | 811 |
| 2 | FC | 30 | 300 | 0.0062 | 1.62 | 2204 |
| 3 | DE | 0 | 369 | 0.0065 | 12.41 | 1177 |
| 4 | WT | 0 | 600 | 0 | 6.50 | 0 |
| 5 | PV | 0 | 250 | 0 | 43.57 | 0 |

In this scenario, the influence of variations in the market price of energy exchanged with the utility on the total operating cost of the MG2 is discussed in the following three cases:

*Case A:* The market price of energy is low;
*Case B:* The market price of energy is in the medium range;
*Case C:* The market price of energy reaches its real value.

The simulation results associated with Case A, Case B, and Case C are shown in Tables 6.8–6.10, respectively.

In Case A, given the low market price of electricity, the utility takes the lead in supplying the load inside the MG2, particularly during the periods with low and mean load levels.

In Case B, the market price of energy is in the medium range and significantly higher in relation to Case A. Most of the load portion is supplied from MT and FC.

*Table 6.7 The forecast load demands and market prices for MG2*

| Time (h) | Load (kW) | MP (€ct/kWh) | | |
|---|---|---|---|---|
| | | Case A | Case B | Case C |
| 1 | 480 | 1.4 | 2.8 | 3.5 |
| 2 | 480 | 1.4 | 2.8 | 3.5 |
| 3 | 600 | 1.4 | 2.8 | 3.5 |
| 4 | 600 | 1.4 | 2.8 | 3.5 |
| 5 | 720 | 2.8 | 5.6 | 7 |
| 6 | 720 | 2.8 | 5.6 | 7 |
| 7 | 840 | 2.8 | 5.6 | 7 |
| 8 | 840 | 2.8 | 5.6 | 7 |
| 9 | 960 | 2.8 | 5.6 | 7 |
| 10 | 960 | 2.8 | 5.6 | 7 |
| 11 | 1080 | 4.2 | 11.2 | 12.6 |
| 12 | 1080 | 4.2 | 11.2 | 12.6 |
| 13 | 1200 | 4.2 | 11.2 | 12.6 |
| 14 | 1200 | 4.2 | 11.2 | 12.6 |
| 15 | 1080 | 4.2 | 11.2 | 12.6 |
| 16 | 840 | 2.8 | 5.6 | 7 |
| 17 | 840 | 2.8 | 5.6 | 7 |
| 18 | 960 | 2.8 | 5.6 | 7 |
| 19 | 1080 | 4.2 | 11.2 | 12.6 |
| 20 | 1080 | 4.2 | 11.2 | 12.6 |
| 21 | 960 | 2.8 | 5.6 | 7 |
| 22 | 840 | 2.8 | 5.6 | 7 |
| 23 | 720 | 2.8 | 5.6 | 7 |
| 24 | 600 | 1.4 | 2.8 | 3.5 |

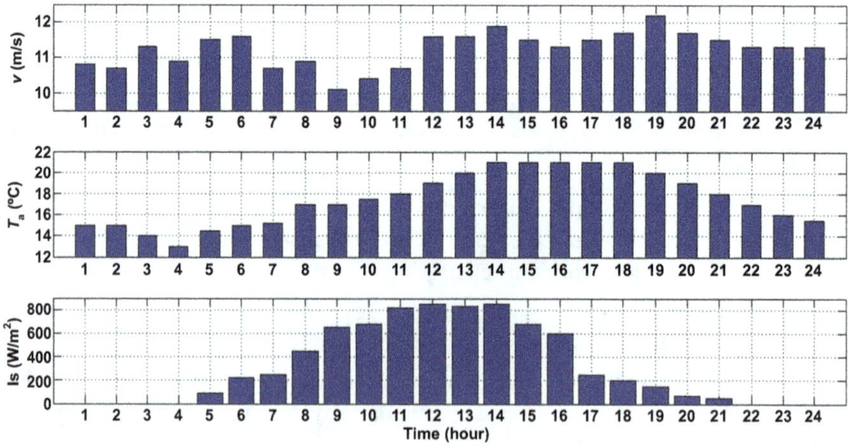


*Figure 6.4 Wind speed, ambient temperature, and solar irradiance data for MG2*

*Table 6.8 Best solutions for Case A in MG2*

| Time (h) | Power (kW) | | | | | | Cost (€/h) |
|---|---|---|---|---|---|---|---|
| | MT | FC | DE | WT | PV | Utility | |
| 1 | 90 | 30 | 0 | 252 | 0 | 108 | 62.85 |
| 2 | 90 | 30 | 0 | 246 | 0 | 114 | 62.57 |
| 3 | 90 | 30 | 0 | 279 | 0 | 201 | 65.94 |
| 4 | 90 | 30 | 0 | 257 | 0 | 223 | 64.80 |
| 5 | 90 | 95 | 0 | 291 | 23 | 221 | 81.75 |
| 6 | 90 | 95 | 0 | 296 | 56 | 183 | 95.19 |
| 7 | 90 | 95 | 0 | 246 | 63 | 346 | 99.65 |
| 8 | 90 | 95 | 0 | 257 | 109 | 289 | 118.88 |
| 9 | 90 | 95 | 0 | 215 | 153 | 407 | 138.53 |
| 10 | 90 | 95 | 0 | 230 | 159 | 386 | 141.51 |
| 11 | 300 | 208 | 0 | 246 | 187 | 139 | 160.74 |
| 12 | 300 | 208 | 0 | 296 | 192 | 84 | 163.84 |
| 13 | 300 | 208 | 0 | 296 | 187 | 209 | 166.98 |
| 14 | 300 | 208 | 0 | 314 | 190 | 188 | 168.54 |
| 15 | 300 | 208 | 0 | 291 | 156 | 125 | 149.67 |
| 16 | 90 | 95 | 0 | 279 | 140 | 236 | 132.09 |
| 17 | 90 | 95 | 0 | 291 | 61 | 303 | 100.63 |
| 18 | 90 | 95 | 0 | 302 | 49 | 423 | 99.57 |
| 19 | 300 | 208 | 0 | 332 | 38 | 202 | 103.92 |
| 20 | 300 | 208 | 0 | 302 | 18 | 252 | 95.46 |
| 21 | 90 | 95 | 0 | 291 | 13 | 471 | 84.20 |
| 22 | 90 | 95 | 0 | 279 | 0 | 376 | 75.20 |
| 23 | 90 | 95 | 0 | 279 | 0 | 256 | 71.84 |
| 24 | 90 | 30 | 0 | 279 | 0 | 201 | 65.94 |
| | | | | | Total Cost (€/h) | | 2570.29 |

*Table 6.9 Best solutions for Case B in MG2*

| Time (h) | Power (kW) | | | | | | Cost (€/h) |
|---|---|---|---|---|---|---|---|
| | MT | FC | DE | WT | PV | Utility | |
| 1 | 90 | 95 | 0 | 252 | | 43 | 64.10 |
| 2 | 90 | 95 | 0 | 246 | 0 | 49 | 63.90 |
| 3 | 90 | 95 | 0 | 279 | 0 | 136 | 68.48 |
| 4 | 90 | 95 | 0 | 257 | 0 | 158 | 67.66 |
| 5 | 300 | 300 | 0 | 291 | 0 | −194 | 80.04 |
| 6 | 300 | 300 | 0 | 296 | 23 | −232 | 92.41 |
| 7 | 300 | 300 | 0 | 246 | 56 | −69 | 101.44 |
| 8 | 300 | 300 | 0 | 257 | 63 | −126 | 119.07 |
| 9 | 300 | 300 | 0 | 215 | 109 | −8 | 142.02 |
| 10 | 300 | 300 | 0 | 230 | 153 | −29 | 144.41 |
| 11 | 300 | 300 | 0 | 246 | 159 | 47 | 164.55 |
| 12 | 300 | 300 | 0 | 296 | 187 | −8 | 163.79 |
| 13 | 300 | 300 | 0 | 296 | 192 | 117 | 175.66 |
| 14 | 300 | 300 | 0 | 314 | 187 | 96 | 175.78 |
| 15 | 300 | 300 | 0 | 291 | 190 | 33 | 152.52 |
| 16 | 300 | 300 | 0 | 279 | 156 | −179 | 130.81 |
| 17 | 300 | 300 | 0 | 291 | 140 | −112 | 101.22 |

*(Continues)*

*Table 6.9 (Continued)*

| Time (h) | Power (kW) | | | | | | Cost (€/h) |
|---|---|---|---|---|---|---|---|
| | MT | FC | DE | WT | PV | Utility | |
| 18 | 300 | 300 | 0 | 302 | 61 | 8 | 103.52 |
| 19 | 300 | 300 | 0 | 332 | 49 | 110 | 112.17 |
| 20 | 300 | 300 | 0 | 302 | 38 | 160 | 107.19 |
| 21 | 300 | 300 | 0 | 291 | 18 | 57 | 89.51 |
| 22 | 300 | 300 | 0 | 279 | 13 | −39 | 77.82 |
| 23 | 300 | 300 | 0 | 279 | 0 | −159 | 71.10 |
| 24 | 90 | 95 | 0 | 279 | 0 | 136 | 68.48 |
| | | | | | | Total cost (€/h) | 2637.65 |

*Table 6.10 Best solutions for Case C in MG2*

| Time (h) | Power (kW) | | | | | | Cost (€/h) |
|---|---|---|---|---|---|---|---|
| | MT | FC | DE | WT | PV | Utility | |
| 1 | 242 | 152 | 0 | 252 | 0 | −165 | 63.79 |
| 2 | 242 | 152 | 0 | 246 | 0 | −160 | 63.63 |
| 3 | 242 | 152 | 0 | 279 | 0 | −73 | 68.82 |
| 4 | 242 | 152 | 0 | 257 | 0 | −50 | 68.15 |
| 5 | 300 | 300 | 0 | 291 | 23 | −194 | 77.33 |
| 6 | 300 | 300 | 0 | 296 | 56 | −232 | 89.16 |
| 7 | 300 | 300 | 0 | 246 | 63 | −69 | 100.47 |
| 8 | 300 | 300 | 0 | 257 | 109 | −126 | 117.30 |
| 9 | 300 | 300 | 0 | 215 | 153 | −8 | 141.91 |
| 10 | 300 | 300 | 0 | 230 | 159 | −29 | 144.00 |
| 11 | 300 | 300 | 15 | 246 | 187 | 32 | 165.19 |
| 12 | 300 | 300 | 15 | 296 | 192 | −23 | 163.66 |
| 13 | 300 | 300 | 15 | 296 | 187 | 102 | 177.28 |
| 14 | 300 | 300 | 15 | 314 | 190 | 81 | 177.12 |
| 15 | 300 | 300 | 15 | 291 | 156 | 19 | 152.97 |
| 16 | 300 | 300 | 0 | 279 | 140 | −179 | 128.30 |
| 17 | 300 | 300 | 0 | 291 | 61 | −112 | 99.65 |
| 18 | 300 | 300 | 0 | 302 | 49 | 8 | 103.64 |
| 19 | 300 | 300 | 15 | 332 | 38 | 96 | 113.70 |
| 20 | 300 | 300 | 15 | 302 | 18 | 145 | 109.41 |
| 21 | 300 | 300 | 0 | 291 | 13 | 57 | 90.30 |
| 22 | 300 | 300 | 0 | 279 | 0 | −39 | 77.27 |
| 23 | 300 | 300 | 0 | 279 | 0 | −159 | 68.87 |
| 24 | 242 | 152 | 0 | 279 | 0 | −73 | 68.82 |
| | | | | | | Total cost (€/h) | 2630.75 |

During the periods with mean load levels, corresponding with the mean market price (5.4 €ct/kWh), the surplus of energy is exported from the MG to the utility.

In the third case, the market price of energy is increased to its real market value. The MT and FC are employed with their maximum power in periods with mean and high load levels. Moreover, the surplus of energy is exported from the MG2 to the utility in most period of the day.

## 6.5 MATLAB® program *eommg*

An educational MATLAB-based program package, named *eommg*, has been developed to present the efficiency of different metaheuristic optimization methods in solving the EOM problem of MGs. The *eommg* program package is available as the supplementary file to this chapter on the book's homepage at the IET digital library.

The program package offers 20 user-selectable metaheuristic algorithms for solving the EOM problem. In addition to well-established methods such as PSO, GSA, GA, TLBO, and WDO there are several recently published algorithms such as PO, RUN, GBO, TSA, and different hybrid variants of PSO and GSA (PSOGSA, PPSOGSA, PSOSCGSA), etc. These algorithms are given in the form of MATLAB m functions. The user can use the default (defined) algorithmic parameters or enter the desired values for algorithm parameters within the m file of the selected algorithm.

EOM calculation starts by running the main program, *eommg*, as follows:

```
eommg(method,N,max_it)
```

Where input arguments can take the following values:

```
Method: e.g. 'PSO' 'GSA', 'WDO', 'GA', 'PSOGSA', etc.
N - Number of agents (e.g. 50).
max_it - Maximum number of iterations (e.g. 200).
```

Figure 6.5 shows an example of running the *eommg* program. The RUN algorithm was chosen for the EOM calculation in the MG1. The population size and the maximum number of iterations were chosen as 50 and 200, respectively.

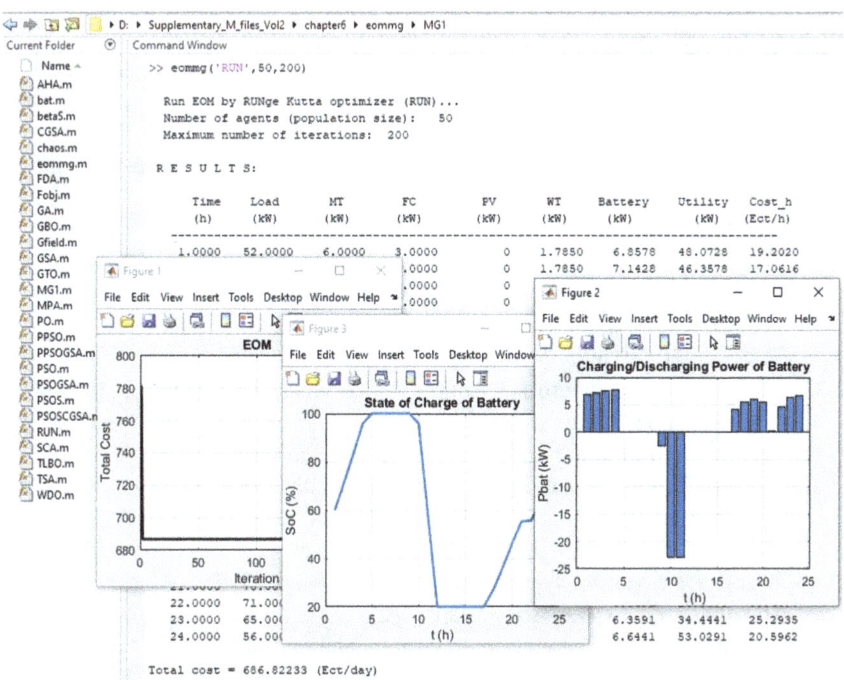


*Figure 6.5 An example of using the eommg program*

## 6.6 Conclusion

In this chapter, the application of metaheuristic optimization algorithms for EOM of MGs is demonstrated.

The proposed approach has been tested and investigated on two grid-connected MGs including different DG units and energy storage. The simulation results show the efficiency of the proposed approach to solve EOM problems under different operational scenarios of the MGs. As such, it can serve as a useful decision-making supporting tool for MG operators and help to find out how different input variables affect the operational characteristics of MGs.

## References

[1] Radosavljević J., Jevtić M., Klimenta D. 'Energy and operation management of a microgrid using particle swarm optimization'. *Engineering Optimization*. 2016;**48**(5):811–30.

[2] Hatziargyriou N. 'MICROGRIDS – large scale integration of micro-generation to low voltage Grids'. Available http://microgrids.eu/micro2000/presentations/16.pdf.

[3] Tsikalakis A.G., Hatziargyriou N.D. 'Centralized control for optimizing microgrids operation'. *IEEE Transactions on Energy Conversion*. 2008;**23**(1): 241–48.

[4] Ustun T.S., Ozansoy C., Zayegh A. 'Recent developments in microgrids and example cases around the world – a review'. *Renewable and Sustainable Energy Reviews'*. 2011;**15**:4030–41.

[5] Jiayi H., Chuanwen J., Rong X. 'A review on distributed energy resources and MicroGrid'. *Renewable and Sustainable Energy Reviews'*. 2008; **12**:2472–83.

[6] Lopes J.A.P., Moreira C.L., Madureira A.G. 'Defining control strategies for microgrids islanded operation'. *IEEE Transactions on Power Systems*. 2006; **21**(2):916–24.

[7] Basak P., Chowdhury S., nee Dey S.H., Chowdhury S.P. 'A literature review on integration of distributed energy resources in the perspective of control, protection and stability of microgrid'. *Renewable and Sustainable Energy Reviews*. 2012;**16**:5545–56.

[8] Planas E., Gil de Muro A., Andreu J., Kortabarria I., Martinez del Algria I. 'General aspects, hierarchical controls and droop methods in microgrids: a review'. *Renewable and Sustainable Energy Reviews*. 2013;**17**:147–59.

[9] Niknam T., Golestaneh F., Malekpour A. 'Probabilistic energy and operation management of a microgrid containing wind/photovoltaic/fuel cell generation and energy storage devices based on point estimate method and self-adaptive gravitational search algorithm' *Energy*. 2012;**43**:427–37.

[10] Hatziargyriou N.D., Anastasiadis A.G., Tsikalakis A.G., Vasiljevska J. 'Quantification of economic, environmental and operational benefits due to

significant penetration of Microgrids in a typical LV and MV Greek network'. *European Transaction on Electrical Power*. 2011;**21**(2):1217–37.

[11] Zhang Y., Gatsis N., Giannakis G.B. 'Robust energy management for microgrids with high-penetration renewables'. *IEEE Transactions on Sustainable Energy*. 2013;**4**(4):944–53.

[12] Moghaddam A.A., Seifi A., Niknam T., Pahlavani M.R.A. 'Multi-objective operation management of a renewable MG (micro-grid) with back-up microturbine/fuel cell/battery hybrid power source'. *Energy*. 2011;**36**:6490–507.

[13] Radosavljević J., Milovanović M., Arsić N., Jovanović A., Perović B., Vukašinović J. 'Optimal power dispatch in distribution networks with PV generation and battery storage'. *In Proceedings of the IX International Conference on Electrical, Electronic and Computing Engineering IcETRAN 2022*, Novi Pazar, Serbia, 6–9 June 2022.

[14] Mohamed F.A., Koivo H.N. 'System modeling and online optimal management of MicroGrid using mesh adaptive direct search'. *Electrical Power and Energy Systems*. 2010;**32**:398–407.

[15] Villanueva D., Pazos J.L., Feijo A. 'Probablistic load flow including wind power generation'. *IEEE Transactions on Power Systems*. 2011;**26**(3):1659–67.

[16] Atwa Y.M., El-Saadany E. F., Salama M. M. A., Seethapathy R., Assam M., Conti S. 'Adequacy evaluation of distribution system including wind/solar DG during different modes of operation'. *IEEE Transactions on Power Systems*. 2011;**26**(4):1945–52.

[17] Tiwari G.N., Dubey S. *Fundamentals of Photovoltaic Modules and Their Applications*. London: RSC Energy Series Cambridge; 2010. p. 423.

[18] Wu H., Liu X., Ding M. 'Dynamic economic dispatch of a microgrid: mathematical models and solution algorithm'. *Electrical Power and Energy Systems*. 2014;**63**:336–46.

[19] Campanari S., Macchi E. 'Technical and tariff scenarios effect on microturbine trigenerative applications'. *Journal of Engineering for Gas Turbines and Power*. 2004;**126**: 581–89.

[20] Hardy Diesel Generators. Model: HDJW 225 T6 [online]. 2012. Available from http://www.hardydiesel.com/john-deere-generators/dl/john-deere-generator-hdjw-225-t6.pdf (Accessed September 2017).

# Chapter 7
# Optimal placement and sizing of distributed generation

## 7.1 Introduction

Distributed generation (DG) is made up of various electrical energy sources with relatively low power that are integrated into a local distribution network. Over the past few decades, DG units have been widely used as a result of increasing consumption, deregulation and liberalization of the electrical power market, and pollution control measures. In general, integrating DG units into a distribution network has good technical implications. The DG units can have an impact on reducing power loss, improving voltage quality, increasing reliability and efficiency of consumer electricity supply, and lowering pollutingemission (by the installation of renewable energy sources).

The degree and quality of a DG's impact on the distribution network performance is determined by its power, type, and location. Optimal placement and sizing of a DG unit aims to maximize the positive and minimize the negative effects on the distribution network's operation. Accordingly, the selection of the objective function depends on the desired objective, which is usually the following:

- minimizing power and energy losses;
- improving the network voltage profile, that is, minimization of voltage deviations;
- increasing the availability of system elements;
- increasing the reliability of power supply to consumer units;
- minimizing the levels of short-circuit currents;
- minimizing harmonic distortions;
- minimizing total costs pertaining to the operation of a network (installation of equipment, purchase of power, power losses, power supply interruptions);
- minimizing investment, procurement, and costs relating to DG connection.

Based on the possibility of determining locations and sizes, DGs may be classified into two groups. Locations and sizes in the instance of the first group are to the largest extent determined by geographic, hydrological, and meteorological factors (small hydroelectric power plants, wind generators). In this case, optimization of the connection point to the existing distribution network is possible only if a certain set of buses to which a DG may be connected is specified. Optimal solutions may be looked for during the stage of planning of an entirely new or reconstruction of the existing network in a specified (geographically and meteorologically) permissible area. The second group includes those DGs with predictable power outputs (micro turbines, fuel cells, diesel generators, and so on), which may be connected to any point in a distribution network.

Several approaches to resolving the problem of selecting an optimal location and size of DGs are represented in scientific literature. Starting from analytical techniques, through methods of linear and quadratic programming, all the way to recently rather popular metaheuristic optimization methods that mimic natural processes and patterns.

An analytical approach to determining optimal locations of DGs in networks with radial and loop topology, which does not include computations of power flows, is presented in [1]. For simple radial systems, the procedure is based on the assumption that the feeder's parameters are constant (with the same cross-section in all sections). In contrast, load is uniformly distributed along the feeder. According to this assumption, an analytic function of change in the current is derived along the feeder with the change in the distance from the feed point. The optimal point of connection of a DG with a defined power is determined based on the condition of minimum power losses. In large networks with looped topology, the procedure is based on modifying the bus admittance matrix when adding DGs to network buses and examining the values of the objective function after the modification that has occurred. Assessing the favorability of buses for connecting a DG in terms of improved reliability of load supply, reduction of losses, and an improved voltage profile of the network in [2] is carried out by means of the voltage sensitivity index (*VSI*) and loss sensitivity index (*LSI*) of network buses. According to [3], when characteristics of costs per unit of DG power are known, the optimal location and power of a DG in a distribution network are determined according to the optimal power flow principle. The method is based on computing bus location marginal costs. The aim of such optimization is to generate the maximum profit for DG owners since, in deregulated energy markets, the price of energy is largest in peak (maximum) load periods; therefore, engaging a DG in said period generates the largest profit for the owner. Recently, some interesting analytical approaches have been proposed to solve the problem of optimal siting and sizing of DG in radial networks. In [4], the authors formulated a loss sensitivity factor for the distribution networks and employed it for the determination of the optimum size and location of DG with the aim of minimizing total power loss. Elsaiah *et al.* [5] proposed an analytical method for placement and sizing of DG based on a new formulation for the power flow problem, which is non-iterative, direct, and involves no convergence issues, even for systems with high $r/x$ branch ratios. Naik *et al.* [6] determine the optimal siting and sizing of DG using suitable analytical expressions, which are based on the change in active and reactive components of branch currents caused by the DG placement. A similar approach was proposed by Viral and Khatod [7]. They developed analytical expressions which are based on minimizing the loss associated with the active and reactive components of branch currents by placing the DG at various locations.

The methodology based on the application of linear programming presented in [8] solves the problem of optimal locations for the purpose of maximizing sizes of DGs that may be connected while not impeding technical constraints pertaining to permissible values of bus voltages and limit values of currents/powers of feeders and transformers. An optimization methodology for establishing optimal locations and sizes of DGs is developed in [9] by applying the sequential quadratic programming algorithm with the purpose of reducing losses of power and costs of DGs in a distribution network while respecting voltage constraints.

In recent years, most authors have used some metaheuristic population-based methods for solving the problem of optimal placement and sizing of DG units. A mini-review of optimal placement and sizing of DGs using metaheuristic optimization algorithms is given in Volume 1 of the book (Chapter 6, Section 6.5).

A higher penetration of a DG requires a somewhat altered distribution network planning strategy. In that respect, an optimization model has been developed in [10,11] which, among other things, takes into consideration the optimal location and size of a DG in a network. An optimal solution is employed to minimize investment and operating costs. A method for optimal planning of the location and size of micro-hydropower resources in a catchment area is presented in [12] with the purpose of minimizing investment costs per unit of generated electrical power. This method enables selecting an optimal turbine type for each site, setting its optimal head, flow, and power output.

The problem of optimal DG positioning and size is regarded in [13] from the aspect of mobile reactive compensators, which ensure maximum voltage stability in a distribution network. A quantitative voltage stability index of load buses is proposed. An optimal DG location and size are established based on the mentioned index. The tabu search (TS) algorithm is applied in [14] for concurrent determination of locations and sizes of DGs, as well as of reactive power sources, whereas a multi-objective function of costs is employed, and it consists of power losses, feeder load, and total reactive power capacity of a distribution network.

The algorithms for optimal DG placement and sizing are important supporting tools for the Distribution Network Operators (DNO) since they provide a guideline for proper strategic planning regarding DG penetration. Within the current liberalized energy market, the DG installation is usually ruled by economic incentives related to private investment plans or by environmental and geographical constraints and criteria. Nevertheless, the prioritization regarding the critical installation nodes for DG could allow the DNO to promote DG installation at critical nodes to satisfy the specifications defined by the optimal DG placement planning. Although the DNO might be unable to canalize the funds of private investments, proper planning about the prioritization of such investment actions should be performed and be implemented whenever possible [15].

In order to determine the optimal location and size of a DG, it is necessary to either find or detect the relation between the power injected by a DG at a given point and the change in the size of the value that needs to be optimized. The simplest (most trivial) manner is the successive placement of DGs at each bus and computation of the given value, whereas various sizes and types of DGs must be taken into consideration, as well as different load levels throughout the observed period. When everything is taken into account, along with the fact that real distribution networks contain up to several thousands of buses, the number of possible combinations that should be taken into consideration becomes immense. However, this number may be significantly reduced by taking into account certain information on the configuration and manner in which the network operates. In addition, a preliminary ranking list of favorable locations may be compiled by applying the procedure for computing the bus location coefficients.

A procedure for determining optimal DG locations and sizes in radial balanced symmetrical three-phase distribution networks with the objective of minimizing active power and energy losses is presented in this chapter [16,17]. This is a two-step procedure. Step 1 includes computing bus location coefficients based on the change in active power losses with the change in injecting active and reactive power in the network buses. In this manner, a preliminary ranking list of favourable locations (buses) is compiled for connecting a DG. By taking the initial $M$ buses from the preliminary ranking list out of the total NB buses in the network, the number of combinations to be taken into consideration is significantly decreased. Step 2 includes selecting optimal DG locations and sizes from the narrowed set of possible solutions by means of the selected search technique. To that end, two search techniques are used:

- Partial search of variants;
- Genetic algorithm (GA).

## 7.2 Preliminary locations of DG

The procedure is based on identifying the change in the total power loss of a distributed network caused by the change in the bus injection power. The measure of favourability to connect a DG is expressed by bus location coefficients.

Active and reactive power losses of a network constitute an inevitable consequence of the active and reactive energy flow, the powers of which are $P$ and $Q$, respectively. The power flows, and thereby system power losses, are dependent on voltage. Accordingly, the change in power loss $P_{loss}$ caused by the change in the injection $P$ and $Q$ may be expressed in the following manner [2]:

$$\begin{bmatrix} \dfrac{\partial P_{gub}}{\partial P} \\ \dfrac{\partial P_{gub}}{\partial Q} \end{bmatrix} = \mathbf{J}^{-1} \cdot \begin{bmatrix} \dfrac{\partial P_{gub}}{\partial \theta} \\ \dfrac{\partial P_{gub}}{\partial V} \end{bmatrix} \tag{7.1}$$

where $\mathbf{J}$ is the Jacobian matrix,

$$\mathbf{J} = \begin{bmatrix} \dfrac{\partial \Delta P}{\partial \theta} & \dfrac{\partial \Delta P}{\partial V} \\ \dfrac{\partial \Delta Q}{\partial \theta} & \dfrac{\partial \Delta Q}{\partial V} \end{bmatrix} \tag{7.2}$$

Total power losses may be determined as a sum of injected active powers in all network buses:

$$P_{gub} = \sum_{i=1}^{NB} P_i = \sum_{i=1}^{NB} \sum_{j \in \alpha_i} V_i V_j \left[ G_{ij} \cos\left(\theta_i - \theta_j\right) + B_{ij} \sin\left(\theta_i - \theta_j\right) \right]; \tag{7.3}$$

The derivative of power losses with respect to the angle and magnitude of the voltage at bus $i$ may be determined from (7.3) as follows:

$$\frac{\partial P_{gub}}{\partial \theta_i} = \sum_{i=1}^{NB} \sum_{j \in \alpha_i} V_i V_j \left[ -G_{ij} \sin\left(\theta_i - \theta_j\right) + B_{ij} \cos\left(\theta_i - \theta_j\right) \right] \tag{7.4}$$

$$\frac{\partial P_{gub}}{\partial V_i} = \sum_{i=1}^{NB} \sum_{j \in \alpha_i} V_j \left[ G_{ij} \cos\left(\theta_i - \theta_j\right) + B_{ij} \sin\left(\theta_i - \theta_j\right) \right] \tag{7.5}$$

where $P_i$ and $Q_i$ are active and reactive injecting powers at bus $i$; $V_i$, and $\theta_i$ are the magnitude and angle of the voltage at bus $i$; $V_j$, and $\theta_j$ are the magnitude and angle of the voltage at bus $j$; $\alpha_i$ is the set of buses connected with the bus $i$; $B_{ij}$, and $G_{ij}$ are the susceptance and conductance of the network bus admittance matrix element ($\underline{Y}_{ij} = G_{ij} + jB_{ij}$); and $NB$ is the total number of buses.

The hybrid form of the equations of power debalans in network buses is as follows:

$$\Delta P_i = P_i - V_i^2 G_{ii} - V_i \sum_{\substack{j=1 \\ j \neq i}}^{NB} V_j \left[ G_{ij} \cos\left(\theta_i - \theta_j\right) + B_{ij} \sin\left(\theta_i - \theta_j\right) \right] \tag{7.6}$$

$$\Delta Q_i = Q_i + V_i^2 B_{ii} - V_i \sum_{\substack{j=1 \\ i \neq j}}^{NB} V_j \left[ G_{ij} \sin\left(\theta_i - \theta_j\right) - B_{ij} \cos\left(\theta_i - \theta_j\right) \right] \tag{7.7}$$

Elements of the Jacobian matrix are computed by partial derivatives of (7.6) and (7.7) with respect to the angle and magnitude of the bus voltages:

$$\frac{\partial \Delta P_i}{\partial \theta_i} = V_i \sum_{\substack{j=1 \\ j \neq i}}^{NB} V_j \left[ G_{ij} \sin\left(\theta_i - \theta_j\right) - B_{ij} \cos\left(\theta_i - \theta_j\right) \right] \tag{7.8}$$

$$\frac{\partial \Delta P_i}{\partial \theta_j} = -V_i V_j \left[ G_{ij} \sin\left(\theta_i - \theta_j\right) - B_{ij} \cos\left(\theta_i - \theta_j\right) \right], \quad j \neq i \tag{7.9}$$

$$\frac{\partial \Delta P_i}{\partial V_i} = -2V_i G_{ii} - \sum_{\substack{j=1 \\ j \neq i}}^{N} V_j \left[ G_{ij} \cos\left(\theta_i - \theta_j\right) + B_{ij} \sin\left(\theta_i - \theta_j\right) \right] \tag{7.10}$$

$$\frac{\partial \Delta P_i}{\partial V_j} = -V_i \left[ G_{ij} \cos\left(\theta_i - \theta_j\right) + B_{ij} \sin\left(\theta_i - \theta_j\right) \right], \quad j \neq i \tag{7.11}$$

$$\frac{\partial \Delta Q_i}{\partial \theta_i} = -V_i \sum_{\substack{j=1 \\ i \neq j}}^{N} V_j\left[G_{ij}\cos\left(\theta_i - \theta_j\right) + B_{ij}\sin\left(\theta_i - \theta_j\right)\right] \tag{7.12}$$

$$\frac{\partial \Delta Q_i}{\partial \theta_j} = V_i V_j\left[G_{ij}\cos\left(\theta_i - \theta_j\right) + B_{ij}\sin\left(\theta_i - \theta_j\right)\right], j \neq i \tag{7.13}$$

$$\frac{\partial \Delta Q_i}{\partial V_i} = 2V_i B_{ii} - \sum_{\substack{j=1 \\ i \neq j}}^{N} V_j\left[G_{ij}\sin\left(\theta_i - \theta_j\right) - B_{ij}\cos\left(\theta_i - \theta_j\right)\right] \tag{7.14}$$

$$\frac{\partial \Delta Q_i}{\partial V_j} = -V_i\left[G_{ij}\sin\left(\theta_i - \theta_j\right) - B_{ij}\cos\left(\theta_i - \theta_j\right)\right], \quad j \neq i \tag{7.15}$$

Values of the magnitudes and angles of the bus voltages are determined by computing the power flows. This enables computing the Jacobian matrix and derivative of the power loss with respect to the bus voltage magnitude and phase. Equation (7.1) enables determining the changes in active power losses in a distribution network caused by the change in injection power in the nodes. As a measure of favourability for DG connection, values of the loss sensitivity index (*LSI*) are taken as the bus location coefficients (*BLC*s), which are, according to [2], determined in the following manner:

$$BLC_i = LSI_i = w\frac{\partial P_{gub}}{\partial P_i} + (1 - w)\frac{\partial P_{gub}}{\partial Q_i} \tag{7.16}$$

In (7.16), $w$ is the weight factor dependent on the network $r/x$ ratio. The ratio of resistance and reactance ($r/x$) of feeders and transformers in distribution networks varies broadly. The approach adopted herein implies that in distribution networks with a vast range of the $r/x$ ratio, the weight factor $w$ is computed for each bus individually. Due to the radial topology of the network and relatively small powers, the change in injection power at a given bus has the largest effect on power losses at the correspondent branch. Given the manner of enumeration of buses and branches in the radial distribution network, the computation of weight factors is rather simple:

$$w_i = 1 - \frac{1}{\frac{r_i}{x_i} + 1} \tag{7.17}$$

where $w_i$ is the weight factor for bus $i$ and $r_i$ and $x_i$ are the resistance and reactance of the $i$th branch, respectively.

The need for such an approach is particularly justified for lateral branches of a radial distribution network, since ratios of resistances and reactances of distribution transformers differ significantly from the $r/x$ ratio of medium voltage feeders to which they are connected, particularly when smaller distribution transformers are considered in a network. If a distribution network is homogenous (one voltage level), the feeders have approximately the same parameters; therefore, the $r/x$ ratio is within narrow limits. In that instance, a unique weight factor is employed for the entire system, and it is determined as a mean value of weight factors of all buses computed according to (7.17).

When power is successively injected in two buses with different *BLC* values, the relation between the *BLC* values and power losses in a network depends on the ratio between the DG power and power load. In other words, *BLC* values as a criterion for assessment of favourability of a bus for DG connection should be construed while taking into account the ratio between the DG power and the total network load power. This ratio is important since it defines the power distribution in a network and consequentially losses as well. In general, if DG power is smaller than 50% of the power load, buses with larger *BLC* values are more suitable for connecting, and vice versa: when DG power is larger than 50% of the power load, buses with smaller *BLC* values are considered more suitable for DG connecting.

On the basis of *BLC* values, a set of preliminary buses is defined as a starting point for determining an optimal location or several optimal locations for DG connecting for the purpose of minimizing power and energy losses in a distribution network. To that end, a certain optimization procedure or a search technique is applied. The algorithm for creating preliminary DG locations based on the *BLC*s may be summarized in the following six steps:

(1) Defining the topology (configuration) of the network and parameters of elements. The network is defined by enumerating buses by branches, i.e. buses on the longest branch are marked first, followed by marking of buses on other branches, starting from the branch most distant from the root bus.

(2) Defining the power load in a given distribution network. It is difficult to determine the exact values, but in practice, minimum and maximum power of each bus over a certain period may be familiar. On this occasion, power load is determined by multiplying rated values ($P_{ni}$, $Q_{ni}$) with the load coefficient ($k_{load}$), the value of which ranges between 0 and 1

$$P_{Pi} = k_{load} \cdot P_{ni} \tag{7.18}$$

$$Q_{Pi} = k_{load} \cdot Q_{ni} \tag{7.19}$$

(3) Preliminary computation of power flows and bus voltages.

(4) Using the magnitudes and angles of network bus voltages computed in Step 3, computing the elements of Jacobian matrix (7.8)–(7.15) and derivatives of the power loss function [(7.4) and (7.5)].

(5) Computing the *BLC* values for all network buses (7.16).

(6) Selection of preliminary DG locations based on the *BLC* values.

If there are DGs in a specific distribution network, when determining *BLC* buses, these are treated as negative constant power loads (*PQ*).

## 7.3 Partial search of variants

Variant search includes computation of the selected objective function for each combination of the DG size, type, and location. The optimal solution is obtained by comparing the values of objective function. To evaluate a variant, the power flow computation is performed. As the number of possible combinations may be rather large, they are reduced by taking into account certain information on the real operation of the network. Accordingly, partial search algorithms appear as a separate group of methods. In such a manner, better or worse local optima are found with a smaller or larger extent of efficiency. The goal is to have them as close to the global optimum as possible, whereas it is insisted that this is performed in as short a time as possible, that is, to search for as many combinations in as efficiently formulated manner as possible. This procedure directly implies (includes):

- selection of a favorable set of buses for searching;
- adequate DG modeling;
- taking into account various levels of load over a given period;
- defining the objective function;
- power flow computation;
- taking into account the operating constraints.

### *7.3.1 Optimal DG placement by using partial search of variants*

The basic idea is to narrow the starting set of possible solutions by employing the results of previous research while starting from characteristics of the specific network and thereby obtain a solution faster. Criteria for narrowing the set may vary. For instance, it was demonstrated that connecting a DG close to the root bus has a minimal impact on the loss reduction and voltage improvement, that is, the best effects in radial networks are achieved by placing a DG in peripheral parts (buses), or on the basis of bus location coefficients determined according to the algorithm referred to in Section 7.2.

The number of preliminary locations may be selected arbitrarily. A larger number ensures the reliability of the optimal solution, but the number of combinations and time of computation is consequentially increased. Conversely, a smaller number of preliminary locations implies a smaller number of search variants, but also a larger probability that the optimal solution found has a local character. In this manner, the number of search variants is reduced significantly. For other practical reasons, sometimes a specific set of buses is given to which a DG may be connected. Therefore, taken as a starting point herein are $M_p$ buses ($M_p$ < NB), which constitute a preliminary set of buses for DG connecting. A single combination of locations (buses) for connecting the given number of DGs constitutes a single variant. Thereby, it is adopted that only one DG may be connected to a single bus. If the number of preliminary locations is $M_p$, and the number of DGs for which optimal locations are to be established is $N_{DG}$, then the number of variants to be explored is equal to the number of variations without repetition:

$$N_{\text{var}} = N_{DG}!\binom{M_p}{N_{DG}} = M_p(M_p - 1)(M_p - 2)...(M_p - N_{DG} + 1) \tag{7.20}$$

For each possible variant, it is necessary to perform at least one power flow computation in order to determine its quality according to the adopted optimization criterion. Therefore, the efficiency of the power flow computation method is very

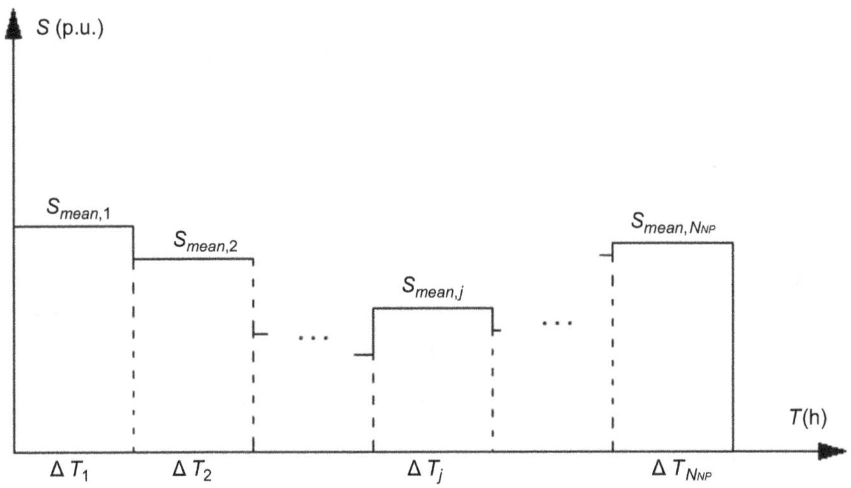


*Figure 7.1 Mean values of load over a period (year)*

important. A DG must be adequately modeled based on the output power and voltage regulation ability. The DGs are modeled as PQ and PV buses herein. The relevance of the DG placement and sizing results is directly related to the accuracy of the model and the level of load over the observed period. The term period employed herein implies a single year (8760h). The annual load diagram is approximated by a step curve, as shown in Figure 7.1. The larger the number of segments in the step curve, the more accurate the approximation.

The symbols in Figure 7.1 have the following meaning: $S_{\text{mean,j}}$ is the mean value of the power of a typical daily load diagram in the segment $j$ of the annual load diagram; $\Delta T_j$ is the duration of the segment $j$ of the annual load diagram; and $N_{NP}$ is the total number of annual load diagram segments.

It is assumed that the form of the annual load diagram is the same for all consumer units in the considered distribution network. It means that it is necessary to define only one diagram in which the size is expressed in relative units ranging from 0 to 1. The actual annual load diagram of a consumer unit is obtained by multiplying the rated or maximum power of the consumer unit with the relative diagram.

The optimal solution (optimal DG locations) depends on the adopted optimization criterion. The most commonly used optimization criterion is the effect of a DG on the loss reduction. In order to encompass the effect of the change in load during the course of a year, total annual active energy loss in the distribution network is taken as the objective function. Therefore, the aim to be achieved by optimization is minimizing annual active energy loss:

$$\min \Delta W_v = \min \sum_{j=1}^{j=N_{NP}} P^j_{loss,v} \cdot \Delta T_j \tag{7.21}$$

where $\Delta W_v$ denotes total annual active energy loss in the network for the $v$th variant of DG locations, $v = 1, \ldots, N_{var}$; $P^j_{loss,v}$ is the power loss in the network in the segment $j$ of the annual load diagram for the $v$th variant of DG locations; $\Delta T_j$ is the duration of the segment $j$ in the annual loss diagram; and $N_{NP}$ is the total number of annual load diagram segments.

In addition, operational constraints must be satisfied:

$$V_{i\ \min} \leq V_i^j \leq V_{i\ \max} \tag{7.22}$$

$$J_i^j \leq J_{i\ \max} \tag{7.23}$$

where $V_i^j$ is the voltage at bus $i$ for the $j$th load level; $V_{i\ \min}$, and $V_{i\ \max}$ are minimum and maximum permissible values of the voltage at bus $i$, respectively; $J_i^j$ is the current in branch $i$ for the $j$th load level; and $J_{i\ \max}$ is the maximum permissible RMS current of branch $i$ of the network.

The algorithm for determining optimal DG locations is quite simple, and it consists of several steps as follows:

(1) Defining the network configuration, data on parameters of network elements and on loads;
(2) Defining the annual load diagram parameters: $S_{\text{mean},j}$ and $\Delta T_j$, $j = 1, ..., N_{NP}$ (according to Figure 7.1);
(3) Defining preliminary locations (buses) – based on bus location coefficients, $BLC_i$, $i = 1, ..., M_p$;
(4) Defining the number $N_{DG}$, type, and size of DGs to be allocated;
(5) Defining possible DG location variants, $v = 1, ..., N_{var}$;
(6) Computing the objective function $\Delta W_v$ (7.21) and examining the constraints (7.22) and (7.23). If any of the constraints are violated, the current variant $v$ is excluded.
(7) Comparing the criterion function values obtained and determining the optimal combination of locations for connecting DGs of the defined types and sizes.

### 7.3.2 *Optimal DG sizing by using partial search of variants*

This problem may be characterized as both operation-related and planning-related. An operation-related problem refers to establishing an optimal distribution of power among several DGs, that is, establishing the optimal DG operation in a given stationary network mode aimed at loss reduction. A planning-related problem pertains to the selection of installed power of DGs to be connected to pre-defined locations, which are not necessarily optimal. Both cases refer to the selection of optimal variant of operation of DGs connected to given network buses. A single variant is a single possible solution that constitutes a combination of DG output powers from the defined DG power ranges. Various DGs may operate within various power ranges. Owing to the combinatorial nature of the problem, it is convenient to represent the mentioned range as a segment function in the following form:

$$P_{DGk} = a \cdot \Delta P_{DG} \tag{7.24}$$

where $a$ is a coefficient ($a = 1, 2, ..., N_s$); $N_s$ is the total number of segments within the DG power range; $\Delta P_{DG}$ is a power change step within the defined range.

In a general case, DGs may operate within various power ranges. However, all of them have the same number of segments, the width of which depends on the width of the range of the specific DG. If $N_{DG}$ is the number of DG locations, and if the power range of all DGs is divided into $N_s$ segments, then the number of possible solutions is equal to the number of variations with the repetition $N_{\text{var}}^{p} = N_s^{N_{DG}}$. As an illustration, if: $N_{DG} = 3$ and $N_s = 50$, then the number of variants to be examined is $N_{\text{var}}^{p} = 125000$. If we assume that a computer requires 0.1 s to examine one solution variant for a network of several dozens of buses, then it would take about 3.5 h to examine all 125,000 possible variants in this example. Therefore, it is necessary to resort to a drastic decrease in the number of variants to be examined. The approach applied herein includes simultaneous repetition of search. For instance, if the initial DG power range from the previous example is divided into ten segments, the number of possible solutions is 1000. The first search results in a rough estimate of optimal DG size. Then, in the following step new DG size ranges are estimated close to the optima obtained in the previous step. The search process is repeated on the newly formed size ranges with the same number of segments. The entire cycle is repeated several times. In this manner, a satisfactory solution is obtained with a drastic decrease in the number of variants examined. Figure 7.2 shows the described approach of reducing the number of variants in the partial search process.

The objective is to minimize power losses of a network for the given load power level. The process of partial search of variants is performed for any level of consumption over a given period. The result is optimal DG size for all load levels over a given period. For each DG size variant from within the defined ranges, constraints (7.22) and (7.23) are examined. Violation of any constraint means eliminating the current variant. The criterion function constitutes power losses of a network $P^{j}_{loss,v}$, where $v$ denotes (is the number of) the variant and $j$ is the segment over the period (year). The algorithm for optimal DG sizes at the given locations by applying a partial search of variants includes the following steps:

(1) Defining the network configuration, data on parameters of network elements and on loads;

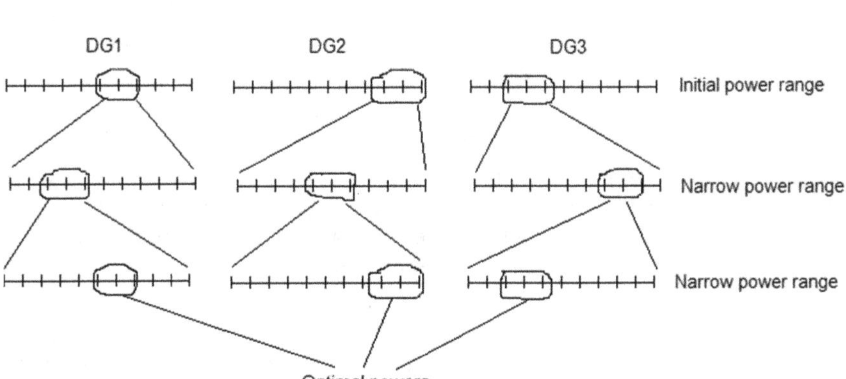


*Figure 7.2 Reducing the number of variants by narrowing the DG power range*

(2) Defining the load diagram parameters in a given period: $S_{\mathrm{mean},j}$ and $\Delta T_j$, $j = 1, \ldots, N_{NP}$;
(3) Defining locations and DG power ranges;
(4) Defining the number of $N_s$ segments for the DG size range;
(5) Defining the number of repetitions so as to narrow the range;
(6) Dividing the size ranges of all DGs into a uniform number of segments $N_s$;
(7) Defining possible variants of DG sizes – solutions, $v = 1, \ldots, N_{\mathrm{var}}$;
(8) Computing power losses $P^j_{loss,v}$ for various variants of DG sizes ($v$) for a given level of load power ($j$);
(9) Verifying whether conditions (7.22) and (7.23) have been met. If any of the given conditions have been breached, the current variant is excluded;
(10) Comparing the obtained values of power losses and establishing optimal DG sizes;
(11) Narrowing the DG size range close to optimal values and returning to Step 6.

The algorithm is performed for any power load level $j = 1, \ldots, N_{NP}$.

## 7.4 Genetic algorithm

Distribution network optimization involves solving the problem of optimal locations and DG powers. Distribution networks of real dimensions have up to a few hundred nodes. The nature of variables whose values need to be optimized is twofold. In addition to continuous variables such as voltage and power, there are discrete and integer variables such as locations (nodes) and the number of DG units. For this case, the application of classical optimization methods based on the differentiation of an objective function can be difficult. Therefore, the so-called evolutionary optimization techniques are used to overcome this problem. They are based on searching for a set of possible solutions using the rules of natural evolution. Due to the large number of possible solutions, the problems of optimizing the distribution network with DG can be considered combinatorial optimization problems.

One of well-established evolutionary methods that is very effective in solving these combinatorial optimization problems is the genetic algorithm (GA). The GA as an optimization method was created by Holland [18]. He elaborated an idea of how to transform the characteristics of natural evolution into a computer program.

The GA operates with a population of individuals. Each individual constitutes a potential solution to a given optimization problem. An individual can be described as a set of variable conditions, the values of which are being optimized. The quality of an individual is quantified over the value of the fitness function or the function of goodness. A population of individuals constitutes a set of solutions to the given optimization problem. One generation has a population with a certain number of individuals with better or worse fitness function values. Through a series of GA operations, a new population is obtained, and its individuals are created by the individuals from the previous population according to the natural evolution principles: the selection, crossover, and mutation.

The basic steps of the GA are

*Step* 1. Representation of individuals. All data (variables) that make an individual are written in a string. A string is composed of substrings. Each substring represents a binary encoded variable on which the process of optimization is carried out.

*Step* 2. Initialization. Individuals with random strings are generated that set up the initial population.

*Step* 3. Fitness function calculation. It is used to rate the quality of an individual, and it represents an equivalent of the function that should be optimized, that is, objective function.

*Step* 4. Selection. During the selection process the individuals that will participate in the reproduction (parents) are selected. The point of the selection is to store and transfer good individuals to the next generation.

*Step* 5. Crossover. The way in which coded column parts (substrings) are crossed over actually makes a GA. Crossover is an exchange process of column parts between two individuals, that is, "parents." One or two new individuals engender by the crossover, that is, a "child." The possibility of inheriting the first parent's characteristics by a child is introduced during this process.

*Step* 6. Mutation. Mutation is a way to give a new piece of information to an individual. It represents an accidental bit variation of an individual, generally with a constant probability for each bit within a population.

*Step* 7. Ending conditions. The GA is an iterative process that ends when a maximum number of generations is achieved or when another criterion is fulfilled, such as a minimum offset from the best fitness value and medium fitness value of all individuals in a current population. Otherwise, return to Step 3.

The main advantages of GA over the classical optimization methods are:

- The objective function to be optimized is entirely optional, that is, there are no special requirements such as continuity, differentiability, etc.;
- Reliability of the results obtained may be increased by simple repetition of the process;
- They solve any problem that may be presented as optimization-related, regardless of whether the state variables are real numbers, bits, or characters;
- Availability of program support. There are ready-made software packages that may be applied to solving concrete problems. For instance, the *globaloptim* toolbox in the MATLAB® software package.

The main weakness of GA are as follows:

- Parameters exert a considerable impact on efficiency. There is no universal rule for parameter setting;
- It is impossible to obtain a solution that is 100% reliable.

### 7.4.1 Optimal DG placement and sizing by using GA

In this optimization problem, the total energy losses of the distribution network are considered an objective function, whereas the control variables are locations (i.e. indexes of connecting buses) and power outputs of the DG units. Therefore, a potential solution to the problem can be represented by a vector (string) consisting of a combination of locations and powers of DG units, as shown in Figure 7.3.

Both DG types, *PQ* and *PV* are considered. The total energy losses are computed based on the system power losses and time during which the mentioned losses occur. In order to obtain as realistic a model as possible, the load curve is approximated by a step curve with $N_{NP}$ levels, as shown in Figure 7.1. An objective function of energy losses, in a general case, may be formulated in the following manner:

$$\min \Delta W = \min \sum_{j=1}^{j=N_{NP}} P^j_{gub} \cdot \Delta T_j \tag{7.25}$$

where $P^j_{loss}$ is the power loss at the *j*th load level and $\Delta T_j$ is the duration of the *j*th load level.

At the same time, the following operating constraints must be satisfied:

(1) Power flow equations:

$$f^j\left(x_j, P^j_{DGi}\right) = 0 \tag{7.26}$$

where $P^j_{DGi}$ is the power of DG at bus *i* in case the *j*th load level and $x_j$ is the state variable (bus voltages) for the *j*th load level.

(2) The bus voltages and branch currents must be within the permissible limits:

$$V_{i\,\min} \le V^j_i \le V_{i\,\max} \tag{7.27}$$

$$J^j_i \le J_{i\,\max} \tag{7.28}$$

where $V^j_i$ is the voltage at bus *i* for the load level *j*, $V_{i\,\min}$, and $V_{i\,\max}$ are minimum and maximum permissible values of voltage at the bus *i*, $J^j_i$ is current of the *i*th branch for the load level *j*, and $J_{i\,\max}$ is the maximally permissible current of the branch *i*.

(3) Constraints pertaining to the size or number of DGs that may be connected to certain distribution network buses:

$$P^j_{DGi\,\min} \le P^j_{DGi} \le P^j_{DGi\,\max} \tag{7.29}$$

| 1 | | 2 | | | $N_{DG}$ | |
|---|---|---|---|---|---|---|
| Bus | $P_{DG}$ | Bus | $P_{DG}$ | ... | Bus | $P_{DG}$ |

*Figure 7.3 An individual (possible solution) for optimal location and size of DG*

where: $P^j_{DGi}$ is the size of the DG at the bus $i$ for the $j$th load level; $P^j_{DGi\,\min}$, and $P^j_{DGi\,\max}$ are minimum and maximum sizes of a DG that may be connected to the bus $i$ for the $j$th load level.

A general flowchart that briefly describes the application of GA in determining the optimal locations and size of DGs in distribution networks is presented in Figure 7.4. The MATLAB program realization of the GA within the toolbox/gads module is applied herein.

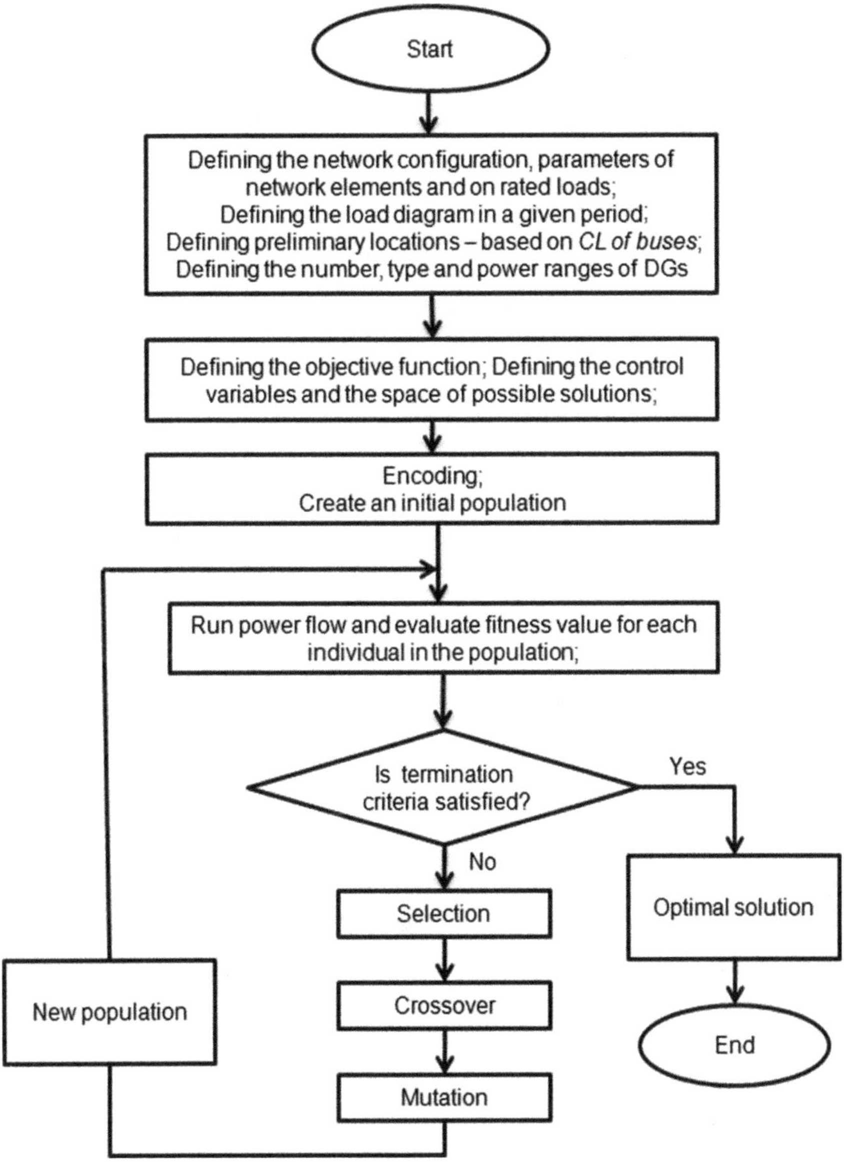


*Figure 7.4. General flow chart of DG placement and sizing using GA*

Starting from the assumption that the total number of DGs to be connected to a distribution network is $N_{DG}$, a single set of buses is associated with each DG of the defined size type and range. Classification of any bus into a set depends on the given or adopted criterion of favourability for connecting the DG to that bus. Criteria may vary, as already mentioned above. The *BLC* values determined in the manner described in Section 7.2 have been taken herein as a criterion for defining the set of preliminary buses.

## 7.5 Simulation results

Testing of the algorithms presented in Sections 7.2–7.4 was performed on two distribution networks. The first one is the IEEE 31-bus system, for which data were taken from [19]. The other test network is an actual distribution network from the ED Zajecar area, Serbia.

### 7.5.1 IEEE 31-bus system

The network is shown in Figure 7.5. The data on the network elements data and load data are given in Table 7.1. In all the computations that follow, the root bus voltage is constant and is equal to 1 p.u.

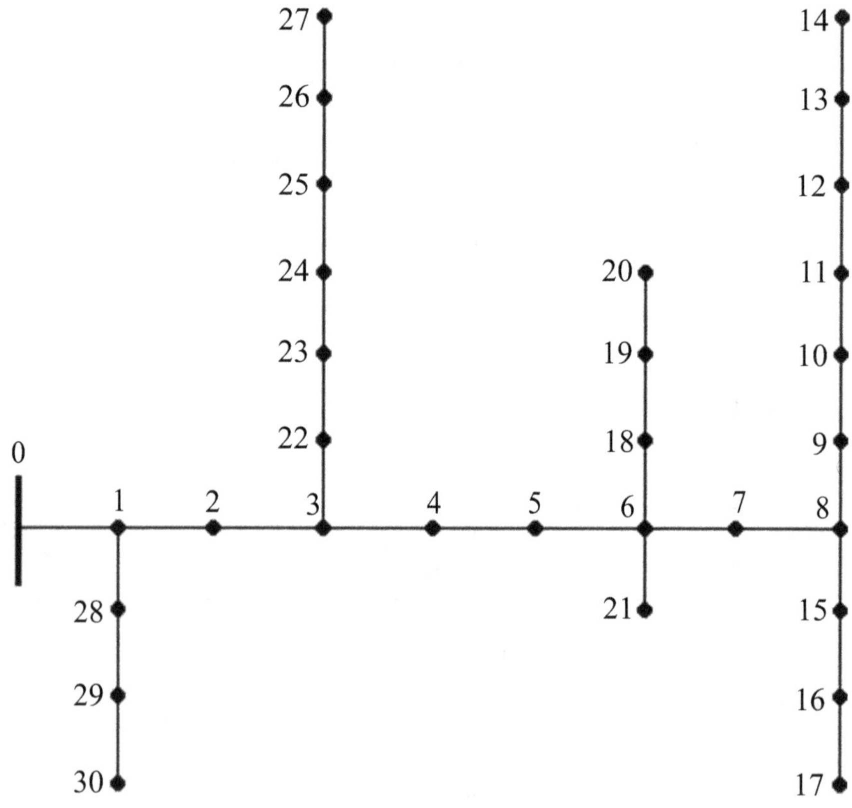


*Figure 7.5 IEEE 31-bus test system*

*Table 7.1 Data for IEEE 31-bus test system*

| Branch | $V_n$ (kV) | R (Ω) | X (Ω) | R (p.u.) | X (p.u.) | $P_p$ (kW) | $Q_P$ (kVAr) | $P_p$ (p.u.) | $Q_P$ (p.u.) |
|---|---|---|---|---|---|---|---|---|---|
| 0–1 | 23 | 0.5096 | 1.7030 | 0.000963 | 0.003219 | – | – | – | – |
| 1–2 | 23 | 0.2191 | 0.0118 | 0.000414 | 0.000022 | 522 | 174 | 0.522 | 0.174 |
| 2–3 | 23 | 0.3485 | 0.3446 | 0.000659 | 0.000651 | – | – | – | – |
| 3–4 | 23 | 1.1750 | 1.0214 | 0.002221 | 0.001931 | 936 | 312 | 0.936 | 0.312 |
| 4–5 | 23 | 0.5530 | 0.4806 | 0.001045 | 0.000908 | – | – | – | – |
| 5–6 | 23 | 1.6625 | 0.9365 | 0.003143 | 0.001770 | – | – | – | – |
| 6–7 | 23 | 1.3506 | 0.7608 | 0.002553 | 0.001438 | – | – | – | – |
| 7–8 | 23 | 1.3506 | 0.7608 | 0.002553 | 0.001438 | – | – | – | – |
| 8–9 | 23 | 1.3259 | 0.7469 | 0.002506 | 0.001412 | 189 | 63 | 0.189 | 0.063 |
| 9–10 | 23 | 1.3259 | 0.7469 | 0.002506 | 0.001412 | – | – | – | – |
| 10–11 | 23 | 3.9709 | 2.2369 | 0.007506 | 0.004229 | 336 | 112 | 0.336 | 0.112 |
| 11–12 | 23 | 1.8549 | 1.0449 | 0.003506 | 0.001975 | 657 | 219 | 0.657 | 0.219 |
| 12–13 | 23 | 0.7557 | 0.4257 | 0.001429 | 0.000805 | 783 | 261 | 0.783 | 0.261 |
| 13–14 | 23 | 1.5389 | 0.8669 | 0.002909 | 0.001639 | 729 | 243 | 0.729 | 0.243 |
| 8–15 | 23 | 0.4752 | 0.4131 | 0.000898 | 0.000781 | 477 | 159 | 0.477 | 0.159 |
| 15–16 | 23 | 0.7282 | 0.4102 | 0.001377 | 0.000775 | 549 | 183 | 0.549 | 0.183 |
| 16–17 | 23 | 1.3053 | 0.7353 | 0.002467 | 0.001390 | 477 | 159 | 0.477 | 0.159 |
| 6–18 | 23 | 0.4838 | 0.4206 | 0.000915 | 0.000795 | 432 | 144 | 0.432 | 0.144 |
| 18–19 | 23 | 1.5898 | 1.3818 | 0.003005 | 0.002612 | 672 | 224 | 0.672 | 0.224 |
| 19–20 | 23 | 1.5389 | 0.8669 | 0.002909 | 0.001639 | 495 | 165 | 0.495 | 0.165 |
| 6–21 | 23 | 0.6048 | 0.5257 | 0.001143 | 0.000994 | 207 | 69 | 0.207 | 0.069 |
| 3–22 | 23 | 0.5639 | 0.5575 | 0.001066 | 0.001054 | 522 | 174 | 0.522 | 0.174 |
| 22–23 | 23 | 0.3432 | 0.3393 | 0.000649 | 0.000641 | 1917 | 639 | 1.917 | 0.639 |
| 23–24 | 23 | 0.5728 | 0.4979 | 0.001083 | 0.000941 | – | – | – | – |
| 24–25 | 23 | 1.4602 | 1.2692 | 0.002760 | 0.002399 | 1116 | 372 | 1.116 | 0.372 |
| 25–26 | 23 | 1.0627 | 0.9237 | 0.002009 | 0.001746 | 549 | 183 | 0.549 | 0.183 |
| 26–27 | 23 | 1.5114 | 0.8514 | 0.002857 | 0.001609 | 792 | 264 | 0.792 | 0.264 |
| 1–28 | 23 | 0.4659 | 0.0251 | 0.000881 | 0.000047 | 882 | 294 | 0.882 | 0.294 |
| 28–29 | 23 | 1.6351 | 0.9211 | 0.003091 | 0.001741 | 882 | 294 | 0.882 | 0.294 |
| 29–30 | 23 | 1.1143 | 0.6277 | 0.002106 | 0.001187 | 882 | 294 | 0.882 | 0.294 |

#### 7.5.1.1 Preliminary locations – BLCs

The *BLC*s, adopted as criteria for determining the preliminary locations of DGs in the network, were computed for two levels of power load: 80% and 40% of the rated power. The results are shown in Figure 7.6. Although the *BLC*s were obtained for two different levels of power load, the shape of the *BLC* diagram in Figure 7.6 is the same in both cases. Somewhat higher values are obtained at a lower level of power load. In both instances, the maximum *BLC* value is at bus 14, whereas the minimum value is at bus 28. The order of buses according to *BLC* values is obviously not dependent on the level of power load for which these are computed. The form or trend in the diagram shown in Figure 7.6 clearly indicates that the buses at peripheral parts of the network have higher *BLC* values. There is a clear pattern in terms of an increase in the *BLC* values for buses, which are part of certain feeders. Thus, the trend of increase in the *BLC* values for the main feeder from bus

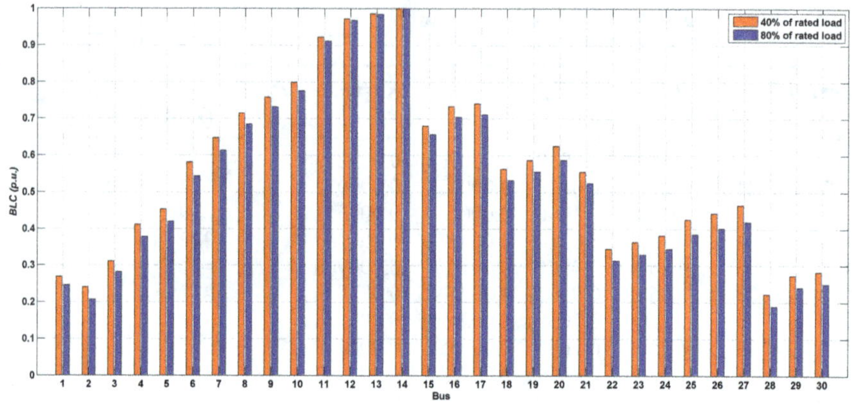


*Figure 7.6 BLC values for the IEEE 31-bus test system for two load levels*

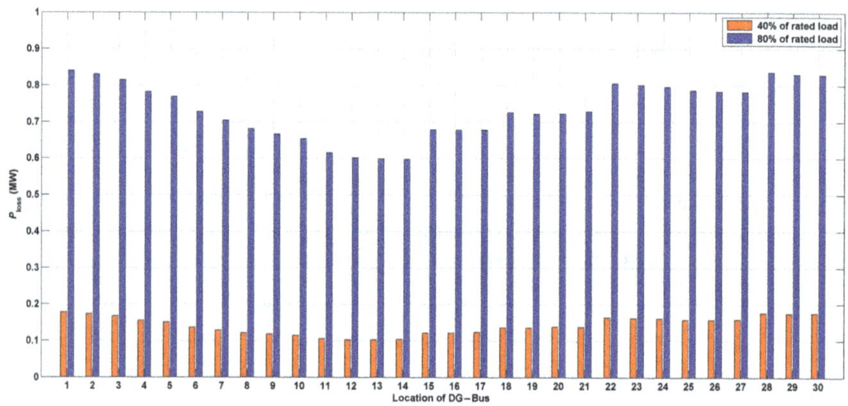


*Figure 7.7 Active power losses in the IEEE 31-bus network at successive connecting of a 1 MW DG to buses for two load levels*

2 to bus 14 is approximately of a linear character. A similar pattern in terms of increase in the *BLC* values is observed in auxiliary feeders between buses 15–17, 18–20, and 22–27. It may be concluded that the network configuration has a crucial impact on *BLC* values of buses.

Power and energy losses were computed at successive connecting of DGs of various sizes at various load levels for the purpose of observing the relation between *BLC* values and the suitability of buses for connecting DGs in terms of power loss reduction in a network. The results are shown in Figures 7.7 and 7.8.

When the diagrams in Figures 7.6 and 7.7 are compared, a relation between the *BLC* values and power losses in the network becomes evident. When DG power is injected in buses with the largest *BLC* values, the loss values are minimal. It is apparent that the trend of increase in *BLC* values follows the trend of power loss

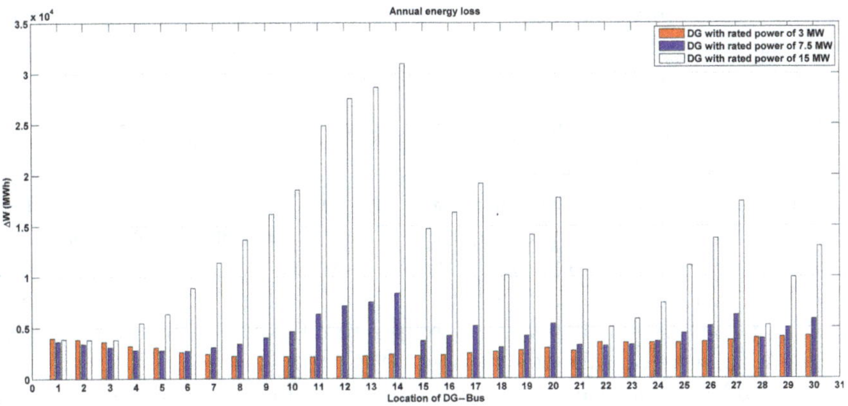


*Figure 7.8 Active energy losses in the IEEE 31-bus network at successive connecting of DGs of various sizes to buses at the annual load levels given in Table 7.2*

reduction. However, annual energy losses of a network at successive connecting of DGs of various sizes are given in Figure 7.8. Three typical cases were selected: 3 MW, which is 30% of the total rated power load; 7.5 MW, which is 50% of the total rated power load; and 15 MW, which accounts for 100% of the total rated power load of the network. The results obtained clearly indicate that the optimal location of DGs depends on the ratio between the DG size and distribution network load. For a DG of a smaller size, optimal locations are in peripheral parts of the network. As the size of a DG increases, the optimal location is shifted toward the network root. When the diagrams in Figures 7.8 and 7.6 are compared for the 7.5 and 15 MW DGs, which account for 50% and 100% of the network power load, the trend of change in energy losses follows the trend of *BLC* bus values, that is, the diagram shapes become equal to the increase in the DG size. It means that the *BLC* values of buses, as a criterion for forming preliminary locations, must be taken conditionally. In that respect, a tentative framework is given herein for defining preliminary locations based on the *BLC* values depending on the ratio between the DG size and a distribution network power load:

- If the DG size is smaller than 30% of the total network power load, buses with larger *BLC* values are more suitable for connecting a DG compared with buses with smaller *BLC* values. The initial $M_n$ buses are selected for preliminary locations, sorted according to the descending order of *BLC* values, starting from the bus with the maximum *BLC* value.
- If the DG size is larger than 70% of the total network power load, buses with smaller *BLC* values are more suitable for connecting a DG compared with buses with larger *BLC* values. Preliminary locations now are $M_n$ buses sorted according to the ascending order of *BLC* values starting from the bus with the minimum *BLC* value;

- In instances of DGs the sizes of which fall between 30% and 70% of the power load, *BLC* values may not be employed as clear criteria for assessing the favorability of buses for connecting a DG.

#### 7.5.1.2 Optimal DG placement and sizing by using partial search of variants

The test is performed for DG sizes smaller than 30% of the total power load. Accordingly, ten buses with the largest *BLC* values determined in Section 7.5.1.1 were taken as preliminary optimal locations. These are the following buses: 14, 13, 12, 11, 10, 9, 17, 16, 8, and 15. A one-year period is observed. The data on power load levels and the duration thereof are given in Table 7.2. Three different scenarios (tests) were considered in this case.

*Test B1: Optimal location of DGs of various sizes*. The procedure of determining an optimal location is carried out for three values of the rated power of the DG: 3 MW, 2 MW, and 1 MW, whereby the DG was modeled as a PQ bus with the unit power factor. The results obtained are given in Figure 7.9 in the form of a diagram of annual energy losses. According to the results shown in Figure 7.9, optimal buses vary for various values of installed power of DGs. For a 3 MW DG, the optimal is bus 11, whereas for a 2 MW DG optimal is bus 12, and for a 1 MW DG, the bus in question is 13. As the power of a DG increases, the optimal bus is shifted toward the network root. A 1 MW DG covers the power load at bus 13 and

*Table 7.2 Annual load levels and their durations*

| Load level (p.u.) | 1.0 | 0.8 | 0.6 | 0.4 |
|---|---|---|---|---|
| Duration (h) | 1215 | 1190 | 1515 | 4840 |

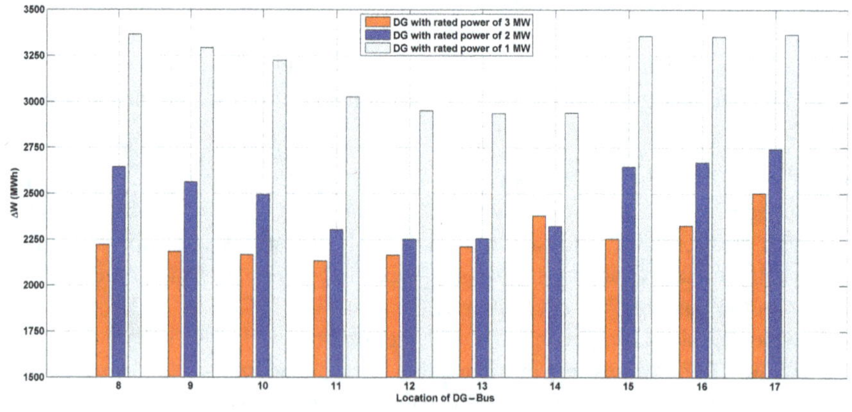


*Figure 7.9 Annual energy losses in IEEE 31-bus system with DGs of different sizes located at preliminary buses*

partially at bus 14. In this manner, preceding feeders are unloaded, and losses are reduced. As the DG power consumption increases, its capacity to supply more load downstream is increased concurrently, thus unloading the feeders upstream (toward the network root). That is the underlying reason for shifting the optimal point in this specific instance. The same conclusion may be drawn when the diagram in Figure 7.8 is observed.

*Test B2: Optimal locations of several DGs with specified sizes.* In the instance of a number of DGs of various sizes and types that must not be connected to a single bus, the optimal combination of buses should be determined so as to meet the specified optimization criterion – minimization of energy losses. Since the number of preliminary locations (buses) is 10, and the number of DGs to be allocated is 3, the total number of possible combinations of buses for connecting DGs is 720. Data on the DGs is given in Table 7.3, which includes the optimization results presented in the last two columns.

Optimal locations for three DGs were established in three examples different by the DG operating mode. The results differ significantly at first sight. However, if the network topology is regarded, it may be noted that in all three examples the optimal locations are concentrated around bus 8 as a root of that specific part of the network. The results vary due to different manners of modelling and operation of DGs. In Examples 1 and 2, all DGs are represented by the *PQ* model with a constant power factor, whereas in Example 3 DGs are represented by a *PV* model with a variable power factor in the given range. Minimum annual energy losses occur in Example 2, when DGs inject constant active and reactive power. The specified voltage value at all buses is a rated value (1 p.u). In Example 3, DGs inject reactive power in the positive and negative direction with the purpose of achieving and maintaining the specified voltage value. It means that during the optimization process attention must be paid to the values of the voltage magnitude specified at *PV* buses, in accordance with the value of the root bus voltage (network power supply transformer), consumption rate levels, active DG power and possible range

*Table 7.3 Data on the DGs and the results – optimal locations*

| | DG | Type | $P_{DG}$ (MW) | tg$\varphi$ | Results | |
|---|---|---|---|---|---|---|
| | | | | | Optimal locations of DG | Annual energy loss $\Delta W$ (MWh) |
| Example 1 | DG1 | PQ | 3 | 0 | 8 | 2206 |
| | DG2 | PQ | 2 | 0 | 12 | |
| | DG3 | PQ | 1 | 0 | 16 | |
| Example 2 | DG1 | PQ | 3 | 0.4 | 8 | 1820 |
| | DG2 | PQ | 2 | 0.4 | 15 | |
| | DG3 | PQ | 1 | 0.4 | 13 | |
| Example 3 | DG1 | PV | 3 | −0.4–0.4 | 15 | 1991 |
| | DG2 | PV | 2 | −0.4–0.4 | 8 | |
| | DG3 | PV | 1 | −0.4–0.4 | 11 | |

of the change in reactive power of DG. It, however, falls in the domain of exploitation, and therefore the optimization of the specified values of the voltage magnitude at *PV* buses is specifically dealt with in the chapter on optimal power flows and optimal voltage regulation. If the results of Tests B1 and B2 are compared, it is noticeable that optimal locations closer to the network root correspond to larger DG sizes. It quite certainly depends on the network configuration, level and distribution of load on network buses and the ratio between the power load and the DG size. In the mentioned test examples, DG sizes are constant and selected arbitrarily. In practice, it is necessary to take into account the variability of the DG size over a certain period, a year for instance. In order to obtain relevant results, the most plausible power to operate a DG or the power to operate a DG for the longest time over the considered period should be taken as a starting point when establishing the optimal location. The mentioned power depends on the rated power, technology of distributed generation and type of the primary energy source.

*Test B3: Optimal power outputs of DGs at the specified locations.* This test example includes determining optimal power outputs of DGs connected at predefined buses. The aim of this task is optimal operation of DGs for a given level of power load. Optimal DG power outputs are to be determined at their simultaneous operation with the purpose of minimizing power losses of the distribution network. The test was conducted with the input data given in Table 7.4. Optimal DG locations were selected according to the results of Test B2 presented in Table 7.3. In the procedure of successive reduction of variants the DG power ranges were divided into 10 equal segments. The results are shown in Table 7.5.

The obtained values in Table 7.5. constitute local optima simply because the specified ranges of possible DG powers in the procedure of determining optimal power outputs were divided into a finite number of segments (see Figure 7.2).

Reduction of annual energy losses of the network when DGs operate at optimal power outputs relative to operating with rated power outputs is as follows:

Example 1: 2206–1357 MWh = 849 MWh
Example 2: 1820–978 MWh = 842 MWh
Example 3: 1991–1134 MWh = 857 MWh

*Table 7.4 Data on DGs for which optimal power outputs need to be determined*

| | DG | Type | $P_{DG}$ (MW) | tg$\varphi$ |
|---|---|---|---|---|
| Example 1 | DG1 at bus 8 | PQ | 0–3 | 0 |
| | DG2 at bus 12 | PQ | 0–2 | 0 |
| | DG3 at bus 16 | PQ | 0–1 | 0 |
| Example 2 | DG1 at bus 8 | PQ | 0–3 | 0.4 |
| | DG2 at bus 15 | PQ | 0–2 | 0.4 |
| | DG3 at bus 13 | PQ | 0–1 | 0.4 |
| Example 3 | DG1 at bus 15 | PV | 0–3 | −0.4–0.4 |
| | DG2 at bus 8 | PV | 0–2 | −0.4–0.4 |
| | DG3 at bus 11 | PV | 0–1 | −0.4–0.4 |

*Table 7.5 Optimal power outputs of DGs*

| | | | | | | |
|---|---|---|---|---|---|---|
| Example 1 | Load level | | 1.0 | 0.8 | 0.6 | 0.4 |
| | DG1 at bus 8 | $P$ (MW) | 3 | 2.4 | 1.8 | 1.2 |
| | DG2 at bus 12 | $P$ (MW) | 2 | 2 | 1.5 | 1 |
| | DG3 at bus 16 | $P$ (MW) | 1 | 1 | 0.7 | 0.5 |
| | Active power loss, $P_{loss}$ (MW) | | 0.4334 | 0.2618 | 0.1436 | 0.0623 |
| | Annual energy loss, $\Delta W$ (MWh) | | 1357 | | | |
| Example 2 | Load level | | 1.0 | 0.8 | 0.6 | 0.4 |
| | DG1 at bus 8 | $P$ (MW) | 3 | 3 | 2 | 1.2 |
| | DG2 at bus 15 | $P$ (MW) | 2 | 1.4 | 1 | 0.6 |
| | DG3 at bus 13 | $P$ (MW) | 1 | 1 | 1 | 0.9 |
| | Active power loss, $P_{loss}$ (MW) | | 0.3279 | 0.1898 | 0.0994 | 0.0419 |
| | Annual energy loss, $\Delta W$ (MWh) | | 978 | | | |
| Example 3 | Load level | | 1.0 | 0.8 | 0.6 | 0.4 |
| | DG1 at bus 15 | $P$ (MW) | 2.9 | 3 | 2.7 | 1.6 |
| | DG2 at bus 8 | $P$ (MW) | 1.3 | 1 | 0.2 | 0 |
| | DG3 at bus 11 | $P$ (MW) | 1 | 1 | 1 | 1 |
| | Active power loss, $P_{loss}$ (MW) | | 0.3857 | 0.2275 | 0.1123 | 0.0463 |
| | Annual energy loss, $\Delta W$ (MWh) | | 1134 | | | |

Energy saving is apparent when a DG operates at optimal power output. However, the underlying motive (aim, interest) of distribution system owners is to generate as much energy as possible and supply it to the network, that is, to consumers. From the point of individual (private) owner of a distribution system, energy losses of a distribution network are not too relevant and therefore, optimization of a DG size becomes senseless. Nevertheless, if an owner of a distribution system is a distribution company operating a distribution network, it must pay attention to various technical and financial aspects of connecting and operating DGs. In this case, the optimization of the DG size becomes very significant. In addition, optimal operation of DGs also includes coordinating the manner in which DGs operate along with other regulation resources in a distribution network.

#### 7.5.1.3 Optimal DG placement and sizing by using GA

*Test C1: Optimal DG location and size.* An optimal DG location and size was determined for the possible DG size range between 0 and 5 MW and for the list of preliminary buses (8, 9, 10, 11, 12, 13, 14, 15, 16, and 17) according to the criterion of minimizing annual active energy loses. The test was conducted for two DG models:

- $PQ$ model with $tg\phi = 0$;
- $PV$ model with $tg\phi = (-0.4 - 0.4)$.

The MATLAB realization of the GA within the additional gads module was used to solve the problem, with parameters and options given in Table 7.6. The results are shown in Table 7.7. Applying the GA ensures certainly approximation to the global optimum; however, owing to the stochastic nature of the solutions, these are nevertheless local optima. For the purpose of illustration, values of two local

*Table 7.6 GA parameters and options*

| | |
|---|---|
| Generation | 50 |
| Population | 20 |
| Initial population | Feasible population |
| Scaling function | Top, quantity: 0.4 |
| Selection | Tournament, Tournament K: 4 |
| Crossover | Heuristic, ratio: 1.2 |
| Mutation | Adaptive feasible |
| Ending conditions | Max generation: 50 Termination tolerance on the function value: $10^{-6}$ |

*Table 7.7 Optimal DG location and size obtained by using GA*

| DG | Optimal location – Bus | Optimal power (MW) | Annual energy loss $\Delta W$ (MWh) |
|---|---|---|---|
| PQ | **8** | **4.0736** | **2072.8** |
| | 11 | 2.8383 | 2127.4 |
| PV | 8 | 3.5251 | 1717.8 |
| | **8** | **3.5392** | **1715.5** |

*Note*: The bold values indicate the best results.

optima are given, one of which was selected as a global optimum and final solution (in bold). The global optimal solution is best solution obtained from 10 consecutive runs of GA. Figures 7.10 and 7.11 contain the best and mean values of the fitness function during the GA runs for these two examples.

*Test C2: Optimal locations of several DGs with specified sizes.* Optimal DG locations were determined for Example 2 in Table 7.3. Owing to the stochastic nature of the GA, the solutions obtained are local optima. For that reason, the procedure was repeated several times. Table 7.8 presents the best results of 10 consecutive runs of GA. Since a unique optimal solution was determined by applying the partial search of variants in Test B2, it may be said with certainty that the local optimum specified in Table 7.8 is concurrently a global optimum.

*Test C3: Optimal power outputs of DGs at the specified locations.* In this case, the GA was used to determine the optimal power outputs of DGs for Examples 1 and 3 as given in Table 7.4. The results are given in Table 7.9. When the results shown in Table 7.9 are compared with the corresponding results from Table 7.5, it is noticeable that they are rather similar. However, in neither instance can it be said that unique optimal solutions have been achieved (global optimum). The accuracy in the case using partial search of variants depends on the number of DG power range segments considered. The larger the number of segments, the more accurate the obtained solution. In the instance when the GA is applied, the reason for the above-mentioned lies in the stochastic nature of the process. In addition to said reasons, there is often more than one global optimum, that is, there are a larger number of solutions – combinations of variables (DG size) for which the minimum of the criterion function is achieved.

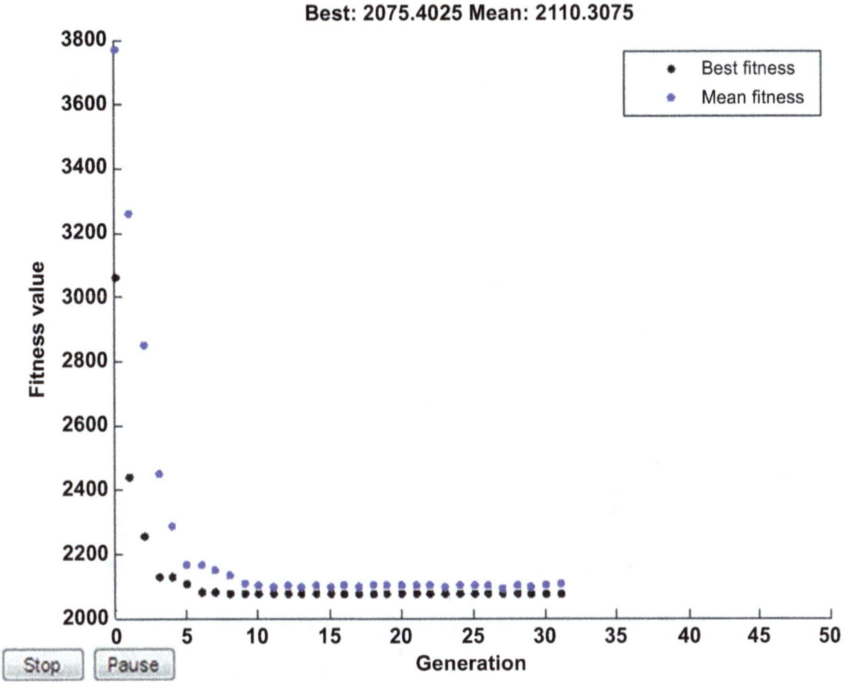


*Figure 7.10* *Convergence of GA in the case of a PQ-type DG*

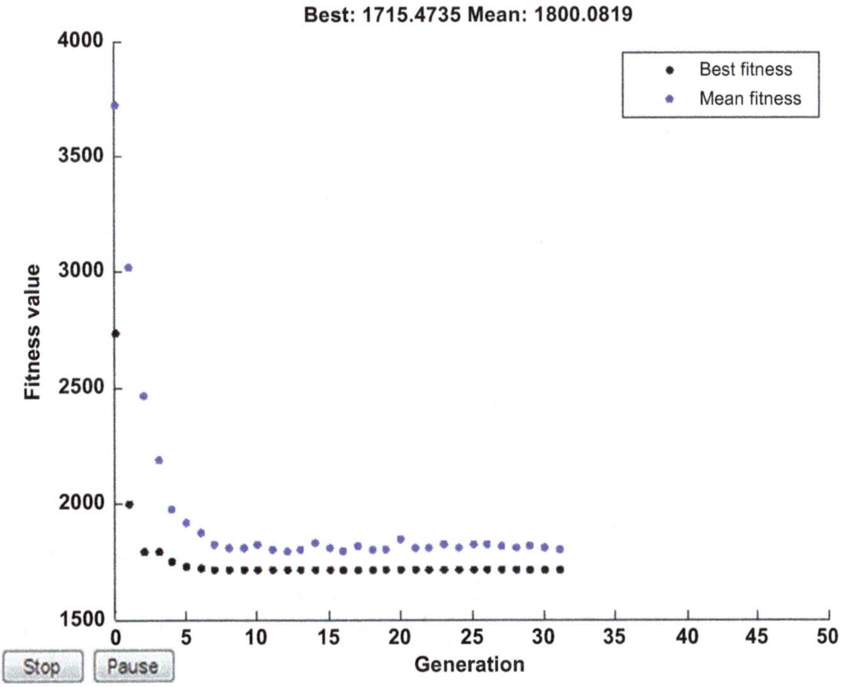


*Figure 7.11* *Convergence of GA in the case of a PV-type DG*

Table 7.8 Optimal locations of DG using GA

| Locations – Buses | | | Objective function ΔW (MWh) |
|---|---|---|---|
| DG1 | DG2 | DG3 | |
| 8 | 15 | 13 | 1819.7 |

Table 7.9 Optimal power outputs of DGs obtained by using GA

| | | | | | | |
|---|---|---|---|---|---|---|
| Example 1 | Load level | | 1.0 | 0.8 | 0.6 | 0.4 |
| | DG1 at bus 8 | *P* (MW) | 3 | 2.4137 | 1.8083 | 1.2060 |
| | DG2 at bus 12 | *P* (MW) | 2 | 1.9838 | 1.4822 | 0.9876 |
| | DG3 at bus 16 | *P* (MW) | 1 | 0.9786 | 0.722 | 0.4702 |
| | Active power loss, $P_{loss}$ (MW) | | 0.4334 | 0.2618 | 0.1436 | 0.0623 |
| | Annual energy loss, $\Delta W$ (MWh) | | 1357 | | | |
| Example 3 | Load level | | 1.0 | 0.8 | 0.6 | 0.4 |
| | DG1 at bus 15 | *P* (MW) | 2.9019 | 3 | 2.9 | 1.6059 |
| | DG2 at bus 8 | *P* (MW) | 1.3211 | 0.9373 | 0.0504 | 0.0161 |
| | DG3 at bus 11 | *P* (MW) | 0.9931 | 1 | 1 | 1 |
| | Active power loss, $P_{loss}$ (MW) | | 0.3851 | 0.2274 | 0.1120 | 0.0463 |
| | Annual energy loss, $\Delta W$ (MWh) | | 1132,3 | | | |

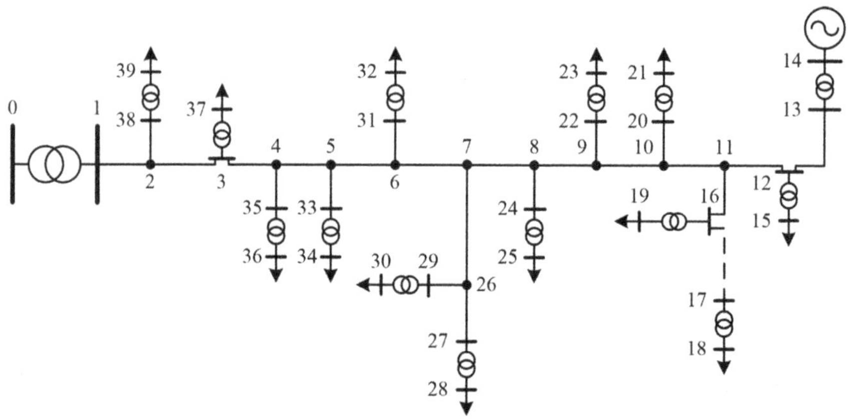


Figure 7.12 Distribution network Zajecar

## 7.5.2 Distribution network Zajecar

The single-line diagram of the test network Zajecar is shown in Figure 7.12. This is a part of a 10 kV distribution network in area of city Zajecar in Serbia. There is already one DG in the network. It is an mini hydropower plant (MHP) at bus 14, which has two synchronous generators the rated power of which is 160 kVA each andcos $\phi = 0.7$. In that area the conditions are favourable for construction of micro and mini hydropower plants, therefore the results below may be of use when selecting an optimal connection point, as well as in considering the effect of MHPs on the distribution network operation.

The network component parameter data and the load data are given in Table 7.10. Rated powers of consumer units at the network buses were estimated according to the rated powers of the associated transformers, whereas it was adopted that the power factor is 0.9. Active transformer resistances and reactances were computed for a higher voltage level, that is, in 0–1 transformers for the

*Table 7.10 Branch data and load data for the test network Zajecar*

| Branch | $V_n$ (kV) | $R$ (Ω) | $X$ (Ω) | $R$ (p.u.) | $X$ (p.u.) | $P_L$ (kW) | $Q_L$ (kVAr) | $P_L$ (p.u.) | $Q_L$ (p.u.) |
|---|---|---|---|---|---|---|---|---|---|
| 0–1 | 35/10.5 | 1.2630 | 9.1000 | 0.00103 | 0.00743 | – | – | – | – |
| 1–2 | 10 | 0.10108 | 0.05967 | 0.0010108 | 0.0005967 | – | – | – | – |
| 2–3 | 10 | 0.10703 | 0.06318 | 0.0010703 | 0.0006318 | – | – | – | – |
| 3–4 | 10 | 0.27350 | 0.16146 | 0.0027350 | 0.0016146 | – | – | – | – |
| 4–5 | 10 | 0.42810 | 0.25272 | 0.0042810 | 0.0025272 | – | – | – | – |
| 5–6 | 10 | 0.10703 | 0.06318 | 0.0010703 | 0.0006318 | – | – | – | – |
| 6–7 | 10 | 0.40433 | 0.23868 | 0.0040433 | 0.0023868 | – | – | – | – |
| 7–8 | 10 | 0.82649 | 0.48789 | 0.0082649 | 0.0048789 | – | – | – | – |
| 8–9 | 10 | 0.29135 | 0.17200 | 0.0029135 | 0.0017200 | – | – | – | – |
| 9–10 | 10 | 0.17838 | 0.10530 | 0.0017838 | 0.0010530 | – | – | – | – |
| 10–11 | 10 | 0.05946 | 0.03510 | 0.0005946 | 0.0003510 | – | – | – | – |
| 11–12 | 10 | 0.07286 | 0.03981 | 0.0007286 | 0.0003981 | – | – | – | – |
| 12–13 | 10 | 0.17989 | 0.10299 | 0.0017989 | 0.0010299 | – | – | – | – |
| 13–14 | 10/6 | 3.4300 | 14.000 | 0.03430 | 0.1400 | – | – | – | – |
| 12–15 | 10/0.4 | 3.4370 | 13.570 | 0.03437 | 0.1357 | 360 | 170 | 0.36 | 0.17 |
| 11–16 | 10 | 0.19212 | 0.08326 | 0.0019212 | 0.0008326 | – | – | – | – |
| 16–17 | 10 | 0.16217 | 0.02542 | 0.0016217 | 0.0002542 | – | – | – | – |
| 16–19 | 10/0.4 | 6.8800 | 15.326 | 0.0688 | 0.1533 | 225 | 109 | 0.225 | 0.109 |
| 17–18 | 10/0.4 | 11.719 | 22.080 | 0.1172 | 0.2208 | 144 | 70 | 0.144 | 0.070 |
| 10–20 | 10 | 0.03557 | 0.02106 | 0.0003557 | 0.0002106 | – | – | – | – |
| 20–21 | 10/0.4 | 20.000 | 34.6400 | 0.2000 | 0.3464 | 90 | 43 | 0.090 | 0.043 |
| 9–22 | 10 | 0.11892 | 0.07020 | 0.0011892 | 0.0007020 | – | – | – | – |
| 22–23 | 10/0.4 | 20.000 | 34.6400 | 0.2000 | 0.3464 | 90 | 43 | 0.090 | 0.043 |
| 8–24 | 10 | 0.14865 | 0.08775 | 0.0014865 | 0.0008775 | – | – | – | – |
| 24–25 | 10/0.4 | 3.4370 | 13.570 | 0.03437 | 0.1357 | – | – | – | – |
| 7–26 | 10 | 1.05211 | 0.33460 | 0.0105211 | 0.0033460 | – | – | – | – |
| 26–27 | 10 | 0.06014 | 0.01865 | 0.0006014 | 0.0001865 | – | – | – | – |
| 27–28 | 10/0.4 | 20.000 | 34.640 | 0.2000 | 0.3464 | 90 | 43 | 0.090 | 0.043 |
| 26–29 | 10 | 0.76848 | 0.33304 | 0.0076848 | 0.003304 | – | – | – | – |
| 29–30 | 10/0.4 | 48.000 | 64.000 | 0.4800 | 0.6400 | 90 | 43 | 0.090 | 0.043 |
| 6–31 | 10 | 0.72168 | 0.22380 | 0.0072168 | 0.002238 | – | – | – | – |
| 31–32 | 10/0.4 | 48.000 | 64.000 | 0.4800 | 0.6400 | 90 | 43 | 0.090 | 0.043 |
| 5–33 | 10 | 0.10703 | 0.06318 | 0.0010703 | 0.0006318 | – | – | – | – |
| 33–34 | 10/0.4 | 48.000 | 64.000 | 0.4800 | 0.6400 | 90 | 43 | 0.090 | 0.043 |
| 4–35 | 10 | 0.48112 | 0.14920 | 0.0048112 | 0.0014920 | – | – | – | – |
| 35–36 | 10/0.4 | 20.000 | 34.640 | 0.2000 | 0.3464 | 90 | 43 | 0.090 | 0.043 |
| 2–38 | 10 | 0.20811 | 0.12285 | 0.0020811 | 0.0012285 | – | – | – | – |
| 3–37 | 10/0.4 | 3.4370 | 13.570 | 0.03437 | 0.1357 | 360 | 170 | 0.360 | 0.170 |
| 38–39 | 10/0.4 | 3.4370 | 13.570 | 0.03437 | 0.1357 | 360 | 170 | 0.360 | 0.170 |

voltage level of 35 kV, and in other transformers for the voltage level of 10 kV. Transforming the variables and parameters into per unit system was carried out for the base power of 1 MVA and base voltages:

- 35 kV for the 0–1 transformer;
- 10 kV for other distribution transformers and feeders;
- 6 kV for the MHP.

### 7.5.2.1 Preliminary locations – *BLCs*

The procedure of determining the *BLCs* was performed for two values of load level and power output of DG in bus 14, obtained by random selection:

- 45.43% of the rated power and
- 95.72% of the rated power.

Figure 7.13 shows the *BLC* values in both cases. The largest values are at low-voltage buses of MV/LV transformers to which the loads are connected. The reason is a large transformer impedance relative to MV feeders, as well as the fact that they are located on the periphery of the radial network. The *BLC* diagram form is approximately the same for both levels of the load power. It means that the network configuration and values of parameters of elements (resistance and reactance) have a decisive effect on the *BLC* values. There is already a DG (MHP) on bus 14; therefore, the *BLC* value of that bus is markedly different from other buses. This is yet another indirect indicator that *BLC* values can be used to assess the favourability of buses for connecting a DG to it (a bus to which a DG is already connected may not be included in the list of preliminary locations for connecting new DGs, which is clearly indicated by the *BLC* value of the bus in question).

For the purpose of identifying a relation between the *BLCs* and favorability of buses for connecting DGs in terms of power loss reduction in the network, power losses when 100 kW DGs are successively connected were computed for all levels

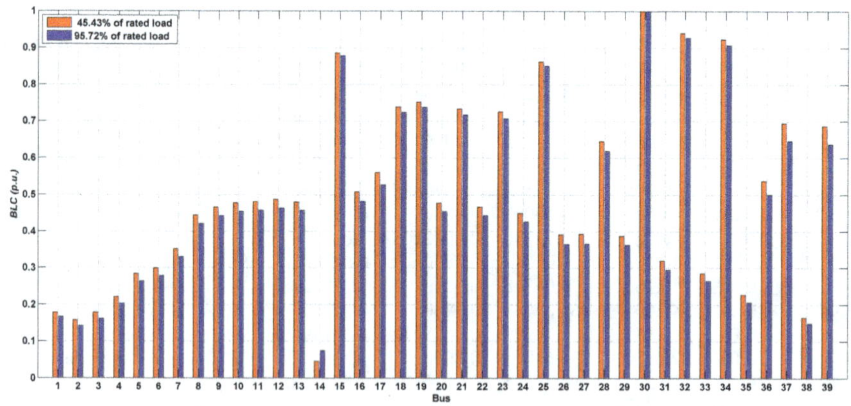


*Figure 7.13 BLCs for test network Zajecar*

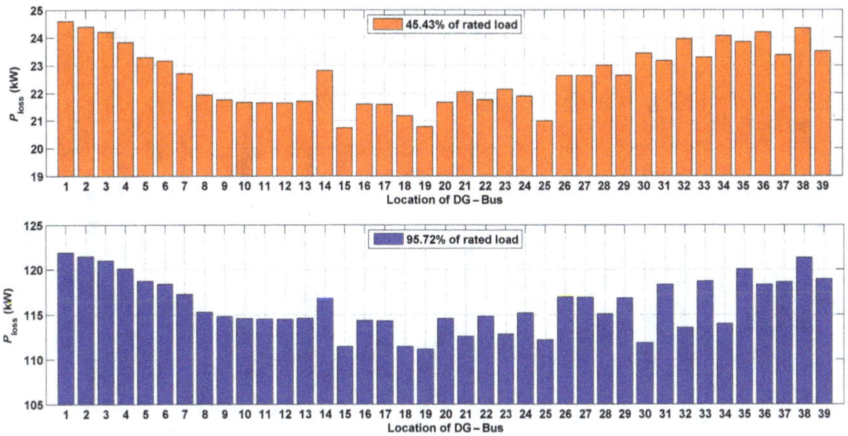


*Figure 7.14 Active power losses in the distribution network Zajecar with DG of 100 kW in different buses*

of power load in the same manner as the *BLC* values. It was assumed that DGs operate at a unit power factor. Figure 7.14 shows the values of network losses when DG power is injected in certain network buses. When the *BLC* diagrams in Figure 7.13 are compared with the loss diagrams in Figure 7.14, the relation between the *BLC* values and power losses in the network is clearly visible. Minimum loss values are detected when DG power is injected in buses with large *BLC* values. Minimum power losses for both the load power levels correspond to the case of connecting DGs to buses 15, 18, 19, and 25. These buses may be selected as preliminary locations in the process of determining optimal locations of DGs.

### 7.5.2.2 Optimal DG placement and sizing

Buses with the largest *BLC* values, determined in Section 7.5.2.1, were taken as preliminary locations. These buses are as follows: 15, 18, 19, 21, 23, 25, 28, 30, 32, 34, 36, 37, and 39. A 1-year period is observed. The annual power load diagram has four levels relative to the rated power of distribution transformers (1, 0.8, 0.6, 0.4) with a power factor of 0.9. The output power of the DG at bus 14 is changed according to the same pattern as the load (relative to its rated power). It is represented by the model of constant power as a negative load with a rated power factor of 0.7. All load levels have the same duration of 2190 h.

*Test D1: Optimal location of DG with various rated powers*. The procedure of determining an optimal location was carried out for the following DG-rated power values: 300, 200, and 100 kW, whereas DGs were modeled as PQ buses with the unit power factor. Since only one DG is treated, the process is rather simple and it includes determining annual power losses of the network when DGs of the specified size are successively connected to the defined preliminary buses.

Figure 7.15 includes the results – annual power losses of the network for certain values of DG-rated power. Minimum annual power losses for 300 and 200 kW

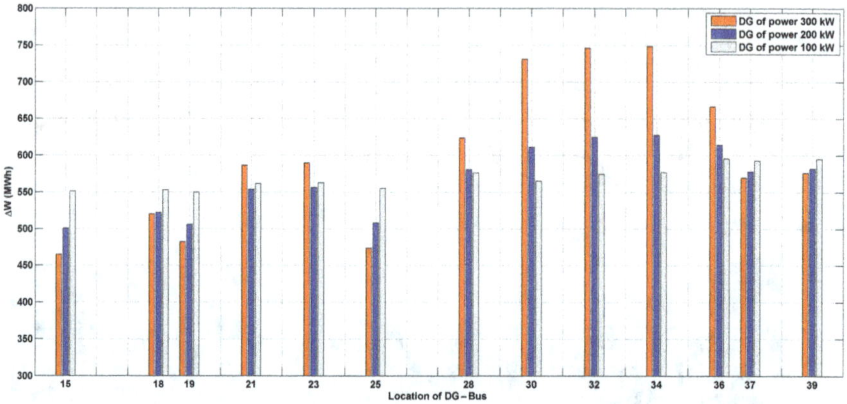


*Figure 7.15 Annual active energy losses for various DG sizes in preliminary buses*

*Table 7.11 Data on the DGs for Test D2*

| DG | Type | $P_{DG}$ (kW) | tg$\varphi$ |
|---|---|---|---|
| DG1 | PV | 300 | −0.4–0.4 |
| DG2 | PQ | 200 | 0 |
| DG3 | PQ | 100 | 0 |

DGs are observed at bus 15. Accordingly, bus 15 is the optimal bus for connecting DGs of the mentioned installed power. If a 100 kW DG is to be installed, the optimal bus would be bus 19, although the difference compared to bus 15 in that instance is practically negligible, therefore, it would be no mistake if bus 15 was also selected as optimal. According to the results in Figure 7.13, bus 15 is not the one with the largest BLC value but bus 30. Therefore, BLC values of buses should not be taken as absolute a priori indicators of optimality but solely as a guideline or a good initial guess for a starting point in the further process of final establishing of an optimal location or locations for DG connecting. As in the case of the IEEE 31 test system, in this example optimal buses are also located on the periphery, that is, further away from the network root relative to other preliminary locations, which is the result of the relation between the load power and DGs.

*Test D2: Optimal locations for a number of DGs with specified rated power.* Table 7.12 contains the results of optimization for the DG on which the data is given in Table 7.11. This table also contains the results of applying both procedures – partial search of variants and GA application. The results are identical.

Given the number of preliminary locations and the number of DGs, there are 1716 possible variants – combinations examined when performing a partial search of variants. Since all possible solution variants are encompassed according to the defined criteria, the obtained solution constitutes a global optimum.

*Table 7.12 Results – optimal DG locations*

| **Partial search of variants** | | **GA** | | |
|---|---|---|---|---|
| **DG** | **Optimal location** | **DG** | **Optimal location** | **ΔW (MWh)** |
| DG1 | 15 | DG1 | 15 | 326 |
| DG2 | 25 | DG2 | 25 | |
| DG3 | 19 | DG3 | 19 | |

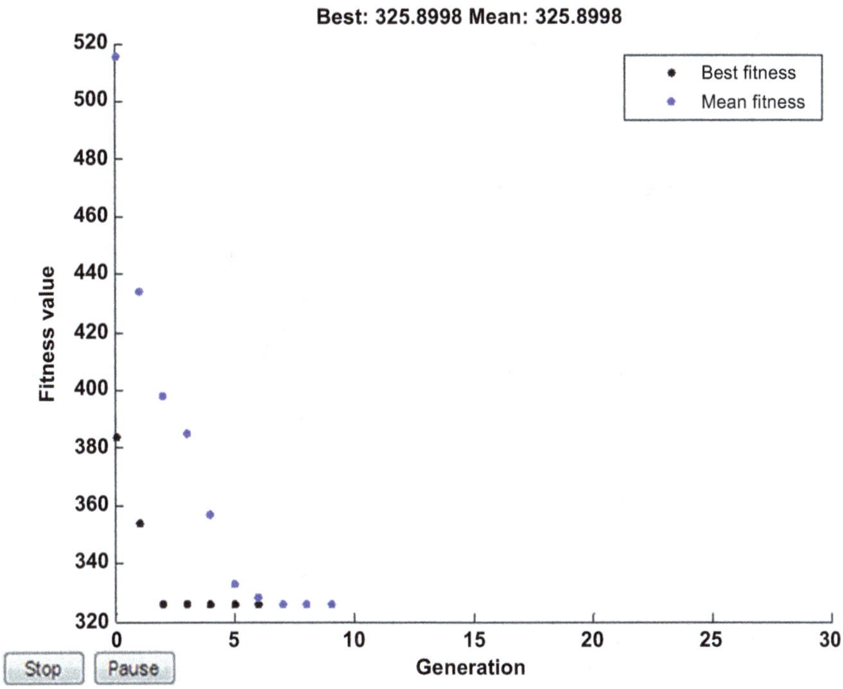


*Figure 7.16 Convergence characteristic of GA in Test D2*

Given a small number of variables of the same nature (integers), when applying a GA the solution is obtained rather quickly, after only nine generations. In addition, a global solution is obtained in spite of the stochastic character of the GA and owing to the aforementioned nature of variables and relatively narrow search set (preliminary locations). This was confirmed by several repeated GA runs, whereas identical results as those specified in Table 7.12 are obtained in each run. Figure 7.16 shows the convergence characteristic of GA for this test case.

*Test D3: Optimal power outputs of DGs at specified locations*. The test was performed for the data given in Table 7.13. The results of a partial search of

*Table 7.13 Data on the DGs for Test D3*

| DG | Type | $P_{DG}$ (kW) | tg$\varphi$ |
|---|---|---|---|
| DG1 at bus 15 | PV | 0–300 | −0.4–0.4 |
| DG2 at bus 25 | PQ | 0–200 | 0 |
| DG3 at bus 19 | PQ | 0–100 | 0 |

*Table 7.14 Optimal results for Test D3 obtained by applying partial search of variants*

| Load level | Optimal power of DG1 (kW) | Optimal power of DG2 (kW) | Optimal power of DG3 (kW) | Power loss, $P_{loss}$ (kW) |
|---|---|---|---|---|
| 1.0 | 300 | 200 | 100 | 76.6 |
| 0.8 | 300 | 200 | 100 | 42.5 |
| 0.6 | 300 | 200 | 100 | 20.5 |
| 0.4 | 210 | 180 | 100 | 8.5 |

*Table 7.15 Optimal results for Test D3 obtained by applying GA*

| Load level | Optimal power of DG1 (kW) | Optimal power of DG2 (kW) | Optimal power of DG3 (kW) | Power loss, $P_{loss}$ (kW) |
|---|---|---|---|---|
| 1.0 | 300 | 200 | 100 | 76.6 |
| 0.8 | 300 | 200 | 100 | 42.5 |
| 0.6 | 300 | 200 | 100 | 20.5 |
| 0.4 | 202.3 | 184 | 100 | 8.5 |

variants are given in Table 7.14. Table 7.15 presents the solutions obtained by applying the GA.

Since maximum powers (rated powers) were obtained as optimal power outputs of DGs for all load levels (excluding 0.4), it is evident that the objective function is decreasing and that in order to achieve its minimum it is necessary to increase the DG sizes at given locations. However, optimization includes taking into account realistic conditions, which in turn imply that DGs are of limited power capacities. Accordingly, in terms of the defined DG size range, the results obtained constitute global optima for all levels of power load other than for 0.4.

When partial search of variants was applied, DG size ranges were divided into ten equal segments. For all levels of power load other than for 0.4, already during the first search the results obtained were the same as those specified in Table 7.14. However, the procedure was conducted once again on the narrowed DG power ranges close to the previously obtained optima (for DG1 200–300 kW, for DG2 100–200 kW, and for DG3 the same range as in the first search was used). The results obtained are identical.

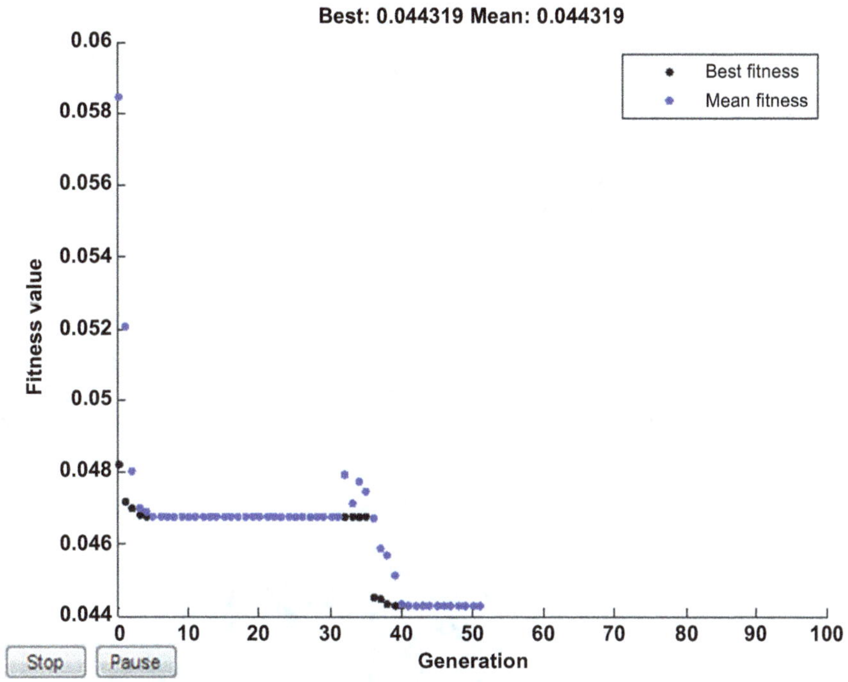


*Figure 7.17 Convergence characteristic of GA in Test D3 for the load level of 0.8*

When GA was applied in this test example, the optimal solution was obtained after 50 generations, as shown in Figure 7.17. This procedure is several times longer than in the example from Test D2. The reason for this is that the variables are not integers and have a much larger range of possible values and thereby a much larger number of possible solutions. If these results are compared with the results in Table 7.14, it is apparent that global optima were obtained for all levels of power load other than for 0.4 when the obtained solutions have a character of a local optimum relative to the defined DG power ranges.

## 7.6 MATLAB® program *opsdgm*

Based on the methods described above, the *opsdgm* program package was developed. The *opsdgm* is an educational program package of MATLAB m-files for determining optimal placement and sizing of DG units in distribution networks. The program package is available as the supplementary file to this chapter on the book's homepage at the IET digital library. The simulation results presented in Section 7.5 are obtained by the *opsdgm* program package. An example of the usage of the program package is shown in Figure 7.18.

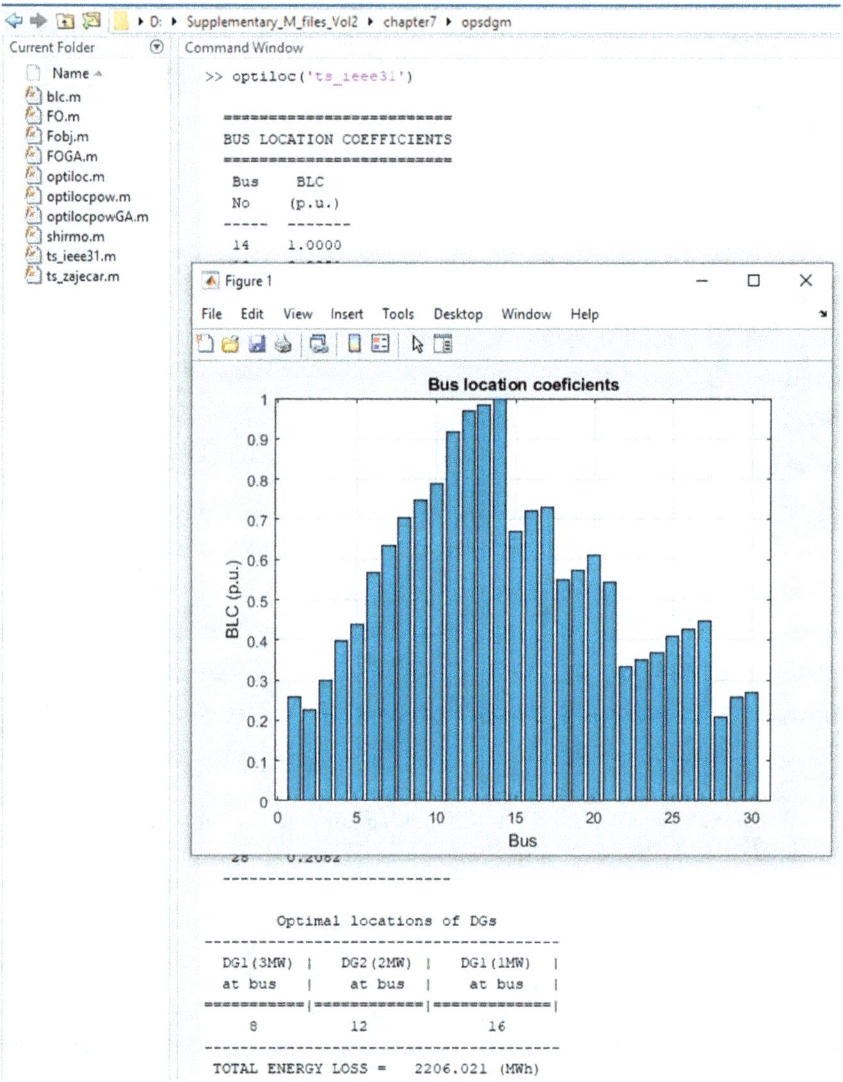


*Figure 7.18 Main window of the opsfgm program package*

## 7.7 Conclusion

The procedure for optimal siting and sizing of DG units requires two levels. The first level aims to reduce the solution space by introducing a list of preliminary locations based on the sensitivity of power losses with power injections into the network buses. This significantly reduces the number of possible variants of the solution, which are searched in the second stage of the procedure.

Two search techniques were used for searching: partial search of variants and metaheuristic optimization methods – GA.

Simulation results on two test networks confirm the efficiency and robustness of this approach for both search techniques. However, it should be noted that the partial search technique is suitable for small distribution networks, while the GA also provides high-quality solutions for real systems with a large number of buses and potential locations for DG units.

## References


[1] Wang C., and Nehrir M.H. 'Analytical approaches for optimal placement of distributed generation sources in power systems'. *IEEE Transactions on Power Systems*. 2004;**19**(4):2068–76.

[2] Popović D.H., Greatbanks J.A., Begović M., and Pregelj A. 'Placement of distributed generators and reclosers for distribution network security and reliability'. *Electrical Power and Energy Systems*. 2005;**27**:398–408.

[3] Gautam D., and Mithulananthan N. 'Optimal DG placement in deregulated electricity market'. *Electric Power Systems Research*. 2007;**77**:1627–36.

[4] Gözel T., and Hocaoglu M.H. 'An analytical method for the sizing and siting of distributed generatorsin radial systems'. *Electric Power Systems Research*. 2009;**79**:912–18.

[5] Elsaiah S., Benidris M., and Mitra J. 'Analytical approach for placement and sizing of distributed generation on distribution systems'. *IET Generation, Transmission & Distribution*. 2014;**8**(6):1039–49.

[6] Naik S.N.G., Khatod D.K., and Sharma M.P. 'Analytical approach for optimal siting and sizing of distributed generation in radial distribution networks'. *IET Generation, Transmission & Distribution*. 2015;**9**(3):209–20.

[7] Viral R., and Khatod D.K. 'An analytical approach for sizing and siting of DGs in balanced radial distribution networks for loss minimization'. *Electrical Power and Energy Systems*. 2015;**67**:191–201.

[8] Keane A., and O'Malley M. 'Optimal allocation of embedded generation on distribution networks'. *IEEE Transactions on Power Systems*. 2005;**20**(3): 1640–46.

[9] Le A.D.T., Kashem M.A., Negnevitsky M., and Ledwich G. 'Optimal distributed generation parameters for reducing losses with economic consideration'. *Proceedings of IEEE Power Engineering Society General Meeting*; Tampa, FL, USA, June 2007. IEEE; 2007. pp. 1–8.

[10] El-Khattam W., Hegazy Y.G., and Salama M.M.A. 'An integrated distributed generation optimization model for distribution system planning'. *IEEE Transactions on Power Systems*. 2005;**20**(2):1158–65.

[11] Mori H., and Iimura Y. 'An improved tabu search approach to distribution network expansion planning under new environment'. *Proceedings of International Conference on Power System Technology*; Singapore, Singapore, Nov 2004. IEEE; 2005. pp. 981–86.

[12] Roy S. 'Optimal planning of generating units over micro-hydro resources within a catchment area'. *IEEE Transactions on Energy Conversion*. 2005; **20**(1):231–36.

[13] Yuan Y., Qian K., and Zhou C. 'The optimal location and penetration level of distributed generation'. *Proceedings of 2nd International Universities Power Engineering Conference*; Brighton, UK, Sep 2007. IEEE; 2008. pp. 917–23.

[14] Golshan M.E.H., and Arefifar S.A. 'Optimal allocation of distributed generation and reactive sources considering tap positions of voltage regulators as control variables'. *European Transactions on Electrical Power*. 2007;**17**(3): 219–39.

[15] A mini-review of optimal placement and sizing of DGs using metaheuristic optimization algorithms is given in Chapter 6 (Section 6.4) of this book.

[16] Radosavljević J., Jevtić M., and Klimenta D. 'Determining optimal location of generators in radial distribution networks'. *Elektroprivreda.* 2006;**59**(1): 63–71.

[17] Radosavljević J., and Jevtić M. 'Optimal location and sizing of distributed generators in radial distribution networks using genetic algorithm'. *Proceedings of 28th Conference JUKO CIGRE*; Vrnjačka Banja, Serbia, Sep 2007. R C6-02.

[18] Holland J.H. *Adaptation in Natural and Artificial Systems*. Cambridge: MIT Press; 1975.

[19] Santoso N.I., and Tan O.T. 'Neural-net based real-time control of capacitors installed on distribution systems'. *IEEE Transactions on Power Delivery*. 1990;**5**(1):266–72.

# Chapter 8

# Optimal placement and sizing of renewable-based distributed generation

## 8.1 Introduction

The continuous increase in consumption, the need to reduce greenhouse gas emissions, the deregulation and liberalization of the electricity market, and the privileged prices of green energy have led to the rapid growth of renewable energy sources in the last two decades. It seems that wind and solar energy are the best alternatives to fossil fuels for power generation. The rapid growth of wind and photovoltaic (PV) power installation has been enabled by the technological improvement in wind turbine (WT) and PV generation systems as well as the reduction in total installation costs [1]. Moreover, it can be argued that the WT and PV generation systems are well-established and standardized technologies. As reported in [2], electricity from renewable distributed generation (RDG) units such as WT and PV will soon be consistently cheaper than from fossil fuels.

In principle, there are three ways to use the WT and PV sources:

1. as large wind farms and photovoltaic arrays integrated into the power system;
2. as RDG units constituting an essential part of active distribution networks (DNs) and microgrids;
3. as power resources in small stand-alone hybrid power systems.

The integration of RDG units such as WT and PV leads to major challenges due to their uncertain power generation characteristics. In general, the task of optimal planning of RDG is to determine its optimal location and rated power in order to minimize or maximize a desired objective function, considering different technical, economic, and environmental constraints. In mathematical formulation, this is a large-scale, non-linear, probabilistic-constrained optimization problem with both continuous and discrete control variables. The general framework for defining and solving this problem must include the following aspects:

- the significance and context of this issue;
- the modeling of RDG units, e.g. the modeling of WT and PV output power due to uncertain characteristics of wind speed and solar irradiation;
- the modeling of load uncertainties;
- the choice of objective functions;

- the definition of technical constraints, control variables, and dependent variables;
- the method for solving the optimization problem.

So far, many studies have dealt with the problem of optimal placement and sizing (OPS) of RDG focusing on some of the above tasks. References [3–6] gave comprehensive state-of-the-art reviews in this area.

References [7–10] used Weibull and Beta probability distribution functions (PDFs) to model the stochastic behavior of wind speed and solar irradiance at a specific location. References [11,12] took into account different types of loads such as linear and non-linear loads as well as load growth.

As noted in [5], the objectives for OPS of distributed generation (DG) can be summarized in three groups: technical objectives, financial objectives, and multi-objectives. Technical objectives are the base objectives and include the energy loss minimization [8–10,13], and the improvement of voltage conditions [14–17] in DNs. Financial objectives are related to the minimization of investment cost, operation cost, and maintenance cost for DGs [11], as well as the maximization of the profit of the DG owner [18]. Multi-objective functions with various combinations of objectives have also been implemented. One such case is the simultaneous minimization of annual average power loss, maximization of the voltage stability index, and minimization of the DN security index [7]. In [19], the multi-objective optimization consists of the minimization of power loss, the total electrical energy cost, the pollutant emission, and the improvement of voltage stability. References [12,20] define a multi-objective function based on the total active power loss and the voltage deviation. The multi-objective optimization in [18] is not only considered from the operation aspects such as improving voltage profile and power loss reduction, but also an economic analysis is performed based on the viewpoints of distribution companies and DG owners.

Any optimal solution, which implies the OPS of a DG, must meet different technical and economic constraints to ensure standardized operation or design conditions, regardless of the type of objective function [5]. These constraints can be classified into:

1. The power system conservation constraints, which include power flow balance, bus voltage limits, branch current limits, short circuit level, power quality limits, etc.
2. The DG limitations, which include the power factor of DG units, penetration level of DG, number of DG units, size of DG units, type of DG units, contract price limits between the DG owner and the distribution company, capitalization constraint of DG owner, etc.

Different approaches for solving the problem of OPS of DG can be classified into three main groups: analytical techniques, classical optimization methods, and metaheuristic optimization algorithms. References [3–5] present a comprehensive overview of these approaches.

In recent research, some metaheuristic population-based methods have been used to solve this problem. In [7], a weighted aggregation particle swarm optimization (PSO) is proposed to find the optimal mix of RDG units in DNs with multi-objectives such as the minimization of power loss and the improvement of voltage stability and network security. In [11,13], the PSO is used to determine the OPS of RDG units in a DN considering technical, economic, and environmental constraints. In [18], a multi-objective PSO algorithm is used to find the OPS of RDG units, in addition to determining their optimal prices of generated electricity in a competitive market. In [8], an evolutionary programming-based approach is used for solving the problem of optimal locations of RDG such as PV and WT units in a DN. In [14,15], the application of the ant lion optimization algorithm to solve the OPS of RDG units in DNs is proposed. In [16], the differential evolution is proposed to evaluate the optimal RDG capacity for minimizing power losses in sub-transmission systems. In [17,20], a well-established meta-heuristic optimization method, namely genetic algorithm, is used to solve the problem of optimal planning of RDG in DNs considering multiple aspects of DN operation. In [19], a hybrid optimization algorithm consisting of an ant colony optimization algorithm and an artificial bee colony algorithm is proposed for solving probabilistic OPS of RDG units in DNs.

This chapter deals with the application of metaheuristic algorithms to solve the problem of OPS of RDG in DNs, observing the problem from the viewpoints of the DN operator and the RDG owners, as proposed in [21].

The rest of the chapter is organized as follows: Section 8.2 presents the probabilistic models of RDG and loads. The problem of OPS of RDG is mathematically formulated in Section 8.3. The proposed algorithm and its application are explained in Section 8.4. The simulation results are discussed in Section 8.5, and the MATLAB® program *opsrdgm,* which is used to obtain these results, is described in Section 8.6. Conclusions are listed in Section 8.7.

## 8.2 Modeling of RDG and load

### *8.2.1 WT generation modeling*

The output power of a WT for a given wind speed ($v$) is calculated using the power characteristic of the WT, which is a non-linear function of wind speed [9]:

$$P_{WT}(v) = \begin{cases} 0 & v \le v_{ci} \\ \dfrac{v^2 - v_{ci}^2}{v_n^2 - v_{ci}^2} P_{wtn} & v_{ci} < v \le v_n \\ P_{wtn} & v_n < v \le v_{co} \\ 0 & v > v_{co} \end{cases} \tag{8.1}$$

where $P_{wtn}$, $v_n$, $v_{ci}$, and $v_{co}$ are the nominal power, nominal wind speed, cut-in wind speed, and cut-out wind speed of the WT, respectively. These data as well as the experimentally determined power curve are given by the manufacturers of WTs.

The stochastic nature of wind speed in a predefined time period $t$ at a certain location can be generally described by Weibull PDF [7,9]:

$$f_v(v) = \frac{k}{C} \cdot \left(\frac{v}{C}\right)^{k-1} \cdot e^{-\left(\frac{v}{C}\right)^k} \tag{8.2}$$

where $f_v(v)$ is the Weibull PDF for wind speed data collected during time period $t$, and $C$ and $k$ are the scale and shape parameters of the Weibull distribution at time period $t$.

The cumulative density function (CDF) for the Weibull distribution is

$$F_v(v) = 1 - \cdot e^{-\left(\frac{v}{C}\right)^k} \tag{8.3}$$

The CDF with its inverse has been utilized to calculate the wind speed:

$$v = C(-\ln(r))^{\frac{1}{k}} \tag{8.4}$$

where $r$ is a random number uniformly distributed on [0, 1].

In practice, parameters $C$ and $k$ can be calculated approximately using mean value ($\mu_v^t$) and standard deviation ($\sigma_v^t$) of wind speed at time period $t$ [7,9]:

$$k = \left(\frac{\sigma_v^t}{\mu_v^t}\right)^{-1.086} \tag{8.5}$$

$$C = \frac{\mu_v^t}{\Gamma(1 + 1/k^t)} \tag{8.6}$$

where $\Gamma(\cdot)$ is the gamma function. Note that the $\mu_v^t$ and $\sigma_v^t$ are calculated from the wind speed measurements in considered time period $t$.

In the problem of OPS of WTs, it is necessary to collect the wind speed data from the site under study for a time period of at least one year. Based on these historical data, the parameters of Weibull PDF can be calculated.

The yearly measured weather data are classified by seasons, i.e. Spring, Summer, Autumn, and Winter. Each season consists of a number of days corresponding to the months of the season. The days are divided into hours, which are the elemental time segments. For a given season, a typical day is defined as consisting of 24 characteristic hours. The sampling times for wind speed measurements are 1, 5, or 10 min during the entire considered period [22]. This means 6–60 readings of the wind speeds at each hour over the year. Therefore, a characteristic hour in a typical day of a season can be represented by the mean value and the standard deviation of wind speed calculated from measured data corresponding to this hour in all days within the season. By calculating the mean value and the standard deviation of wind speed for each of the 24 hours, a typical day of the considered season is obtained.

Based on the mean value and the standard deviation of wind speed described above, the shape parameter ($k$) and the scale index ($C$) of Weibull PDF can be calculated for each hour $t$ of the typical day by using (8.5) and (8.6).

To realize the Weibull PDF for each hour in discrete form, the hour $t$ is divided into $N_v$ states, where the corresponding wind speed and probability for each state $g$

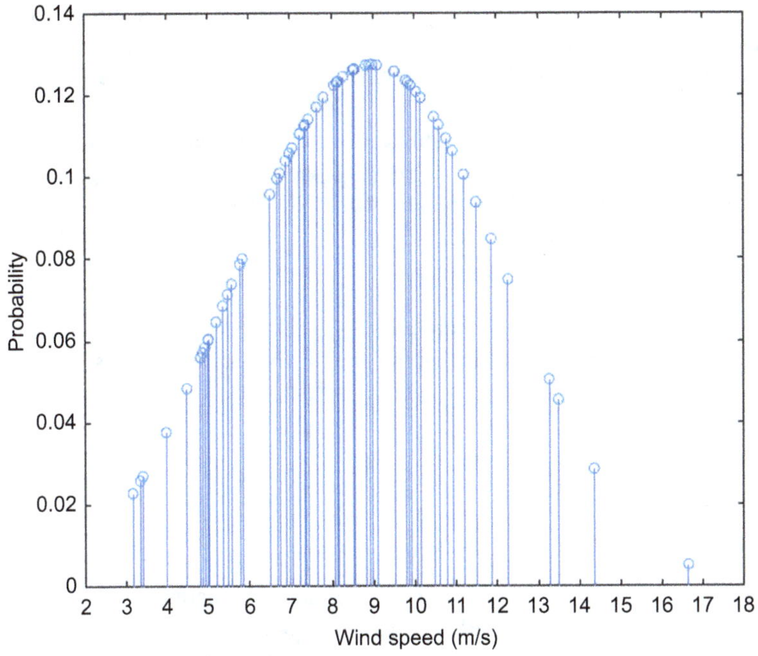


*Figure 8.1 Discrete Weibull PDF of wind speed during an hour*

are calculated by using (8.4) and (8.2), respectively. Figure 8.1 shows the discrete Weibull PDF of wind speed corresponding to an hour with $\mu_v = 9$ m/s, $\sigma_v = 3$ m/s, and $N_v = 60$. The power output of WT is dependent on the probability of all possible states for that hour.

Accordingly, the power generation of WT considering the probability of wind speed for each state during hour $t$ can be calculated as follows:

$$P^t_{WT} = \frac{\sum_{g=1}^{N_v} P_{WTg} f_v\left(v^t_g\right)}{\sum_{g=1}^{N_v} f_v\left(v^t_g\right)} \tag{8.7}$$

where $v^t_g$ is the wind speed of state $g$ at hour $t$, $P_{WTg}$ is the power generation of WT calculated using (8.1) for $v = v^t_g$, and $f_v(v^t_g)$ is the probability of the wind speed for state $g$ during hour $t$.

### 8.2.2 PV generation modeling

The power output of the PV module with given technical characteristics is dependent on the solar irradiance and ambient temperature [23]:

$$P_{PV}(s, T_c) = P_{NTC}\frac{s}{1000}[1 + \gamma(T_c - 25)] \tag{8.8}$$

where $P_{STC}$ is the maximum power of the PV module at standard test condition (STC), $s$ is the solar irradiance on the PV module surface, $\gamma$ is the temperature coefficient of the PV module for power, and $T_c$ is the temperature of the PV cell (module).

The temperature of the PV module can be calculated as a function of solar irradiance and ambient temperature based on the nominal operating cell temperature (NOCT) of modules. The equation of the NOCT model is [24]

$$T_c = T_a + \frac{s}{800}(T_{NOCT} - 20) \tag{8.9}$$

where $T_a$ is the ambient temperature and $T_{NOCT}$ is the NOCT of the module.

Beta PDF is suitable for describing the stochastic nature of solar irradiance [7,9]:

$$f_s(s) = \begin{cases} \dfrac{\Gamma(\alpha+\beta)}{\Gamma(\alpha)\Gamma(\beta)} s^{\alpha-1}(1-s)^{\beta-1}, \\ \text{for } 0 \le s \le 1, \alpha \ge 0,\ \beta \ge 0 \\ \quad 0, \text{otherwise} \end{cases} \tag{8.10}$$

where $f_s(s)$ is Beta PDF of $s$ and $\alpha$ and $\beta$ are the shape parameters of Beta PDF.

Shape parameters of Beta PDF can be obtained based on the mean value and the standard deviation of solar irradiance for the corresponding time period $t$:

$$\beta = \left(1 - \mu_s^t\right) \cdot \left( \frac{\mu_s^t\left(1 + \mu_s^t\right)}{\left(\sigma_s^t\right)^2} - 1 \right) \tag{8.11}$$

$$\alpha = \frac{\mu_s^t \cdot \beta}{1 - \mu_s^t} \tag{8.12}$$

Based on the mean value and the standard deviation of solar irradiance determined in an analogous manner for the wind speed, the shape parameters of Beta PDF ($\alpha$ and $\beta$) can be calculated for each hour $t$ of typical days using (8.11) and (8.12). To realize Beta PDF for each hour in discrete form, the hour $t$ is divided into $N_s$ states, where the corresponding solar irradiance and probability for each state $g$ are calculated using (8.10). Figure 8.2 shows the discrete Beta PDF of solar irradiance related to an hour with $\mu_s = 436$ W/m$^2$, $\sigma_s = 295$ W/m$^2$, and $N_s = 60$. The power output of the PV module is dependent on the probability of all possible states for that hour.

Accordingly, the power generation of the PV module considering the probability of solar irradiance for each state during hour $t$ can be calculated as follows:

$$P_{PV}^t = \frac{\sum_{g=1}^{N_s} P_{PVg} f_s\left(s_g^t\right)}{\sum_{g=1}^{N_s} f_s\left(s_g^t\right)} \tag{8.13}$$

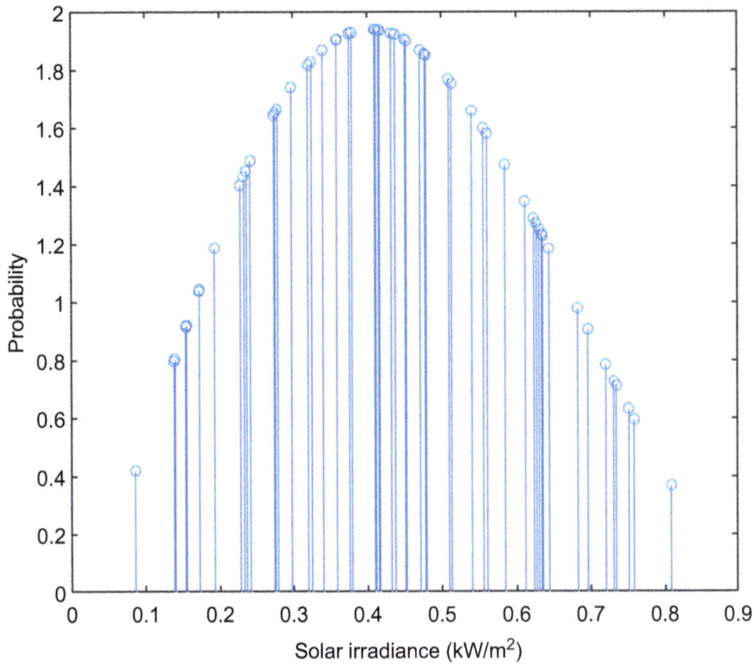


*Figure 8.2 Discrete Beta PDF of solar irradiance during an hour*

where $s_g^t$ is the solar irradiance state $g$ at hour $t$, $P_{PVg}$ is the power generation of PV module calculated using (8.8) for $s = s_g^t$, and $f_s(s_g^t)$ is the probability of the solar irradiance for state $g$ during hour $t$.

### 8.2.3 Load modeling

It is assumed that the load profiles are the same for both active and reactive power. The random nature of the load change is modeled by the normal PDF. Generally, the load is assumed to be a random variable ($L$) following the same PDF within each hour of a given daily load diagram.

$$f_s(L) = \frac{1}{\sigma_L\sqrt{2\pi}} e^{-\frac{(L-\mu_L)}{2\sigma_L^2}} \tag{8.14}$$

$$L = \mu_L + \sqrt{2}\sigma_L \mathrm{erf}^{-1}(2r - 1) \tag{8.15}$$

where $\mu_L$ and $\sigma_L$ are the mean value and standard deviation of $L$, respectively; $r$ is a random number in [0, 1]; and $\mathrm{erf}^{-1}(\cdot)$ is the inverse error function.

To realize the normal PDF for each hourly load in discrete form, hour $t$ is divided into $N_L$ states, where the corresponding load and probability for each state $g$ are calculated using (8.15) and (8.14), respectively. Figure 8.3 shows the discrete normal PDF for the hourly load level with $\mu_L$ = 0.7 p.u., $\sigma_L$ = 5 %, and $N_L$ = 60.

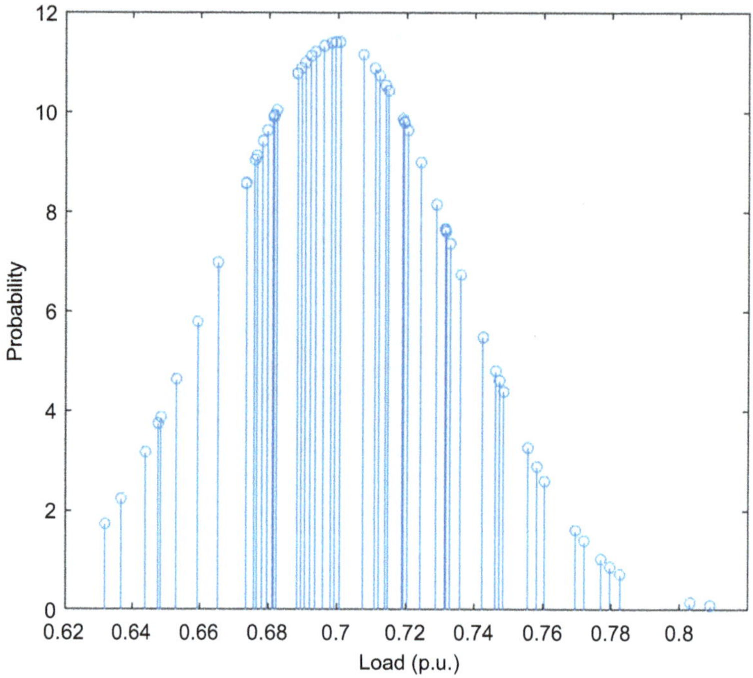


*Figure 8.3 Discrete normal PDF of load during an hour*

The load level related to a time segment is determined by the probability of all possible states for that hour. Accordingly, considering the probability of load for each state during hour $t$, the load level can be calculated as follows:

$$L^t = \frac{\sum_{g=1}^{N_L} L_g f_L\left(L_g^t\right)}{\sum_{g=1}^{N_L} f_L\left(L_g^t\right)} \tag{8.16}$$

where $L_g^t$ is the load of state $g$ at hour $t$, $L_g$ is the load level calculated using (8.15), and $f_L\left(L_g^t\right)$ is the probability of the load level for state $g$ during hour $t$.

## 8.3 Problem formulation

The problem of OPS of RDG is considered a constrained non-linear combinatorial optimization planning problem with two objectives:

1. minimizing the total energy loss in DNs;
2. maximizing the profit of RDG owners.

Several practical assumptions have been adopted which are necessary for the proper definition of this problem. The same or similar assumptions used by most authors to deal with this problem are as follows:

1. No geographic limitations to installing various RDG technologies within DNs [18].
2. All buses in the DN under study are subjected to the same wind profile and solar irradiance [9,10,13].
3. Only one type of RDG can be connected to the same bus in DNs [9,10].
4. All the RDG units are modeled as negative loads with unity power factor, i.e. producing active power only, as recommended by IEEE 1547 [9,13,25].
5. The maximum penetration of RGD is assumed to be equal to the maximum load of DNs [26].

### 8.3.1 Objective function

Two conflicting objectives are considered: the minimization of total energy loss of DNs and the maximization of profit of RDG owners in a given planning horizon of $N_y$ years. The multi-objective optimization problem can be converted to a single-objective optimization problem by the weighted aggregation method. Therefore, the multi-objective function for simultaneously minimizing the total energy loss and maximizing the profit can be formulated as follows:

$$\min(F) = \min\left(w_1F_1 + w_2\frac{1}{F_2}\right) \tag{8.17}$$

where $w_1$ and $w_2$ are the weight coefficients. The total energy loss is calculated as follows:

$$F_1 = \frac{365}{4}\sum_{y=1}^{N_y}\sum_{t=1}^{96} P_{loss,t}^{y} \tag{8.18}$$

where $P_{loss,t}^{y}$ is the power loss for hour $t$ in year $y$ of the considered time period of $N_y$ years.

The objective function $F_2$ can be defined as the difference between the incomes and costs of RDG owners:

$$F_2 = IN_{RDG} - C_{investment} - C_{oper,maint} \tag{8.19}$$

where $IN_{RDG}$ is the income of DG owners; $C_{investment}$ is the investment cost; and $C_{oper,maint}$ is the operation and maintenance cost.

The investment cost $C_{investment}$ contains different initial costs such as the amount of money spent on unit construction, installation, and essential equipment for each RDG unit. This cost can be formulated as [18]:

$$C_{investment} = \sum_{i=1}^{N_{RGD}} P_{RDGi}C_{inv} \tag{8.20}$$

where $P_{RDGi}$ is the rated power of RDG unit $i$, $N_{RDG}$ is the number of RDG units, and $C_{inv}$ is the investment cost of RDG unit $i$.

The operation and maintenance cost $C_{oper,maint}$ includes cost of generation, renewing, repairing, and restoring unit equipment in case of necessity. The equation for modeling the present worth of this cost is as follows:

$$C_{oper,maint} = \sum_{y=1}^{N_y} \sum_{i=1}^{N_{RDG}} P_{RDGi} C_{OM} \left( \frac{1 + INFR}{1 + INTR} \right)^y \tag{8.21}$$

where $C_{OM}$ is the operation and maintenance cost of RDG units per year, and $INFR$ and $INTR$ are the inflation rate and interest rate, respectively.

The RDG owner earns a profit by selling the generated energy to the distribution company at the contract price. The present worth of the income of DG owners $IN_{RDG}$ is [18]

$$IN_{RDG} = \frac{365}{4} \sum_{y=1}^{N_y} \sum_{i=1}^{N_{RDG}} \sum_{t=1}^{96} P^y_{RDGi,t} \cdot CP_{RDGi} \cdot \left( \frac{1 + INFR}{1 + INTR} \right)^y \tag{8.22}$$

where $P^y_{RDGi,t}$ is the generated active power of RDG unit $i$ at hour $t$ of year $y$ and $CP_{RDGi}$ is the contract price of electricity selling between the RDG owner and the distribution company.

### 8.3.2 Control variables

The control variables in this optimization problem are locations, i.e. indexes of connecting buses and numbers of elementary RDG units which should be connected at these buses. Thus, the optimal rated power of the RDG farms can be obtained as

$$P_{RDGF} = N_{RDG} P_{RDG1} \tag{8.23}$$

where $P_{RDGF}$ is the total rated power of the RDG farms, $N_{RDG}$ is the number of elementary RDG units that form an RDG farm (WT farm or PV farm), and $P_{RDG1}$ is the rated power of an elementary RDG unit.

### 8.3.3 Constraints

The optimization problem is subjected to various technical constraints, which are described below.

1. *Power flow constraints*
   The power flow constraints in DN with RDG units operating with unity power factor are the equality constraints represented by the power balance equations as follows:

$$P_{grid} = \sum_{i=1}^{N_B} P_{Li} - \sum_{i=1}^{N_{RDGF}} P_{RDGFi} + \sum_{i=1}^{N_{Br}} P_{lossi} \tag{8.24}$$

$$Q_{grid} = \sum_{i=1}^{N_B} Q_{Li} + \sum_{i=1}^{N_{Br}} Q_{lossi} \tag{8.25}$$

where $N_B$ is the number of busses in the network; $N_{RDGF}$ is the number of RDG farms; $N_{Br}$ is the number of branches in the network; $P_{grid}$ is the active power injected to the substation; $Q_{grid}$ is the reactive power injected to the substation; $P_{RDGFi}$ is the active power generation of RDG farm $i$; $P_{Li}$ is the active power of load at bus $i$; $Q_{Li}$ is the reactive power of load at bus $i$; and $P_{lossi}$ and $Q_{lossi}$ are the active and reactive power losses in branch $i$, respectively.

The backward/forward sweep algorithm presented in Chapter 1 is suitable to solve the above power balance equations for radial DNs.

2. *Bus voltage and branch load constraints*
   The OPS of RDG should be determined in such a way that bus voltages and branch loads remain in standard intervals during all normal operation states of DNs. These constraints can be defined as follows:

$$V_i^{\min} \leq V_i \leq V_i^{\max} \quad i = 1, 2, \ldots, N_B \tag{8.26}$$

$$S_{li} \leq S_{li}^{\max} \quad i = 1, 2, \ldots, N_{Br} \tag{8.27}$$

where $V_i^{\min}$ and $V_i^{\max}$ are the minimum and maximum allowable values of voltage magnitude of bus $i$ and $S_{li}^{\max}$ is the maximum load of branch $i$ of the network.

3. *RDG capacity constraints*
   The active power capacity of each RDG farm is limited to a specific maximum $P_{RDGFi}^{\max}$ as follows:

$$P_{RDGFi} \leq P_{RDGFi}^{\max} i = 1, 2, \ldots, N_{RDGF} \tag{8.28}$$

According to (8.23), the limit of RDG capacity can be expressed as

$$N_{RDGi} P_{RDGi1} \leq N_{RDGi}^{\max} P_{RDGi1} \tag{8.29}$$

where $N_{RDGi}$ is the number of elementary RDG units, which consists of the RGD farm at location $i$, $P_{RDGi1}$ is the rated power of elementary RDG unit at location $i$, and $N_{RDGi}^{\max}$ is the maximum number of elementary RDG units at location $i$.

## 8.4 Solution method

Metaheuristic optimization methods are population-based stochastic search techniques. In general, a search agent can be represented as a vector ($x_i$) whose elements are the values of the control variables of the optimization problem. The number of control variables ($n$) is the search space dimension of the optimization problem. At time (iteration) $t$, the agent $x_i(t)$ can be represented as $\mathbf{x}_i(t) = \left[x_i^1(t), \ldots, x_i^d(t), \ldots, x_i^n(t)\right]$, where $x_i^d(t)$ is the position of the agent $i$ with respect to the dimension $d$, i.e. the values of the control variable $d$ in the candidate solution $i$. The population (**POP**) is defined as a set of search agents that represent potential solutions to the optimization problem. The number of agents ($N$) is defined as the size of the population, i.e. $\mathbf{POP}(t) = [\mathbf{x}_1(t), ..., \mathbf{x}_i(t), ..., \mathbf{x}_N(t)]^T$. The essence of metaheuristic methods is an iterative correction of the solution, i.e. generating a new population by applying algorithmic operators with a stochastic search mechanism on agents from the current population.

The general structure of the proposed algorithm can be described as follows.

**Initialization**

*Step 1*: Define the objective function $F(x_i)$ and the space of possible solutions $X$.

*Step 2*: Generate an initial population of $N$ agents, where the initial positions of agents are randomly selected between minimum and maximum values of the control variables. Set the iteration counter $t = 1$.

**Iterative procedure**

*Step 3*: Calculate the fitness value for each agent in the current population **POP**($t$).

*Step 4*: Generate the new population **POP**($t$+1) by applying the algorithmic operators of the selected metaheuristic algorithm on search agents from the current population **POP**($t$).

*Step 5*: Repeat the iterative procedure until the stop criteria is reached.

*Step 6*: Report best solution.

In the case of OPS of RGD, a potential solution can be presented by a vector consisting of a combination of locations and rated power of RDG farms (i.e. the number of elementary RDG units) in these locations. Thus, $\mathbf{x}_i$ can be written as follows:

$$\mathbf{x}_i(t) = \left[Bus_i^1, ..., Bus_{N_{RDGF}}^d, P_1^{d+1} ..., P_{N_{RDGF}}^n\right] \tag{8.30}$$

where $n = 2N_{RDGF}$, $Bus_{N_{RDGF}}^d$ is the position of the $N$th RDGF in the potential solution $i$, and $P_{N_{RDGF}}^n$ is the rated power, i.e. the number of elementary RDG units of the RDGF at this position.

A general procedure of applying the proposed optimization algorithm to solve the problem of OPS of RDG units (WT and PV) in DNs can be described as follows.

*Step 1*: Define the DN configuration, the line data, the transformer data, and the load data.

*Step 2*: Define the technical and commercial data about the elementary RDG units, such as the rated power and other manufacturer specifications, the installation costs, the operation and maintenance costs, the contract power of selling power, the interest rate, the inflation rate, and the total number of years in the planning horizon.

*Step 3*: Define the total number of RDG farms ($N_{RDGF}$) to be connected in DNs, and the maximum number of each type of elementary RDG unit which can be connected at a bus of the network.

*Step 4*: Define the typical daily diagrams of output power for WT and PV, and the typical daily load profiles for each season, as described in Section 8.2.

*Step 5*: Set the algorithmic parameters such as the population size and the maximum number of iterations, and generate an initial random population of $N$ agents.

*Step 6*: Calculate the objective function (8.17) for each agent $x_i(t)$ from the current population $POP(t)$.

*Step 7*: Apply the algorithmic operators of the selected metaheuristic algorithm to create a new population of agents, i.e. the potential solutions to the problem.

*Step 8*: Repeat Steps 6 and 7 until the stop criterion (i.e. the max number of iterations) is reached.

*Step 9*: Report the best $x_i$ from the last iteration, i.e. the optimal locations (list of buses) and rated power (number of elementary RDG units at each of these buses).

## 8.5 Simulation results

The proposed algorithm is applied on the IEEE 69-bus test system with the nominal voltage of 12.66 kV, and the total active and reactive loads of 3791.89 kW and 2694.10 kVAr, respectively. The total power loss in the original system without any DG shown in Figure 8.4 is 224.946 kW. The system data can be found in [27], as well as in the supplementary material for this chapter.

The task is to determine the optimal placement and size for one WT farm and one PV farm in the IEEE 69-bus system. The rated power of elementary RDG units

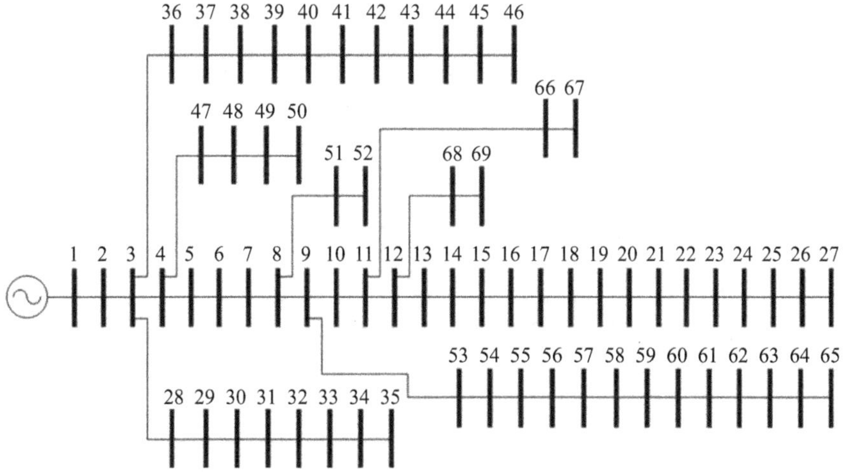


*Figure 8.4 Single-line diagram of IEEE 69-bus test system*

*Table 8.1 Commercial data of RDG units*

| Type | $C_{inv}$ (\$/kW) | $C_{OM}$ (\$/kW) | $CP_{DG}$ (\$/kWh) | *INFR* (%) | *INTR* (%) | $N_y$ (year) |
|---|---|---|---|---|---|---|
| WT | 1100 | 16 | 0.10 | 2 | 1.25 | 10 |
| PV | 1000 | 10 | 0.15 | 2 | 1.25 | 10 |

$P_{RDG1}$ is 200 kW, whereas the maximum size of RDG farms $N_{RDG}^{\max}$ is 10, for both WT and PV generation. The commercial data of the RDG units are given in Table 8.1. The installed cost $C_{inv}$, the operation and management (O&M) cost $C_{OM}$, and the contract price of electricity selling $CP_{DG}$ are adopted based on the report of International Renewable Energy Agency (IRENA) [2]. The planning horizon is $N_y = 10$ years.

The WT units used in this simulation have a rated power of 200 kW, a nominal wind speed of 10 m/s, a cut-in wind speed of 2.7 m/s, and a cut-out wind speed of 25 m/s. The PV has rated power of 200 kW and consists of 800 PV modules with $P_{STC} = 250$ W, $\gamma = -0.0045\ ^\circ\text{C}^{-1}$, and $T_{NOCT} = 46\ ^\circ\text{C}$.

The measured wind speed and solar irradiance data are taken from [22]. The wind speed and solar irradiance data are recorded with sampling periods of 10 and 5 min during the entire year of 2013, respectively. The period of 1 year is divided into four seasons, i.e. Spring (March, April, May), Summer (June, July, August), Autumn (September, October, November), and Winter (December, January, February), and a typical day for each of the seasons is obtained, as explained in Section 8.2. Based on the measured data, the mean values and standard deviations of wind speed and solar irradiance are calculated for each hour of the typical days, as given in Tables 8.2 and 8.3, respectively. Using these data, the discrete PDFs of wind speed and solar irradiance for each hour can be determined, as illustrated in Figures 8.1 and 8.2.

By using the typical day models for seasons, the predicted power of WT and PV is calculated for each year in the planning horizon of 10 years. The normalized output power of WT and PV shown in Figures 8.5 and 8.6 is given relative to their rated power.

A typical daily load profile is assumed for each season according to the IEEE RTS system [28]. Figure 8.7 shows seasonal variations of load levels with a standard deviation of 5%. The mean hourly load levels are given relative to the peak load.

In the system under study, two different cases (Cases 1 and 2) are considered in determining OPS of WT and PV farms, along with an extra reference case (Case 3) for comparison.

Case 1: Consider the simultaneous minimization of total energy loss and the maximization of profit of RDG owner's profit, i.e. (8.17).

Case 2: Consider the minimization of total energy loss only, i.e. (8.18).

Case 3: Compare the total energy loss without RDG integrated into the system.

*Table 8.2 Mean values and standard deviations of wind speed (m/s)*

| Time (hour) | Spring | | Summer | | Autumn | | Winter | |
|---|---|---|---|---|---|---|---|---|
| | $\mu_v$ | $\sigma_v$ | $\mu_v$ | $\sigma_v$ | $\mu_v$ | $\sigma_v$ | $\mu_v$ | $\sigma_v$ |
| 1 | 8.188 | 5.271 | 7.253 | 5.039 | 12.342 | 7.009 | 8.189 | 6.509 |
| 2 | 8.269 | 5.622 | 7.517 | 5.090 | 12.049 | 6.929 | 8.451 | 5.928 |
| 3 | 8.678 | 5.526 | 7.118 | 4.839 | 11.981 | 6.871 | 8.792 | 5.735 |
| 4 | 9.053 | 5.820 | 7.238 | 4.893 | 11.897 | 7.035 | 8.989 | 5.984 |
| 5 | 8.927 | 5.549 | 7.219 | 4.851 | 12.258 | 7.361 | 9.015 | 6.056 |
| 6 | 9.056 | 5.889 | 7.491 | 4.725 | 12.140 | 6.948 | 8.899 | 5.895 |
| 7 | 8.616 | 5.492 | 7.695 | 4.750 | 11.677 | 6.836 | 8.499 | 5.791 |
| 8 | 8.248 | 5.523 | 7.568 | 4.460 | 12.076 | 7.288 | 9.038 | 6.221 |
| 9 | 8.684 | 5.679 | 8.430 | 4.875 | 12.341 | 7.086 | 9.554 | 5.901 |
| 10 | 9.259 | 6.401 | 8.801 | 4.904 | 12.774 | 7.203 | 9.930 | 6.049 |
| 11 | 9.110 | 6.258 | 9.036 | 5.029 | 12.745 | 7.342 | 10.332 | 5.896 |
| 12 | 9.043 | 6.132 | 9.185 | 5.194 | 12.845 | 6.633 | 9.879 | 6.038 |
| 13 | 9.111 | 6.072 | 9.099 | 5.095 | 13.123 | 6.607 | 9.866 | 5.674 |
| 14 | 9.099 | 6.285 | 8.625 | 4.619 | 13.505 | 6.789 | 10.101 | 5.957 |
| 15 | 9.202 | 5.906 | 8.519 | 4.948 | 13.313 | 6.922 | 10.362 | 6.049 |
| 16 | 9.123 | 5.508 | 8.328 | 5.215 | 12.991 | 6.742 | 10.157 | 5.565 |
| 17 | 8.732 | 5.191 | 8.174 | 5.324 | 12.514 | 6.423 | 10.112 | 5.977 |
| 18 | 8.519 | 5.228 | 7.864 | 4.949 | 12.136 | 6.498 | 9.644 | 6.015 |
| 19 | 8.468 | 5.037 | 7.523 | 4.869 | 11.624 | 6.072 | 9.022 | 5.647 |
| 20 | 7.849 | 4.998 | 7.483 | 4.618 | 11.582 | 5.789 | 8.676 | 5.715 |
| 21 | 7.806 | 5.184 | 7.535 | 4.606 | 11.467 | 6.138 | 8.136 | 5.554 |
| 22 | 7.544 | 4.859 | 7.566 | 4.758 | 11.442 | 6.771 | 8.219 | 5.922 |
| 23 | 7.984 | 4.972 | 7.653 | 4.820 | 11.325 | 6.709 | 8.229 | 6.019 |
| 24 | 8.391 | 5.167 | 7.383 | 4.778 | 11.848 | 6.916 | 8.306 | 6.549 |

*Table 8.3 Mean values and standard deviations of solar irradiance ($W/m^2$)*

| Time (hour) | Spring | | Summer | | Autumn | | Winter | |
|---|---|---|---|---|---|---|---|---|
| | $\mu_s$ | $\sigma_s$ | $\mu_s$ | $\sigma_s$ | $\mu_s$ | $\sigma_s$ | $\mu_s$ | $\sigma_s$ |
| 1 | 0.031 | 0.109 | 0.000 | 0.001 | 0.017 | 0.058 | 0.131 | 0.547 |
| 2 | 0.023 | 0.076 | 0.009 | 0.039 | 0.006 | 0.027 | 0.363 | 1.435 |
| 3 | 0.027 | 0.099 | 0.003 | 0.021 | 0.019 | 0.076 | 0.139 | 0.598 |
| 4 | 0.033 | 0.116 | 0.001 | 0.006 | 0.032 | 0.123 | 0.305 | 1.426 |
| 5 | 0.028 | 0.083 | 0.015 | 0.051 | 0.038 | 0.109 | 0.048 | 0.134 |
| 6 | 0.047 | 0.161 | 0.023 | 0.080 | 0.021 | 0.089 | 0.173 | 0.841 |
| 7 | 3.706 | 5.895 | 7.433 | 7.809 | 0.036 | 0.139 | 0.150 | 0.453 |
| 8 | 137.6 | 132.4 | 157.7 | 120.4 | 23.79 | 32.18 | 2.159 | 4.574 |
| 9 | 360.1 | 265.8 | 385.9 | 248.5 | 241.9 | 186.4 | 97.78 | 124.8 |
| 10 | 535.6 | 316.7 | 538.2 | 276.6 | 437.7 | 282.2 | 254.4 | 234.2 |
| 11 | 655.7 | 314.4 | 668.5 | 261.4 | 553.1 | 306.5 | 363.6 | 279.4 |
| 12 | 670.2 | 322.1 | 747.8 | 241.9 | 606.9 | 319.9 | 435.4 | 294.8 |

(Continues)

*Table 8.3 (Continued)*

| Time (hour) | Spring | | Summer | | Autumn | | Winter | |
|---|---|---|---|---|---|---|---|---|
| | $\mu_s$ | $\sigma_s$ | $\mu_s$ | $\sigma_s$ | $\mu_s$ | $\sigma_s$ | $\mu_s$ | $\sigma_s$ |
| 13 | 728.8 | 313.5 | 818.4 | 221.2 | 612.3 | 329.6 | 419.1 | 288.7 |
| 14 | 740.1 | 307.2 | 862.2 | 207.1 | 629.7 | 317.9 | 398.8 | 286.7 |
| 15 | 715.9 | 337.4 | 869.7 | 196.5 | 610.3 | 312.0 | 332.5 | 258.7 |
| 16 | 689.4 | 347.3 | 853.3 | 205.4 | 555.9 | 316.2 | 287.8 | 233.7 |
| 17 | 640.4 | 351.6 | 778.9 | 237.3 | 420.9 | 315.1 | 145.9 | 163.1 |
| 18 | 534.8 | 348.9 | 699.6 | 257.8 | 276.8 | 311.2 | 33.43 | 78.99 |
| 19 | 364.3 | 325.2 | 571.6 | 262.7 | 135.2 | 194.9 | 1.321 | 3.960 |
| 20 | 183.3 | 200.8 | 356.9 | 205.7 | 22.00 | 47.78 | 0.134 | 0.729 |
| 21 | 26.34 | 44.01 | 89.04 | 79.09 | 0.045 | 0.151 | 0.197 | 0.806 |
| 22 | 0.040 | 0.109 | 0.877 | 1.344 | 0.010 | 0.052 | 0.117 | 0.534 |
| 23 | 0.012 | 0.042 | 0.007 | 0.036 | 0.025 | 0.089 | 0.231 | 1.140 |
| 24 | 0.026 | 0.071 | 0.001 | 0.013 | 0.007 | 0.038 | 0.359 | 1.929 |

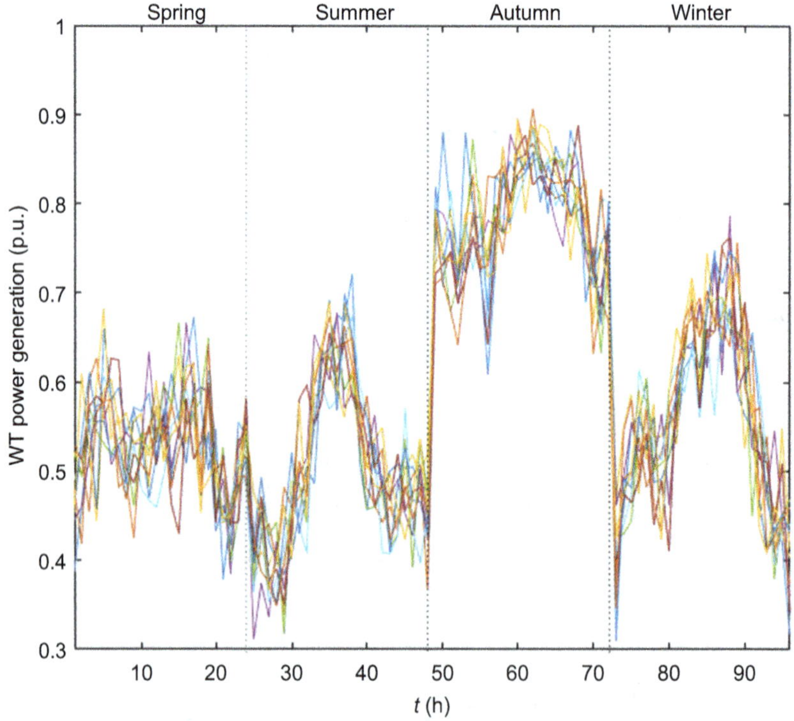


*Figure 8.5 Prediction of WT generation*

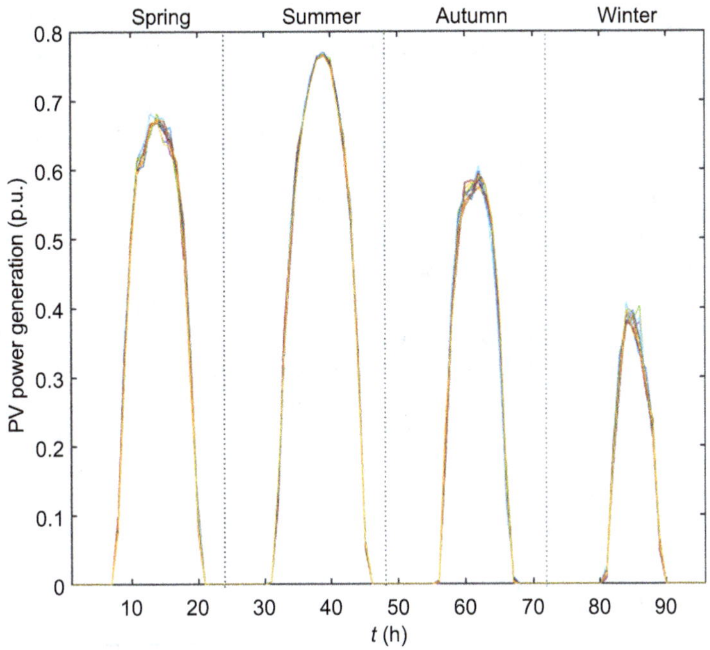


*Figure 8.6 Prediction of PV generation*

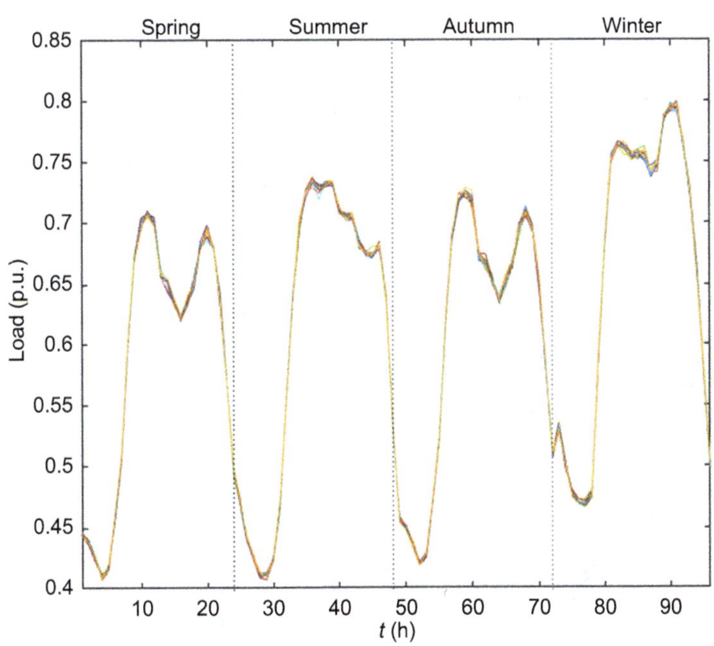


*Figure 8.7 Prediction of load levels*

The optimal results are shown in Tables 8.4 and 8.5. The total energy loss for the considered period of 10 years calculated for Case 3 amounts to 7251.311 MWh.

The optimal solutions in Cases 1 and 2 indicate a huge reduction in total energy losses in relation to Case 3. The total energy loss is only 3.5% higher for Case 1 in comparison to Case 2, but the profit of RDG owners is 20.1% higher in Case 1 than that in Case 2. These ratios show that the solution obtained in Case 1 is satisfactory both from the viewpoint of distribution company and the RDG owners. For at least one-third of the year (at night), the energy produced by the PV farm is zero, and thus, the total generation (and profit) from the PV farm is considerably lower than the total generation (and profit) from the WT farm.

The power losses in Cases 2 and 3 are shown in Figure 8.8. The results reveal that the OPS of WT and PV leads to a significant reduction of power losses in the whole period. As shown in Figure 8.9, the power loss reduction is more in periods with higher power generation of WT and PV farms. As expected, the power loss reduction is less in periods without PV generation.

The results shown in the previous section were obtained using the hybrid PPSOGSA method, as suggested in [21]. The convergence profiles of the PPSO,

*Table 8.4 Optimal results of WT and PV farms*

| Case | Farm | Location bus | $N_{RDG}$ | $P_{RDGF}$ (MW) |
|---|---|---|---|---|
| 1 | WT | 61 | 10 | 2.0 |
| | PV | 11 | 8 | 1.6 |
| 2 | WT | 61 | 9 | 1.8 |
| | PV | 16 | 4 | 0.8 |

*Table 8.5 Optimal results of economic and technical indices for considered period of 10 years*

| Index | Case 1 | Case 2 |
|---|---|---|
| Total energy loss $F_1$ (MWh) | 2975.7 | 2871.4 |
| Profit of RDG owners $F_2$ (M$) | 11.168 | 8.923 |
| Profit from WT (M$) | 8.289 | 7.484 |
| Profit from PV (M$) | 2.879 | 1.438 |
| Total investment cost (M$) | 3.800 | 2.780 |
| Investment cost of WT (M$) | 2.200 | 1.980 |
| Investment cost of PV (M$) | 1.600 | 0.800 |
| Total O&M cost (M$) | 0.500 | 0.383 |
| O&M cost of WT (M$) | 0.333 | 0.300 |
| O&M cost of PV (M$) | 0.167 | 0.083 |
| Income of RDG owners (M$) | 15.468 | 12.086 |
| Income from WT (M$) | 10.822 | 9.764 |
| Income from PV (M$) | 4.645 | 2.322 |
| Total generation of WT (MWh) | 103896.5 | 93738.8 |
| Total generation of PV (MWh) | 29730.1 | 14857.7 |

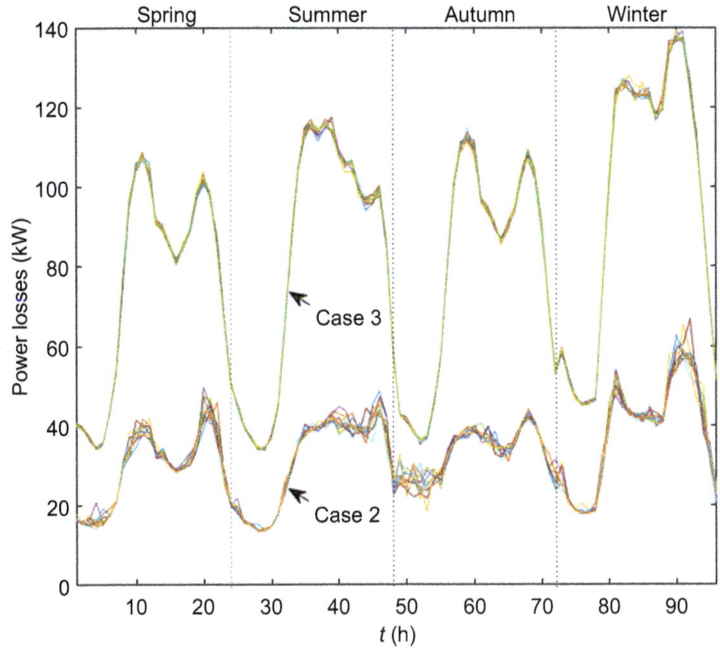


*Figure 8.8 Power losses for Cases 2 and 3*

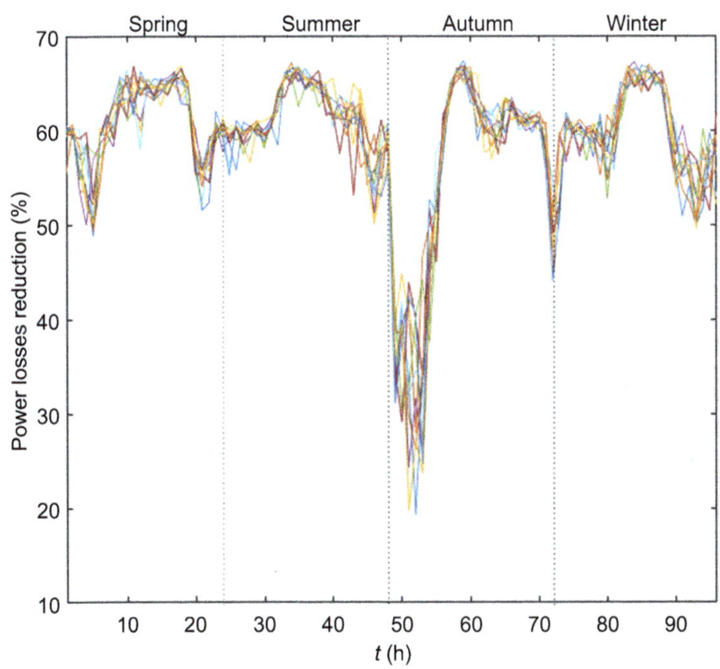


*Figure 8.9 Power losses reduction in Case 2*

GSA, and proposed hybrid algorithm in solving the OPS of WT and PV for Cases 1 and 2 are shown in Figures 8.10 and 8.11, respectively. It is clear that the proposed hybrid algorithm achieves better solutions and converges to an optimal solution with less number of iterations compared to the PPSO and GSA.

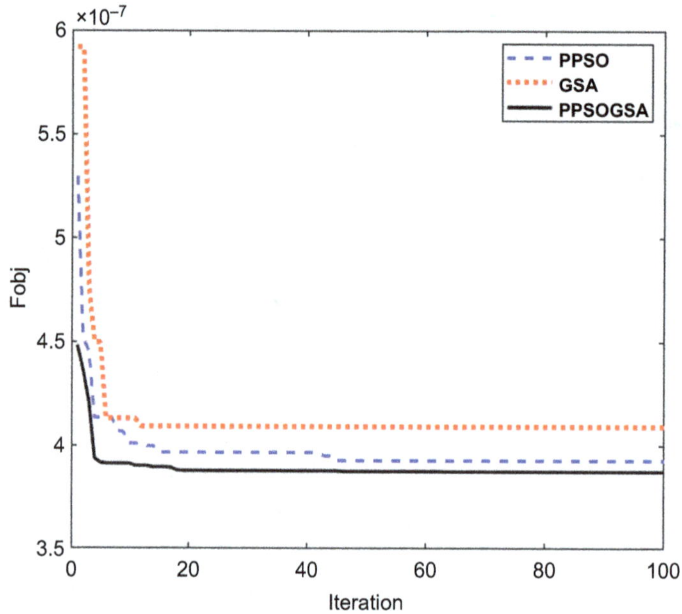


*Figure 8.10 Convergence characteristics in Case 1*

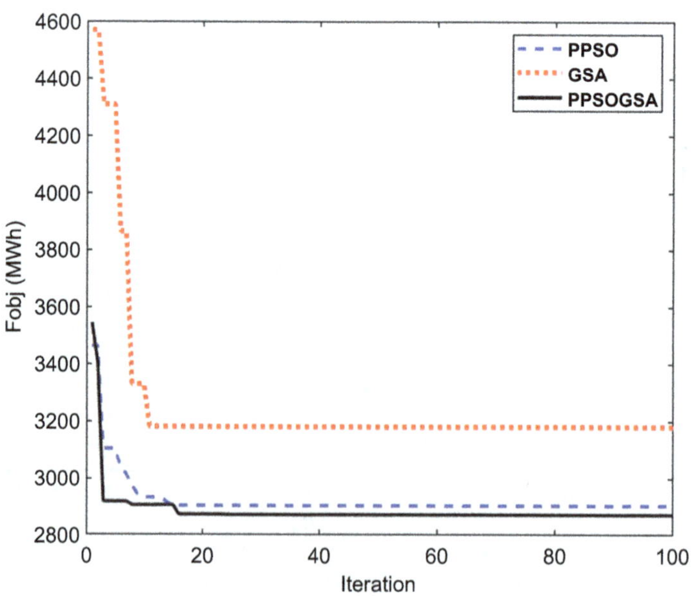


*Figure 8.11 Convergence characteristics in Case 2*

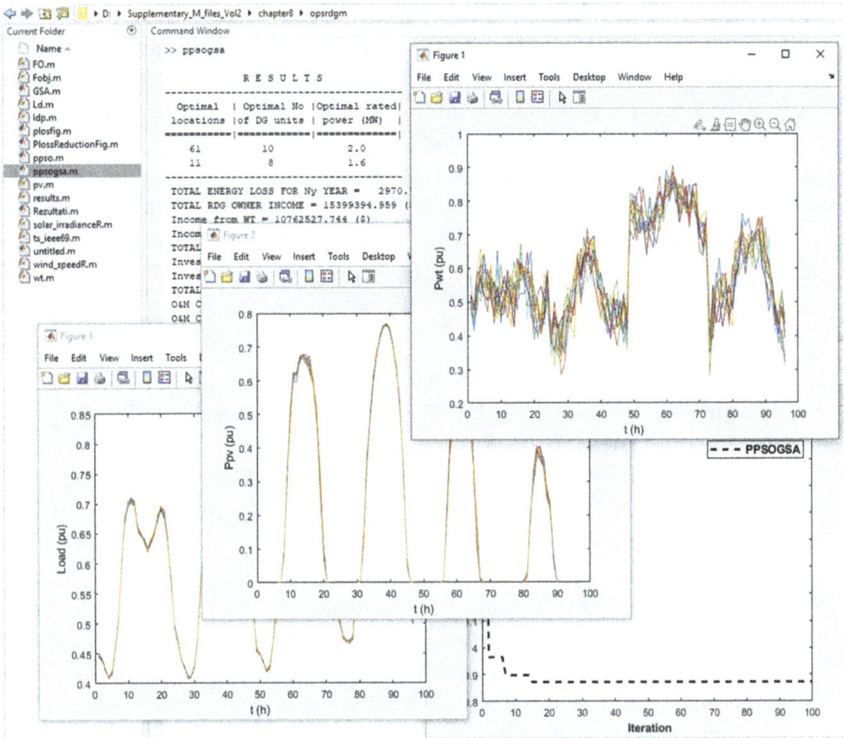


*Figure 8.12 An example of using the opsrdgm program package*

## 8.6 Program package *opsrdgm*

The methodology presented in this chapter is implemented in the MATLAB within an educational program package *opsrdgm*. This program package is available as the supplementary file to Chapter 8 of this book on the book's homepage at the IET digital library.

Figure 8.12 shows an example of running the *opsrdgm* program. The PPSOGSA algorithm was chosen as the solution method.

## 8.7 Conclusion

In this paper, a metaheuristic optimization-based approach is proposed and successfully applied to solve the problem of OPS of RDG with objectives to minimize the total energy loss in DNs and maximizing the profit of RDG owners. The proposed algorithm has been tested on the IEEE 69-bus test system considering the probabilistic models for WT, PV and loads based on the typical daily diagrams representing seasons of years. The conclusions can be summarized as follows.

- The proposed algorithm provides results which are quite satisfactory both from the viewpoint of distribution company and RDG owners. There is a significant reduction of total energy losses both in the case of simultaneous minimization of total energy loss and maximization of profit of RDG owners, and in the case of minimization of total energy loss only.
- The proposed algorithm provides robust and high-quality solutions both in the case of simultaneous minimization of total energy loss and maximization of profit of RDG owners, and in the case of minimization total energy loss.

## References

[1] Xu M., and Zhuan X. 'Identifying the optimum wind capacity for a power system with interconnection lines'. *International Journal of Electrical Power and Energy Systems*. 2013;**51**(1):82–88.

[2] International Renewable Energy Agency. (2018, Jan.). *Renewable Power Generation Costs in 2017*. [online]. 2018. Available from http://www.irena.org/-/media/Files/IRENA/Agency/Publication/2018/Jan/IRENA_2017_Power_Costs_2018.pdf.

[3] Tan W.S., Hassan M.Y., Majid M.S., and Rahman H.A. 'Optimal distributed renewable generation planning: a review of different approaches'. *Renewable and Sustainable Energy Reviews*. 2013;**18**:626–45.

[4] Prakash P., and Khatod D.K. 'Optimal sizing and siting techniques for distributed generation in distribution systems: a review'. *Renewable and Sustainable Energy Reviews*. 2016;**57**:111–30.

[5] Mahmoud P.H.A., Huy P.D., and Ramachandaramurthy V. K. 'A review of the optimal allocation of distributed generation: objectives, constraints, methods, and algorithms'. *Renewable and Sustainable Energy Reviews*. 2017:**75**:293–312.

[6] Adefarati T., and Bansal R.C. 'Integration of renewable distributed generators into the distribution system: a review'. *IET Renewable Power Generation*. 2016;**10**(7):873–84.

[7] Kayal P., and Chanda C.K. 'Optimal mix of solar and wind distributed generations considering performance improvement of electrical distribution network'. *Renewable Energy*. 2015;**75**:173–86.

[8] Khatod D.K., Pant V., and Sharma J. 'Evolutionary programming based optimal placement of renewable distributed generators'. *IEEE Transactions on Power Systems*. 2013;**28**(2):683–95.

[9] Atwa Y.M., El-Saadany E.F., Salama M.M.A., and Seethapathy R. 'Optimal renewable resources mix for distribution system energy loss minimization'. *IEEE Transactions on Power Systems*. 2010;**25**(1):360–70.

[10] Abdelaziz A.Y., Hegazy Y.G., El-Khattam W., and Othman M.M. 'Optimal allocation of stochastically dependent renewable energy based distributed generators in unbalanced distribution networks'. *Electric Power System Research*. 2015;**119**:34–44.

[11] Fard H.H., and Jalilian A. 'Optimal sizing and location of renewable energy based DG units in distribution systems considering load growth'. *Electrical Power and Energy Systems.* 2018;**101**:356–70.

[12] Barik S., and Das D. 'Determining the sizes of renewable DGs considering seasonal variation of generation and load and their impact on system load growth'. *IET Renewable Power Generation*. 2018;**12**(10):1101–10.

[13] Nasri A., Golshan M.E.H., and Najed S.M.S. 'Optimal planning of dispatchable and non-dispatchable distributed generation units for minimizing distribution system's energy loss using particle swarm optimization'. *International Transactions on Electrical Energy Systrems*. 2014;**24**(4):504–19.

[14] Ali E.S., Abd-Elazim S.M., and Abdelaziz A.Y. 'Ant lion optimization algorithm for optimal location and sizing of renewable distributed generations'. *Renewable Energy*. 2017;**101**:1311–24.

[15] Ali E.S., Abd-Elazim S.M., and Abdelaziz A.Y. 'Optimal allocation and sizing of renewable distributed generation using ant lion optimization algorithm'. *Electrical Engineering*. 2016;**100**:99–109.

[16] Arya L.D., Koshti A., and Choube S.C. 'Distributed generation planning using differential evolution accounting voltage stability consideration'. *International Journal of Electrical Power & Energy Systems*. 2012;**42**(1): 196–207.

[17] Biswas S., Goswami S. K., and Chatterjee A. 'Optimal distributed generation placement in shunt capacitor compensated distribution systems considering voltage sag and harmonics distortions'. *IET Generation, Transmission & Distribution*. 2014;**8**(5):783–97.

[18] Ameli A., *et al.* 'A multiobjective particle swarm optimization for sizing and placement of DGs from DG owner's and distribution company's view points'. *IEEE Transactions on Power Delivery*. 2014;**29**(4):1831–40.

[19] Kefayat M., Ara A.L., and Niaki S.A.N. 'A hybrid of ant colony optimization and artificial bee colony algorithm for probabilistic optimal placement and sizing of distributed energy resources'. *Energy Conversion and Management*. 2015;**921**:149–61.

[20] Saric M., Hivziefendic J., Konjic T., and Ktena A. 'Distributed generation allocation considering uncertainties'. *International Transactions on Electrical Energy Systrems*. 2018;**28**(9):e2585.

[21] Radosavljević J., Arsić N., Milovanović M., and Ktena A. 'Optimal placement and sizing of renewable distributed generation using hybrid metaheuristic algorithm'. *Journal of Modern Power Systems and Clean Energy*. 2020; **8**(3):499–510.

[22] IEEE PES. (2018, Oct.). Open data sets. [online]. Availabe from http://sites.ieee.org/pes-iss/data-sets/#weat.

[23] Tiwari G. N., and Dubey S. *Fundamentals of Photovoltaic Modules and Their Applications*. Cambridge, UK: RSC Energy Series, 2010, p. 422.

[24] *Design Qualification Type Approval of Commercial PV Modules*, IEC 61215, 2005.

[25] Kayal P., and Chanda C.K. 'Placement of wind and solar based DGs in distribution system for power loss minimization and voltage stability improvement'. *Electrical Power and Energy Systems.* 2013;**53**:795–809.

[26] Mistry K.D., and Roy R. 'Enhancement of loading capacity of distribution system through distributed generator placement considering techno-economic benefits with load growth'. *International Journal of Electrical Power & Energy Systems.* 2014;**54**:505–15.

[27] Ranjan N., Venkatesh B., and Das D. 'Voltage stability analysis of radial distribution networks'. *Electric Power Components and Systems.* 2001;**23**(2):129–35.

[28] Power Systems Test Case Archive. (2018, August). Reliability test system. [online]. Available from https://www2.ee.washington.edu/research/pstca/rts/pg_tcarts.htm.

*Chapter 9*

# Optimal coordination of directional overcurrent relays

## 9.1 Introduction

Generally, a power system consists of many sections, and each section must be protected against overcurrents. In order to ensure optimum-supply reliability, it is important to obtain selectivity, i.e. to switch off the part of the power system where a fault occurs and to avoid switching off the parts of the system where no fault occurs. In the case that the first overcurrent device fails (although an overcurrent occurs, the overcurrent device does not interrupt this fault current), the second overcurrent device will detect the overcurrent and interrupt the fault current. This backup principle ensures that the system remains safe even if an overcurrent device fails.

When considering a radial system with one feeding point and one directional power flow, it is relatively easy to obtain the desired selectivity by using overcurrent relays with appropriate time delays. When there is bidirectional power flow in a radial system with several source points or a closed ring topology, it is insufficient to use overcurrent relays with an appropriate delay to obtain the desired selectivity property. In these cases, directional overcurrent relays (DOCRs) must be used, which can identify the polarity of the overcurrent.

DOCRs are commonly used for the protection of subtransmission and distribution systems or as a secondary protection to transmission systems [1–7]. The coordination of DOCRs in power systems consists of selecting appropriate settings such that their fundamental protective role is met under the requirements of sensitivity, selectivity, reliability, and speed [8]. These requirements can be translated into conditions such as [7,8]:

1. a relay must get sufficient chance to operate if a fault occurs within the relay jurisdiction;
2. if a primary relay fails, a backup relay should operate;
3. the sequence of primary and backup (P/B) relay operations should be with sufficient coordination margins without excessive time delays.

The goal of DOCR coordination is to find the appropriate time dial setting (TDS) and pickup current setting (PCS) for relays, considering various constraints such as coordination and boundary limits [9,10].

The coordination of DOCRs has been formulated as an optimization problem and solved using different optimization methods, including conventional and heuristic techniques.

In [2,5], optimum coordination has been obtained considering the independent optimization of relay settings. In [10], an analytic approach to optimal coordination of DOCRs was proposed. It has been shown that conventional optimization methods such as linear programming (LP) [1,6,11], sequential quadratic programming (SQP) [12], quadratically constrained quadratic programming [13], gradient-based method [14], and binary integer programming [15] can be applied to optimize DOCR settings. However, conventional optimization methods have a drawback, some time they may trap local minima and fail to give a global optimal solution. In addition, as the size of the system increases, the convergence is also weak [16]. It becomes essential to develop optimization methods able to overcome these drawbacks and handle such difficulties.

Recently, many population-based metaheuristic methods have been used to solve complex constrained optimization problems. Generally, achieving optimal or near optimal solution for a specific problem will require multiple trials as well as appropriate tuning of associated parameters. Combination of several optimization algorithms in hybrid algorithm allows them to exploit the strength of each algorithm. A wide variety of such metaheuristic population-based techniques have been applied in solving the optimal coordination of DOCRs, as outlined in a brief overview given in Volume 1 of the book (Chapter 6, Section 6.7).

This chapter covers the use of a hybrid GSA algorithm with a local search technique based on SQP in solving the DOCRs coordination problem. In this approach, GSA performs global optimization and SQP is a local optimizer to fine-tune the solution, as proposed in [17].

## 9.2 Problem formulation

The DOCR protection zone is a function of the measured fault current. The relay will start when the measured fault current is greater than the predetermined setting of the pickup current. In the coordination of the DOCRs, each relay must offer primary protection for its line (where the relay was installed), and backup protection for all the adjacent lines.

This study takes into account the same assumptions that have been considered in other studies reported in the literature related to the optimal coordination of DOCRs. The assumptions are as follows:

- Each protection zone corresponds to one of the lines.
- DOCRs are placed at both ends of each line.
- The primary relay should only trip for faults within its protection zone (line).
- Only if the primary relay fails to clear the fault, the backup relay will operate after a time delay.

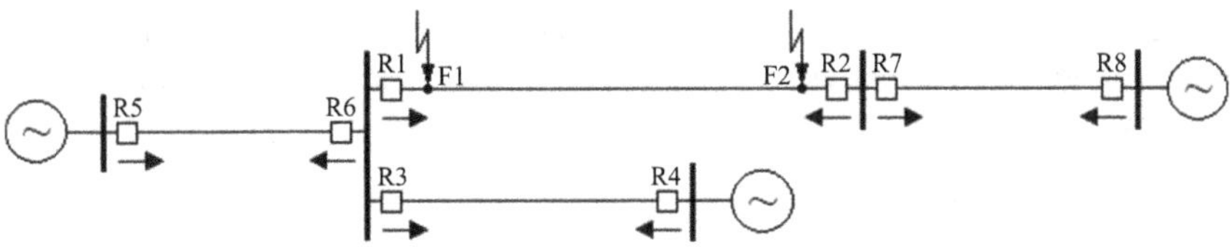


*Figure 9.1 Simple network with primary/backup relay pairs*

- Fault currents for different locations and corresponding primary/backup relay pairs are predetermined. Usually, near-end, far-end, and middle point fault currents are used to coordinate most of the fault situations.
- Network topologies are fixed; the transient configuration of the system is neglected during relay coordination. Thermal limits of lines are not considered.
- The DOCR coordination problem is to determine the sequences of relay operations for each predetermined fault location so that the faulted line is isolated with sufficient coordination margins and without excessive time delays.

Figure 9.1 shows a simple system with the relay pairs for a near-end fault (F1) and a far-end fault (F2). The P/B relay pairs for faults F1 and F2 are R1/R5, R1/R4, and R2/R8.

Obviously, when the network topology is changed, or when the distributed generation (DG) units are added at one or more buses in the network, the short circuit currents seen by the relay are changed. If the same settings of the basic system are used, it is possible that relays see fault currents smaller than their pickup currents at certain fault locations. This will cause some cases of miscoordination between P/B relays. For that, it is necessary to calculate new optimal settings of DOCRs for the system with changed topology and/or added new DG units.

The optimal coordination problem of DOCRs using optimization techniques consists of minimizing an objective function (OF) subject to certain coordination criteria and limits on problem variables [3].

### *9.2.1 Objective function*

The relay, which is supposed to operate first to clear the fault, is called the primary relay. Conventionally, OF in coordination studies is constituted as the summation of operating times (OTs) of all primary relays. Therefore, the OF may be defined as follows [15]:

$$\min_{\mathrm{PCS}_i,\mathrm{TDS}_i} \mathrm{OF} = \sum_{i=1}^{m}\sum_{k} T_{ik} \tag{9.1}$$

where $m$ is the number of relays in the network; $\mathrm{PCS}_i$ and $\mathrm{TDS}_i$ are the pick current and time dial settings of the $i$th relay, respectively; and $T_{ik}$ is the operation time of the $i$th relay for a fault at $k$.

Concerning the operating characteristics of DOCR, the most common and widely used time-current curves, according to IEC 60255-151, are considered as follows:

$$T_{ik} = \frac{\mathrm{TDS}_i \times \alpha}{(I_{ik}/\mathrm{PCS}_i)^n - 1} \tag{9.2}$$

where $I_{ik}$ is the fault current seen by the $i$th relay for a fault at $k$; $\alpha$ and $n$ are constant values, which demonstrate the relay characteristic according to Table 9.1, as illustrated in Figure 9.2.

With each relay, one current transformer (CT) is used to reduce the level of the current so that the relay can withstand it. The primary rating of CT is known in the

*Table 9.1 Characteristics of the overcurrent relays (IEC standard)*

| Characteristic type | $\alpha$ | $n$ |
|---|---|---|
| Long-time inverse | 120 | 1 |
| Extremely inverse | 80 | 2 |
| Very inverse | 13.5 | 1 |
| Normally inverse | 0.14 | 0.02 |

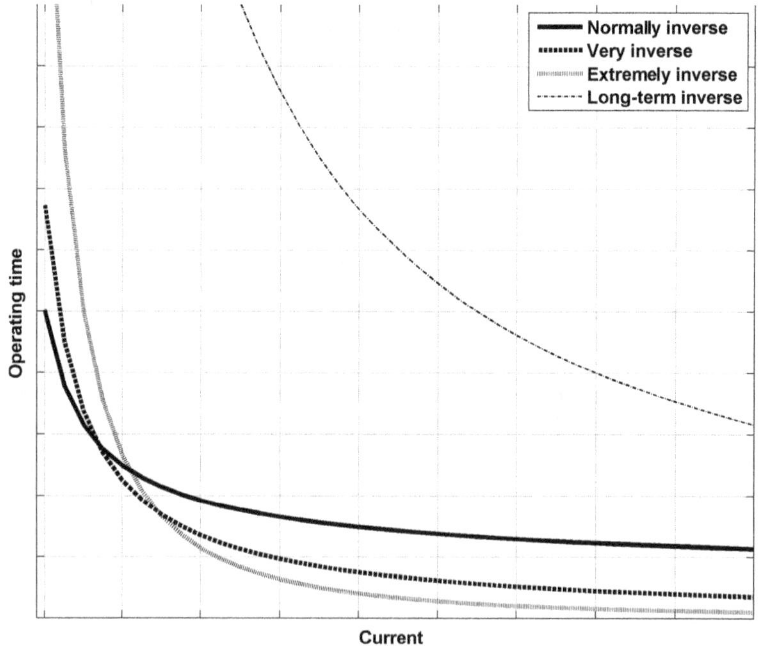


*Figure 9.2 Typical inverse time-current characteristics of DOCRs*

problem. Therefore, the fault current seen by the relay ($I_{ik}$) is obtained as a ratio of the fault current at CT primary terminals ($I_{f,ik}$) and the primary rating of the CT, ($CT_{rating}$), as follows:

$$I_{ik} = \frac{I_{f,ik}}{CT_{rating}} \tag{9.3}$$

### 9.2.2 Limits of the settings

The bounds of the settings can be expressed by

$$\text{TDS}_i^{\min} \le \text{TDS}_i \le \text{TDS}_i^{\max} \quad i = 1, \ldots, m \tag{9.4}$$

$$\text{PCS}_i^{\min} \le \text{PCS}_i \le \text{PCS}_i^{\max} \quad i = 1, \ldots, m \tag{9.5}$$

where $\text{TDS}_i^{\min}$and $\text{TDS}_i^{\max}$ are the lower and upper limits of TDS of the $i$th relay, respectively; and $\text{PCS}_i^{\min}$ and $\text{PCS}_i^{\max}$ are the lower and upper limits of PCS of the $i$th relay, respectively.

The $\text{PCS}_i$ should be less than the minimum fault current and, at the same time, greater than the maximum load current seen by each relay, including a safety margin, which depends on the relay technology and CT errors [10,15].

### 9.2.3 Limits of relay operation time

Minimum and maximum OTs can be established for each relay by

$$T_{ik}^{\min} \le T_{ik} \le T_{ik}^{\max} \quad i = 1, \ldots, m \tag{9.6}$$

where $T_{ik}^{\min}$ and $T_{ik}^{\max}$ are the minimum and maximum operation times of the $i$th relay for a fault $k$, respectively. $T_{ik}^{\min}$ depends on the relay manufacturer, whereas $T_{ik}^{\max}$ depends on the critical clearing time required to prevent equipment damage and preserve system stability [18]. The acceptable range of the primary relay operating time is between values of 0.1 and 1.1 s [13].

### 9.2.4 Coordination criteria

The backup relay operates only when the corresponding primary relay fails. To ensure relay coordination, the OT of the backup relay must be greater than the corresponding primary relay for all the faults. The coordination time interval (CTI) can be defined as follows:

$$\text{CTI} = T_{jk} - T_{ik} \quad i = 1, \ldots, m \tag{9.7}$$

where $T_{jk}$ is the operation time of the $j$th backup relay for a fault at $k$ inside the zone protected by the $i$th primary relay.

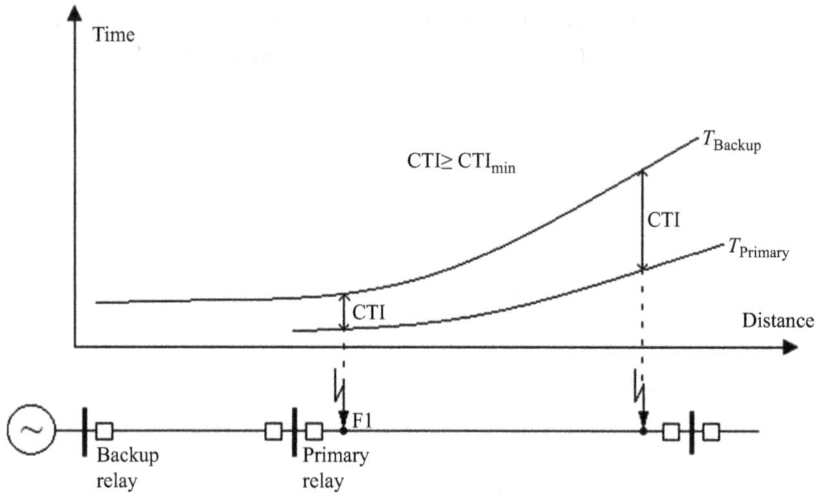


*Figure 9.3 Coordination between P/B relays*

Therefore, the CTI constraint can be stated as

$$\mathrm{CTI} \geq \mathrm{CTI}_{\min} \tag{9.8}$$

where $\mathrm{CTI}_{\min}$ is a minimum coordination time interval. The typical value of $\mathrm{CTI}_{\min}$ is 0.2–0.3 s [19].

This constraint must be satisfied for all primary/backup pairs. Usually, near-end, far-end, and middle point fault currents are used, providing coordination for most of the fault situations. Figure 9.3 shows an example of the CTI constraints for coordination between the P/B relays during the near-end and far-end faults (F1 and F2).

## 9.2.5 Modification of OF for minimization of CTI

Although a minimum CTI between the P/B relay is desirable for proper selectivity, much-delayed operation of backup relays is not preferable from an effective relay coordination point of view. In order to optimize the CTI between P/B relays, the expression for OF is modified as follows [7,20]:

$$\min_{\mathrm{PCS}_i,\mathrm{TDS}_i} \mathrm{MOF} = \alpha_1 \sum_{i=1}^{m} \sum_{k} T_{ik}^2 + \alpha_2 \sum_{p=1}^{m_p} \left[\Delta T_{mbp} - \beta\left(\Delta T_{mbp} - \left|\Delta T_{mbp}\right|\right)\right]^2 \tag{9.9}$$

$$\Delta T_{mbp} = T_{jk} - T_{ik} - \mathrm{CTI} \tag{9.10}$$

where $\Delta T_{mbp}$ is the OT difference with the CTI between $p$th relay pair; $m_p$ is the number of primary/backup relay pairs; $p$ represents each primary/backup relay pair and varies from 1 to $m_p$; $\alpha_1$ and $\alpha_2$ are used to control the weighting of first and second terms of the MOF and $\beta$ is used to consider the miscoordination. By increasing $\beta$, miscoordination is decreased, but the relay OTs are increased. Therefore, the suitable $\beta$ is fit to a value by which the miscoordination is omitted.

## 9.3 Solution method

As seen above, the coordination between DOCRs is formulated as a non-linear and highly constrained optimization problem in which two settings, namely TDS and PCS, are considered control variables. The main objective is to minimize the sum of OTs of all the primary relays, which are expected to operate in order to clear the faults of their corresponding zones. To solve this problem, a hybrid gravitational search algorithm–sequential quadratic programming (GSA–SQP) algorithm is proposed in this study.

### *9.3.1 Overview of GSA*

As explained in Volume 1 of the book, the search agents in GSA are a collection of masses that interact with each other based on Newtonian gravity and the laws of motion. In this algorithm, agents are treated as objects and their performances are measured by their masses. All these objects attract each other by the gravity force, and this force causes a global movement of all objects towards the objects with heavier masses. The position of the mass corresponds to the solution of the problem, and its gravitational and inertial masses are determined using a fitness function. In other words, each mass presents a solution. The algorithm is navigated by properly adjusting the gravitational and inertial masses. By lapse of time, the masses will be attracted by the heaviest mass, which represents an optimum solution in the search space.

The GSA could be considered an isolated system of masses. It is like a small artificial world of masses obeying the Newtonian laws of gravitation and motion.

In a system with $N$ agents (masses), the position and velocity of the $i$th agent is defined by

$$\mathbf{x}_i = \left(x_i^1, \ldots, x_i^d, \ldots, x_i^n\right) \quad \text{for} \quad i = 1, 2, \ldots, N \tag{9.11}$$

$$\mathbf{v}_i = \left(v_i^1, \ldots, v_i^d, \ldots, v_i^n\right) \quad \text{for} \quad i = 1, 2, \ldots, N \tag{9.12}$$

where $n$ is the search space dimension of the problem, $x_i^d$ defines the position of the $i$th agent in the $d$th dimension, and $v_i^d$ is the velocity of the $i$th agent in the $d$th dimension ($d = 1, 2, 3, \ldots, n$).

After evaluating the current population fitness, based on the OF of the optimization problem, the mass of each agent is calculated as follows [21]:

$$M_i(t) = \frac{m_i(t)}{\sum_{j=1}^{N} m_j(t)} \tag{9.13}$$

where

$$m_i(t) = \frac{fit_i(t) - worst(t)}{best(t) - worst(t)} \tag{9.14}$$

where $fit_i(t)$ represents the fitness value of the agent $i$ at time (iteration) $t$; $best(t)$ and $worst(t)$ are the best and worst fitness of all agents, respectively, and defined as follows (for a minimization problem):

$$best(t) = \min_{j \in \{1,\ldots,N\}} fit_j(t) \tag{9.15}$$

$$worst(t) = \max_{j \in \{1,\ldots,N\}} fit_j(t) \tag{9.16}$$

According to the Newton gravitation theory, the total force that acts on the $i$th agent in the $d$th dimension at $t$ time is specified as follows:

$$\mathbf{F}_i(t) = \sum_{j \in Kbest, j \neq i} r \cdot G(t) \frac{M_j(t) \cdot M_i(t)}{R_{i,j}(t) + \varepsilon} \left( \mathbf{x}_j(t) - \mathbf{x}_i(t) \right) \tag{9.17}$$

where $r$ is a random number within the interval [0, 1], $G(t)$ is gravitational constant at iteration $t$, $M_i(t)$ and $M_j(t)$ are masses of agents $i$ and $j$, $\varepsilon$ is a small constant, $R_{ij}(t)$ is the Euclidian distance between the two agents $i$ and $j$, given by the following equation:

$$R_{ij}(t) = \| \mathbf{x}_i(t), \mathbf{x}_i(t) \|_2 \tag{9.18}$$

*Kbest* is the set of first $K$ agents with the best fitness value and biggest mass, which is a function of time, initialized to $K_0$ at the beginning and decreased with time. In such a way, at the beginning, all agents apply the force, and as time passes, *Kbest* is decreased linearly, and at the end, there will be just one agent applying force to the others.

By the law of motion, the acceleration of the $i$th agent, at $t$ time in the $k$th dimension, is given by the following equation:

$$\mathbf{a}_i(t) = \frac{\mathbf{F}_i(t)}{M_i(t)} \tag{9.19}$$

The searching strategy on this notion can be defined to find the next velocity and next position of an agent. Next velocity of an agent is defined as a function of its current velocity added to its current acceleration. Hence, the next velocity and next position of an agent can be computed as follows:

$$\mathbf{v}_i(t+1) = r_i \cdot \mathbf{v}_i(t) + \mathbf{a}_i(t) \tag{9.20}$$

$$\mathbf{x}_i(t+1) = \mathbf{x}_i(t) + \mathbf{v}_i(t+1) \tag{9.21}$$

where $r_i$ is a uniform random variable in the interval [0, 1]. This random number is utilized to give a randomized characteristic to the search.

It must be pointed out that the gravitational constant $G(t)$ given in (9.17) is important in determining the performance of GSA. It is initialized at the beginning and will be reduced with time to control the search accuracy. In other words, the gravitational constant is a function of the initial value $G_0$ and time $t$:

$$G(t) = G_0 \cdot e^{-\alpha \frac{t}{t_{\max}}} \tag{9.22}$$

where $\alpha$ is a user-specified constant, $t$ is the current iteration, and $t_{max}$ is the maximum iteration number. The parameters of maximum iteration $t_{max}$, population size $N$, initial gravitational constant $G_0$, and constant $\alpha$ control the performance of GSA.

### *9.3.2 Overview of SQP*

The SQP method seems to be the best non-linear programming (NLP) method for constrained optimization problems. SQP is a direct method in which the constraints are handled in an explicit manner. It outperforms all other NLP methods in terms of efficiency, accuracy, and percentage of successful solutions over a large number of test problems [4,22]. The method closely resembles Newton's method for constrained optimization, just as is done for unconstrained optimization. At each iteration, Hessian of the Lagrangian function is approximated using the Broyden–Fletcher–Goldfarb–Shanno (BFGS) quasi-Newton updating method. The approximation result is then used to generate a quadratic programming (QP) sub-problem whose solution is used to form a search direction for a line search procedure. Since the OF to be minimized is non-convex, SQP ensures a local minimum for an initial solution.

Each iteration of the SQP algorithm involves three major steps [4,22]: (1) solve the QP subproblem and obtain the search direction; (2) line search and merit function calculation; (3) update the Hessian matrix of the Lagrangian function, using the BFGS formula.

### *9.3.3 Hybrid GSA–SQP algorithm*

GSA, as a probabilistic-based multipoint search technique, has a drawback of, sometimes, converging to the values which may not be optimum. Conversely, GSA searches for a large solution space with excellent convergence, providing approach to the global optimum. NLP methods, such as SQP, being single-point search methods, have the drawback of being trapped in the local optimum point if the initial choice is nearer to the local optimum. The NLP method gives a global optimum solution if a proper initial choice is made [4].

To take advantage of the GSA and SQP methods, and at the same time overcome the drawbacks of these methods, a hybrid GSA with SQP is proposed in [17]. The SQP routine is incorporated in GSA as a local search mechanism to improve the convergence. First, the GSA algorithm is executed normally, and the best fitness per generation is selected in each iteration. From that best fitness, the corresponding agent is set as the initial values of variables to the SQP method. Then, the SQP routine is executed depending on the adopted probability of local search $\alpha_{LS}$, providing improvement of the best fitness obtained from GSA in the current iteration. In this way, the hybrid GSA–SQP algorithm gives the global optimum solution.

### *9.3.4 Implementation of hybrid GSA–SQP algorithm*

The proposed hybrid GSA–SQP approach has been applied to solve the DOCRs coordination problem. The control variables of the DOCRs coordination problem constitute the individual position of several agents that represent a complete solution set. In a system with $N$ agents, the position of the $i$th agent is defined by

$$\mathbf{x}_i = \left[TDS_i^1, \ldots, TDS_i^m, PCS_i^1 \ldots, PCS_i^m\right] \tag{9.23}$$

where $i = 1, 2, \ldots, N$ and $m$ is the number of relays in the network.

Figure 9.4 illustrates the general flowchart for the hybrid GSA–SQP algorithm used to calculate the optimal settings of the DOCRs.

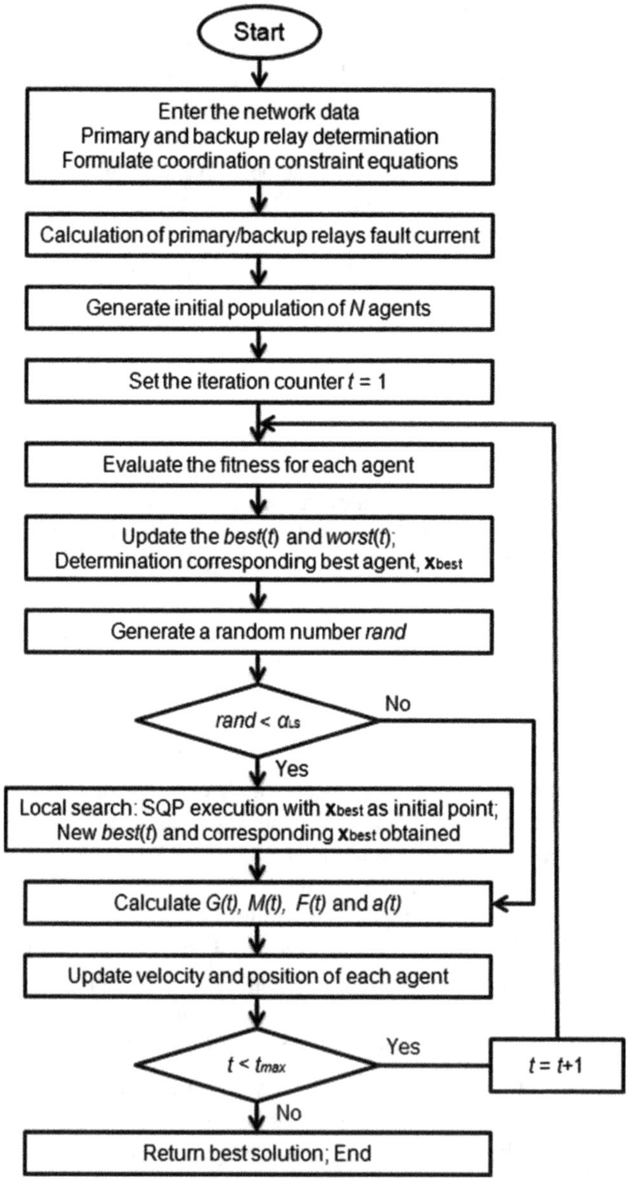


*Figure 9.4 General flowchart of the hybrid GSA–SQP for optimal coordination of DOCRs*

## 9.4 Simulation results

The proposed GSA, SQP, and hybrid GSA–SQP algorithms are simulated over three different test systems with 6, 24, and 38 DOCRs. These test systems are widely used as benchmarks in the power system field for solving the DOCRs coordination problem.

The algorithm parameters used for the simulation are adopted as follows: $\alpha$ is set to 20 and $G_0$ is set to 100; the probability of local search $\alpha_{LS}$ is set to 0.05. The population size ($N$) and maximum iteration number ($t_{max}$) are set to 50 and 200, respectively, for all case studies. The weighting factors in (9.9) are selected as follows: $\alpha_1$ is set to 2, $\alpha_2$ is set to 15, and $\beta$ is set to 200 for test system 1, 500 for test system 2, and 1000 for test system 3.

### 9.4.1 Test system 1

This is the IEEE 3-bus test system, which is shown in Figure 9.5. As can be seen from Figure 9.5, the IEEE 3-bus system contains one synchronous generator, three lines, and six DOCRs with the normally inverse characteristics ($\alpha$ = 0.14 and $n$ = 0.02). The objective is to coordinate the settings of all the six relays responding to clear all near-end and far-end faults. Accordingly, there are 12 decision variables in this problem (two for each relay), i.e. $TDS^1$–$TDS^6$ and $PCS^1$–$PCS^6$. The minimum and maximum limits of TDS are 0.05 and 1.1, respectively. The range of PCS is from 1.25 to 1.50, respectively. $CTI_{min}$ is set to its typical value of 0.3 s. Other data, including the CT ratings and the flowing fault currents for all primary/backup relays, which are calculated for near-end and far-end faults, are given in Tables 9.2 and 9.3.

Table 9.4 shows the optimal settings of TDS and PCS obtained by the proposed hybrid GSA–SQP. The OTs of all primary relays for near-end and far-end faults are given in the last two columns of Table 9.4. The OTs are within an acceptable range of 0.1–1.1 s. The total operating time of the primary relays is 4.78057 s. As can be seen from Table 9.5, the DOCRs operate with no miscoordination pairs. For comparison, the same problem was solved using only the GSA and SQP methods. For

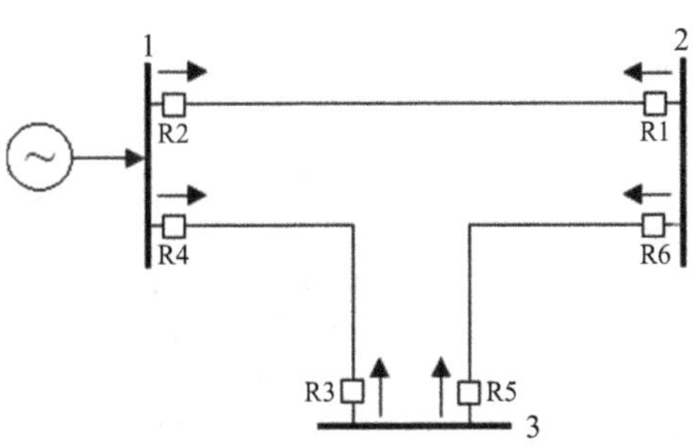


*Figure 9.5 Test system1: IEEE 3-bus system*

*Table 9.2 Fault currents seen by relays for test system 1 [2,23]*

| Relay | Fault current | | Line |
|---|---|---|---|
| | Near-end | Far-end | |
| R1 | 9.46/2.06 | 14.08/2.06 | 1-2 |
| R2 | 29.91/2.06 | 100.63/2.06 | 1-2 |
| R3 | 8.81/2.23 | 12.07/2.23 | 1-3 |
| R4 | 37.68/2.23 | 136.23/2.23 | 1-3 |
| R5 | 17.93/0.80 | 25.90/0.80 | 2-3 |
| R6 | 14.35/0.80 | 19.20/0.80 | 2-3 |

*Table 9.3 Fault currents for P/B relays in test system 1 [2,23]*

| Primary relay | Fault current | | Backup relay | Fault current | |
|---|---|---|---|---|---|
| | Near-end | Far-end | | Near-end | Far-end |
| R1 | 9.46/2.06 | 14.08/2.06 | R5 | 9.46/0.80 | 14.08/0.80 |
| R3 | 8.81/2.23 | 12.07/2.23 | R6 | 8.81/0.80 | 12.07/0.80 |
| R5 | 17.93/0.80 | 25.90/0.80 | R4 | 17.93/2.23 | 25.90/2.23 |
| R6 | 14.35/0.80 | 19.20/0.80 | R2 | 14.35/2.06 | 19.02/2.06 |

*Table 9.4 Optimal TDS and PCS for test system 1*

| Relay | GSA | | SQP | | Hybrid GSA–SQP | | | |
|---|---|---|---|---|---|---|---|---|
| | TDS | PCS | TDS | PCS | TDS | PCS | $OT_{near\_end}$ | $OT_{far\_end}$ |
| R1 | 0.05000 | 1.28160 | 0.05000 | 1.25 | 0.05000 | 1.25 | 0.26549 | 0.20253 |
| R2 | 0.21202 | 1.31711 | 0.19766 | 1.50 | 0.19764 | 1.50 | 0.62549 | 0.38351 |
| R3 | 0.05000 | 1.25000 | 0.05000 | 1.25 | 0.05000 | 1.25 | 0.30066 | 0.23533 |
| R4 | 0.21253 | 1.46965 | 0.20903 | 1.50 | 0.20903 | 1.50 | 0.58970 | 0.38027 |
| R5 | 0.18727 | 1.40548 | 0.18120 | 1.50 | 0.18120 | 1.50 | 0.45649 | 0.40035 |
| R6 | 0.18095 | 1.49486 | 0.18069 | 1.50 | 0.18067 | 1.50 | 0.49712 | 0.44361 |
| OF (s) | 4.82566 | | 4.78081 | | 4.78057 | | | |

the SQP solver, the initial choice of variables was taken as the lower limit of the control variables. It can be seen from Table 9.4 that the hybr GSA–SQP has better performance than the conventional GSA and SQP. The results of the proposed hybrid GSA–SQP algorithm are compared with various algorithms reported in recent years. The comparison is shown in Table 9.6. The algorithm reported in [19] resulted in a lesser minimum OF than the proposed GSA–SQP approach but with

*Table 9.5 CTI for test system 1*

| Fault | Relay | | CTI |
|---|---|---|---|
| | **Backup** | **Primary** | |
| Near-End | 5 | 1 | 0.30000 |
| | 6 | 3 | 0.30000 |
| | 4 | 5 | 0.30000 |
| | 2 | 6 | 0.39008 |
| Far-End | 5 | 1 | 0.33624 |
| | 6 | 3 | 0.32118 |
| | 2 | 6 | 0.40043 |
| | 4 | 5 | 0.30000 |

*Table 9.6 Comparison of the simulation results for test system 1*

| Algorithm | OF (s) | Note |
|---|---|---|
| DE [2] | 4.8421 | Miscoordination |
| MDE5 [2] | 4.7806 | |
| RST2 [2] | 4.8354 | |
| GA [2] | 5.0761 | |
| SOMGA [2] | 4.7898 | |
| ADE [19] | 4.7636 | |
| TLBO [20] | 5.3349 | |
| DE [23] | 4.8422 | |
| OCDE2 [23] | 4.7806 | |
| Hybrid GSA–SQP | 4.78057 | |

miscoordination. It is clear that the proposed GSA–SQP algorithm results in the best setting, compared with the other algorithms.

### *9.4.2 Test system 2*

The 9-bus system shown in Figure 9.6 is considered as test system 2. There are 24 digital DOCRs with the normally inverse characteristics ($\alpha$ = 0.14 and $n$ = 0.02). The range of TDS is from 0.01 to 1.0 and for PCS from 0.5 to 2.5. The $CTI_{min}$ is set to a value of 0.2 s. The CT ratio for each relay is 500/1. Short-circuit faults are generated in the middle of each line. Fault points are marked as A to L. Fault currents seen in each P/B relays are listed in Table 9.7. There are 48 variables that exist in the optimization problem. Those are $TDS^1$–$TDS^{24}$ and $PCS^1$–$PCS^{24}$.

The proposed hybrid GSA–SQP algorithm is used to solve the DOCR coordination problem. The optimized TDS and PCS, as well as the OF value, are given in Table 9.8. The OTs of all primary relays are within an acceptable range of 0.1–1.1 s. The optimized values of decision variables (TDS and PCS) and OF are compared with results obtained using GSA, SQP, and informative differential evolution (IDE)

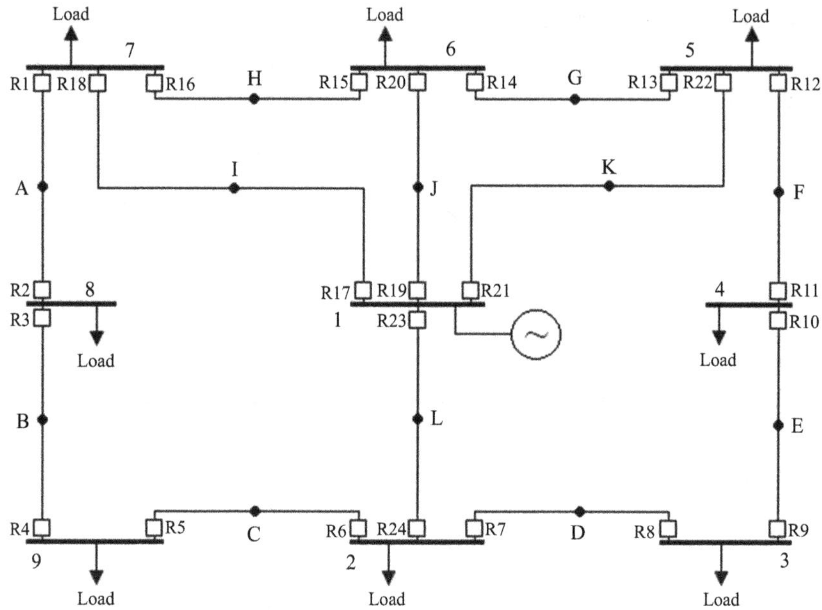


*Figure 9.6 Test system 2: 9-bus system*

*Table 9.7 Fault currents for test system 2 [16]*

| Location | Primary relay | Backup relay | $I_{f,primary}$ (A) | $I_{f,backup}$ (A) |
|---|---|---|---|---|
| A | R1 | R15 | 24,779 | 9150 |
| | R1 | R17 | 24,779 | 15,632 |
| | R2 | R4 | 8327 | 8327 |
| B | R3 | R1 | 16,390 | 16,930 |
| | R4 | R6 | 14,671 | 14,671 |
| C | R5 | R3 | 9454 | 9454 |
| | R6 | R8 | 23,280 | 4777 |
| | R6 | R23 | 23,280 | 18,507 |
| D | R7 | R5 | 23,280 | 4777 |
| | R7 | R23 | 23,280 | 18,507 |
| | R8 | R10 | 9454 | 9454 |
| E | R9 | R7 | 15,304 | 15,304 |
| | R10 | R12 | 16,490 | 16,490 |
| F | R11 | R9 | 8326 | 8327 |
| | R12 | R14 | 24,779 | 9150 |
| | R12 | R21 | 24,779 | 15,631 |
| G | R13 | R11 | 16,087 | 3088 |
| | R13 | R21 | 16,087 | 13,000 |
| | R14 | R16 | 18,213 | 6285 |
| | R14 | R19 | 18,213 | 11,934 |

*(Continues)*

*Table 9.7 (Continued)*

| Location | Primary relay | Backup relay | $I_{f,primary}$ (A) | $I_{f,backup}$ (A) |
|---|---|---|---|---|
| H | R15 | R13 | 18,218 | 6285 |
| | R15 | R19 | 18,218 | 11,935 |
| | R16 | R2 | 16,087 | 3088 |
| | R16 | R17 | 16,087 | 13,000 |
| I | R18 | R2 | 8161 | 2426 |
| | R18 | R15 | 8161 | 5736 |
| J | R20 | R13 | 9286 | 4644 |
| | R20 | R16 | 9286 | 4644 |
| K | R22 | R11 | 8161 | 2426 |
| | R22 | R14 | 8161 | 5736 |
| L | R24 | R5 | 6149 | 3075 |
| | R24 | R8 | 6149 | 3075 |

*Table 9.8 Results for test system 2*

| Relays | IDE [16] | | GSA | | SQP | | Hybrid GSA–SQP | | |
|---|---|---|---|---|---|---|---|---|---|
| | TDS | PCS | TDS | PCS | TDS | PCS | TDS | PCS | $OT_{primary}$ |
| R1 | 0.6352 | 0.5382 | 0.6192 | 1.3782 | 0.3040 | 2.4259 | 0.2765 | 2.5000 | 0.6288 |
| R2 | 0.1895 | 1.3645 | 0.2211 | 1.4531 | 0.1959 | 1.1581 | 0.0918 | 2.5000 | 0.3323 |
| R3 | 0.4605 | 0.7755 | 0.5547 | 1.0784 | 0.2301 | 2.2583 | 0.1975 | 2.5000 | 0.5235 |
| R4 | 0.3814 | 0.5000 | 0.3483 | 1.0006 | 0.2244 | 1.8590 | 0.1470 | 2.5000 | 0.4076 |
| R5 | 0.3974 | 0.5970 | 0.3521 | 1.1756 | 0.1831 | 1.8752 | 0.1385 | 2.5000 | 0.4696 |
| R6 | 0.5101 | 0.5000 | 0.5128 | 0.6301 | 0.2721 | 2.4934 | 0.2191 | 2.5000 | 0.5093 |
| R7 | 0.5123 | 0.5560 | 0.3667 | 2.1981 | 0.2471 | 2.5000 | 0.2209 | 2.5000 | 0.5135 |
| R8 | 0.2403 | 1.6853 | 0.3984 | 0.6316 | 0.1934 | 1.9260 | 0.1377 | 2.5000 | 0.4667 |
| R9 | 0.3418 | 0.9210 | 0.3539 | 1.0966 | 0.1841 | 2.1560 | 0.1475 | 2.5000 | 0.4021 |
| R10 | 0.4740 | 0.5001 | 0.3444 | 1.7680 | 0.2300 | 2.5000 | 0.1967 | 2.5000 | 0.5200 |
| R11 | 0.2012 | 1.6954 | 0.2347 | 1.5582 | 0.1303 | 1.9619 | 0.0923 | 2.5000 | 0.3344 |
| R12 | 0.4190 | 1.6966 | 0.5083 | 1.0618 | 0.3069 | 2.4758 | 0.2723 | 2.5000 | 0.6193 |
| R13 | 0.5414 | 0.5011 | 0.5640 | 0.6511 | 0.2488 | 1.7959 | 0.1903 | 2.5000 | 0.5081 |
| R14 | 0.3947 | 1.3848 | 0.4328 | 1.2291 | 0.4519 | 0.6029 | 0.2377 | 2.5000 | 0.6046 |
| R15 | 0.5640 | 0.6420 | 0.5579 | 1.1417 | 0.3011 | 1.7832 | 0.2404 | 2.5000 | 0.6115 |
| R16 | 0.2634 | 2.1479 | 0.4537 | 0.7078 | 0.2387 | 2.1922 | 0.1887 | 2.5000 | 0.5038 |
| R17 | 0.4808 | 2.1229 | 0.8032 | 1.7951 | 0.5458 | 0.5206 | 0.5001 | 0.5419 | – |
| R18 | 0.1462 | 2.4547 | 0.1359 | 1.2924 | 0.0499 | 1.7434 | 0.0715 | 1.9786 | 0.2323 |
| R19 | 0.4627 | 2.4999 | 0.7782 | 1.1285 | 0.3350 | 2.1148 | 0.3152 | 1.6934 | – |
| R20 | 0.2454 | 0.5016 | 0.0341 | 1.7836 | 0.0717 | 2.0845 | 0.0637 | 2.2012 | 0.2046 |
| R21 | 0.9327 | 0.5851 | 0.6259 | 1.6575 | 0.5190 | 0.6396 | 0.4822 | 0.5967 | – |
| R22 | 0.1667 | 1.8427 | 0.1389 | 0.8907 | 0.0417 | 2.4976 | 0.0719 | 1.9017 | 0.2290 |
| R23 | 0.9995 | 0.5005 | 0.7280 | 0.8679 | 0.3827 | 2.3378 | 0.4238 | 0.6822 | – |
| R24 | 0.1481 | 1.4887 | 0.1113 | 0.8264 | 0.0516 | 2.3987 | 0.1013 | 0.9638 | 0.2714 |
| OF (s) | 14.0448 | | 14.7384 | | 10.2030 | | 8.8923 | | |

algorithms [16]. Table 9.9 shows the corresponding values of CTI. The results show that the total operating time of primary DOCRs is reduced using the proposed hybrid GSA–SQP as compared to IDE [16] and with the original GSA and SQP. The optimal results give no miscoordination, as can be seen in Table 9.9. Moreover, the CTI improves using a hybrid GSA–SQP algorithm, because the sum of CTI values is reduced as compared with other algorithms.

Figure 9.7 shows the comparative convergence profiles of GSA and the proposed hybrid GSA–SQP for test system 2. It may be observed from this figure that the hybrid GSA–SQP explores the global optima, thanks to the exploration and

*Table 9.9 CTI for test system 2*

| Relays | | IDE [16] | GSA | SQP | Hybrid GSA–SQP |
|---|---|---|---|---|---|
| **Backup** | **Primary** | | | | |
| R15 | R1 | 0.200 | 0.2017 | 0.2000 | 0.2000 |
| R17 | R1 | 0.279 | 0.7448 | 0.2110 | 0.2000 |
| R4 | R2 | 0.218 | 0.2235 | 0.2000 | 0.2000 |
| R1 | R3 | 0.200 | 0.2122 | 0.2000 | 0.2000 |
| R6 | R4 | 0.212 | 0.2017 | 0.2000 | 0.2000 |
| R3 | R5 | 0.200 | 0.4543 | 0.2000 | 0.2000 |
| R8 | R6 | 0.200 | 0.2001 | 0.2000 | 0.2000 |
| R23 | R6 | 0.800 | 0.5085 | 0.3114 | 0.2043 |
| R5 | R7 | 0.201 | 0.3367 | 0.2000 | 0.2000 |
| R23 | R7 | 0.782 | 0.4922 | 0.3688 | 0.2000 |
| R10 | R8 | 0.202 | 0.2005 | 0.2006 | 0.2000 |
| R7 | R9 | 0.200 | 0.2295 | 0.2005 | 0.2000 |
| R12 | R10 | 0.200 | 0.2004 | 0.2000 | 0.2000 |
| R9 | R11 | 0.200 | 0.2089 | 0.2000 | 0.2000 |
| R14 | R12 | 0.203 | 0.2010 | 0.2000 | 0.2000 |
| R21 | R12 | 0.736 | 0.5575 | 0.2027 | 0.2000 |
| R11 | R13 | 0.203 | 0.2033 | 0.2002 | 0.2000 |
| R21 | R13 | – | 0.5749 | 0.3585 | 0.3528 |
| R16 | R14 | 0.209 | 0.2086 | 0.2000 | 0.2000 |
| R19 | R14 | 0.586 | 0.8672 | 0.2042 | 0.2075 |
| R13 | R15 | 0.200 | 0.2055 | 0.2000 | 0.2000 |
| R19 | R15 | 0.305 | 0.6421 | 0.2665 | 0.2005 |
| R2 | R16 | 0.202 | 0.2536 | 0.2001 | 0.2000 |
| R17 | R16 | 0.465 | 1.2469 | 0.3337 | 0.3661 |
| R2 | R18 | 0.502 | 0.9027 | 0.7907 | 0.7300 |
| R15 | R18 | 0.780 | 1.2878 | 0.9585 | 0.8557 |
| R13 | R20 | 0.800 | 1.3466 | 0.8181 | 0.7971 |
| R16 | R20 | 0.782 | 1.1026 | 0.9161 | 0.7885 |
| R11 | R22 | 0.800 | 1.1052 | 0.8460 | 0.7392 |
| R14 | R22 | 0.756 | 1.0015 | 0.8900 | 0.8465 |
| R5 | R24 | 0.685 | 1.1842 | 0.8491 | 0.7960 |
| R8 | R24 | 0.800 | 0.9168 | 0.9354 | 0.7896 |
| Sum of CTI (s) | | 13.108 | 18.2230 | 12.4622 | 11.4736 |

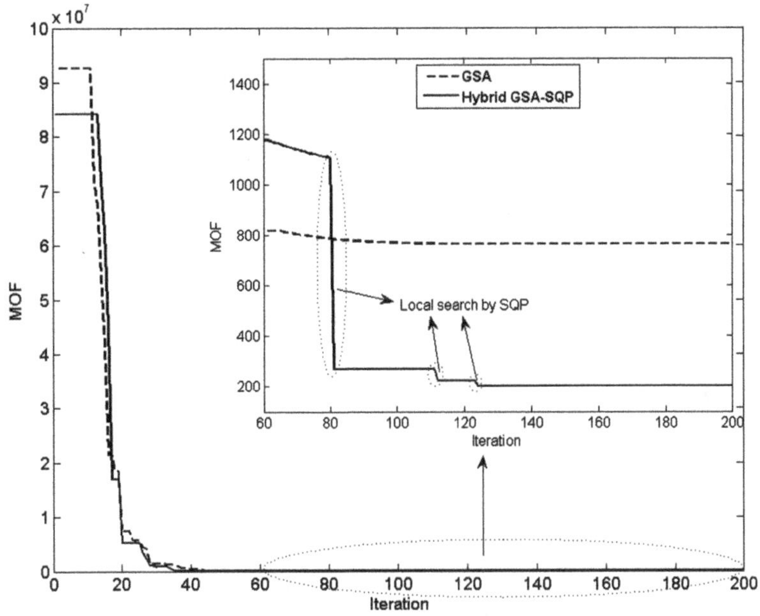


*Figure 9.7 Comparative convergence profiles of GSA and hybrid GSA–SQP for test system 2*

exploitation of GSA in the first part of the iterative process and the local search ability of SQP in subsequent iterations.

### *9.4.3 Test system 3*

To evaluate the effectiveness of the proposed approach in solving a larger power system, the IEEE 30-bus system is used as a test system 3. The system can be considered a meshed subtrasmission/distribution system with DG units, as shown in Figure 9.8. A total of 38 DOCRs, with the normally inverse characteristics ($\alpha$ = 0.14 and $n$ = 0.02), is considered for the system, which are installed on each end of the lines. The short circuit current values for near-end faults are given in Table 9.10. The range of TDS is from 0.1 to 1.1 and for PCS from 1.5 to 6. The CT ratio for each relay was assumed to be equal to 1000/5. The $CTI_{min}$ is set to a value of 0.3 s.

The results obtained using the proposed hybrid GSA–SQP along with GSA and SQP are presented in Table 9.11. Table 9.12 shows the CTI calculated from the optimized values of TDS and PCS, and there is no miscoordination. It can be seen that the results obtained from the hybrid GSA–SQP are better than those obtained from the original GSA and SQP methods. This clearly indicates that the proposed algorithm can be effectively used to solve the DOCRs coordination problem for large-scale power systems.

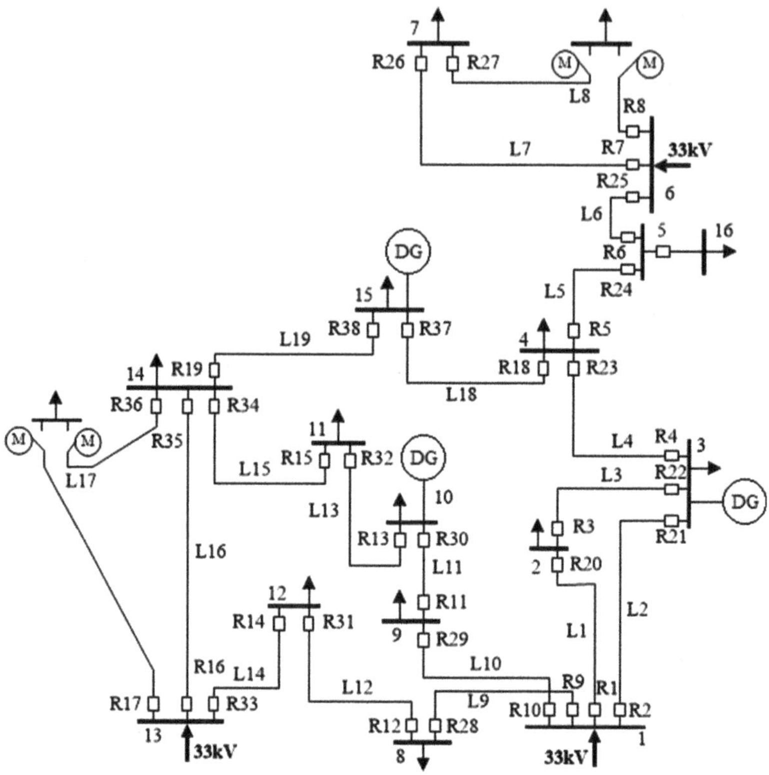

*Figure 9.8 Test system 3: IEEE 30-bus system*

*Table 9.10 Fault currents for test system 3 [24]*

| Primary relay | Backup relay | $I_{f,primary}$ (A) | $I_{f,backup}$ (A) | Primary relay | Backup relay | $I_{f,primary}$ (A) | $I_{f,backup}$ (A) |
|---|---|---|---|---|---|---|---|
| R3 | R1 | 4086.7 | 4086.7 | R9 | R20 | 7212.6 | 1103.5 |
| R4 | R2 | 5411.2 | 2138.8 | R10 | R20 | 7339.3 | 1095.8 |
| R22 | R2 | 4333.0 | 2147.0 | R1 | R21 | 7665.3 | 698.8 |
| R4 | R3 | 5411.2 | 3272.5 | R9 | R21 | 7212.6 | 721.2 |
| R21 | R3 | 5411.8 | 3243.6 | R10 | R21 | 7339.3 | 716.1 |
| R5 | R4 | 4960.8 | 3001.3 | R20 | R22 | 3481.5 | 3481.5 |
| R18 | R4 | 4719.4 | 3002.1 | R21 | R23 | 5411.8 | 2193.5 |
| R6 | R5 | 2416.0 | 2416.0 | R22 | R23 | 4333.0 | 2204.6 |
| R7 | R6 | 5669.0 | 1790.9 | R18 | R24 | 4719.4 | 1717.7 |
| R8 | R6 | 5607.7 | 1774.8 | R23 | R24 | 3689.7 | 1724.2 |
| R27 | R7 | 1472.3 | 1472.3 | R24 | R25 | 2695.0 | 2695.0 |
| R26 | R8 | 1026.8 | 1026.8 | R1 | R28 | 7665.3 | 1552.0 |

*(Continues)*

*Table 9.10 (Continued)*

| Primary relay | Backup relay | $I_{f,primary}$ (A) | $I_{f,backup}$ (A) | Primary relay | Backup relay | $I_{f,primary}$ (A) | $I_{f,backup}$ (A) |
|---|---|---|---|---|---|---|---|
| R12 | R9 | 5034.9 | 5034.9 | R2 | R28 | 7985.7 | 1545.8 |
| R11 | R10 | 3457.1 | 3457.1 | R10 | R28 | 7339.3 | 1538.0 |
| R13 | R11 | 3727.3 | 2875.0 | R1 | R29 | 7665.3 | 1380.6 |
| R14 | R12 | 2906.5 | 2906.5 | R2 | R29 | 7985.7 | 1375.2 |
| R15 | R13 | 2660.5 | 2660.5 | R9 | R29 | 7212.6 | 1379.0 |
| R16 | R14 | 6185.6 | 1668.1 | R29 | R30 | 2518.9 | 2518.9 |
| R17 | R14 | 7492.9 | 1641.1 | R28 | R31 | 2036.8 | 2036.8 |
| R19 | R15 | 5445.2 | 1527.3 | R30 | R32 | 2998.8 | 2149.0 |
| R35 | R15 | 4222.0 | 1533.2 | R31 | R33 | 3263.6 | 3263.6 |
| R36 | R15 | 6420.2 | 1509.7 | R32 | R34 | 2930.4 | 2930.4 |
| R19 | R16 | 5445.2 | 3128.3 | R17 | R35 | 7492.9 | 1885.4 |
| R34 | R16 | 5796.6 | 3123.9 | R33 | R35 | 6456.2 | 1954.5 |
| R36 | R16 | 6420.2 | 3052.4 | R16 | R36 | 6185.6 | 490.9 |
| R19 | R17 | 5445.2 | 801.3 | R33 | R36 | 6456.2 | 500.6 |
| R34 | R17 | 5796.6 | 800.1 | R5 | R37 | 4960.8 | 1961.0 |
| R35 | R17 | 4222.0 | 794.0 | R23 | R37 | 3689.7 | 1968.5 |
| R38 | R18 | 3133.2 | 2292.2 | R34 | R38 | 5796.6 | 1886.8 |
| R37 | R19 | 3788.9 | 2940.9 | R35 | R38 | 4222.0 | 1896.7 |
| R2 | R20 | 7985.7 | 1053.9 | R36 | R38 | 6420.2 | 1867.7 |

*Table 9.11 Results for test system 3*

| Relay | GSA | | SQP | | Hybrid GSA–SQP | | |
|---|---|---|---|---|---|---|---|
| | TDS | PCS | TDS | PCS | TDS | PCS | $OT_{primary}$ |
| R1 | 1.0731 | 1.5393 | 0.4372 | 2.6744 | 0.3038 | 3.8158 | 0.9006 |
| R2 | 0.6752 | 1.5000 | 0.5791 | 1.7575 | 0.3455 | 1.5000 | 0.7131 |
| R3 | 0.5549 | 1.5000 | 0.3472 | 2.6903 | 0.2762 | 2.7562 | 0.9458 |
| R4 | 0.2145 | 6.0000 | 0.3353 | 2.8637 | 0.2584 | 2.6926 | 0.7658 |
| R5 | 0.1670 | 6.0000 | 0.3496 | 2.4616 | 0.2625 | 1.6592 | 0.6611 |
| R6 | 0.1001 | 4.5297 | 0.1769 | 3.2217 | 0.1597 | 1.9809 | 0.6073 |
| R7 | 0.1001 | 4.7929 | 0.2831 | 2.1619 | 0.1732 | 1.8244 | 0.4299 |
| R8 | 0.2648 | 2.3894 | 0.2218 | 1.8887 | 0.1570 | 1.5872 | 0.3718 |
| R9 | 1.0995 | 1.5000 | 0.5090 | 3.4047 | 0.3133 | 3.3528 | 0.9014 |
| R10 | 0.4892 | 5.9977 | 0.4853 | 4.1095 | 0.3123 | 3.5250 | 0.9115 |
| R11 | 0.5000 | 3.4787 | 0.6024 | 2.2426 | 0.2797 | 2.7844 | 1.0530 |
| R12 | 0.4682 | 4.2032 | 0.3627 | 3.1807 | 0.2170 | 3.5937 | 0.7654 |
| R13 | 0.6268 | 1.5000 | 0.3233 | 4.4211 | 0.2494 | 2.6182 | 0.8720 |
| R14 | 0.6755 | 1.5549 | 0.2678 | 2.1714 | 0.2155 | 2.1395 | 0.7723 |
| R15 | 0.4803 | 1.5011 | 0.3311 | 1.7343 | 0.2334 | 1.6058 | 0.7565 |
| R16 | 1.0996 | 1.5000 | 0.4214 | 2.1123 | 0.2678 | 3.0708 | 0.7931 |

*(Continues)*

Table 9.11 (Continued)

| Relay | GSA | | SQP | | Hybrid GSA–SQP | | |
|---|---|---|---|---|---|---|---|
| | TDS | PCS | TDS | PCS | TDS | PCS | $OT_{primary}$ |
| R17 | 0.2929 | 1.5000 | 0.1934 | 1.7176 | 0.1468 | 1.6293 | 0.3176 |
| R18 | 0.2759 | 3.8552 | 0.3870 | 2.0983 | 0.2412 | 2.4930 | 0.7345 |
| R19 | 0.6865 | 1.8317 | 0.4226 | 2.1682 | 0.2405 | 2.8394 | 0.7281 |
| R20 | 0.1069 | 4.1612 | 0.2666 | 2.0309 | 0.1443 | 2.3975 | 0.4993 |
| R21 | 0.3343 | 1.5000 | 0.2160 | 1.7407 | 0.1604 | 1.5000 | 0.3769 |
| R22 | 0.4900 | 1.5000 | 0.3298 | 2.4447 | 0.1702 | 4.0080 | 0.6943 |
| R23 | 0.1558 | 5.4904 | 0.2421 | 3.1497 | 0.2100 | 2.5687 | 0.7311 |
| R24 | 0.3802 | 1.5000 | 0.3247 | 2.1049 | 0.2016 | 2.2364 | 0.7716 |
| R25 | 0.4822 | 1.5000 | 0.3744 | 2.4249 | 0.2523 | 2.6631 | – |
| R26 | 0.3695 | 1.5005 | 0.1681 | 1.9096 | 0.1106 | 1.5095 | 0.6248 |
| R27 | 0.1000 | 4.1890 | 0.2080 | 2.3950 | 0.1215 | 1.6362 | 0.5570 |
| R28 | 0.1039 | 5.8654 | 0.4775 | 1.7040 | 0.1932 | 2.5489 | 0.9631 |
| R29 | 0.4439 | 2.1601 | 0.3064 | 2.1555 | 0.1643 | 2.6733 | 0.7305 |
| R30 | 0.2168 | 6.0000 | 0.3448 | 2.5695 | 0.1837 | 3.6716 | 0.9013 |
| R31 | 0.6293 | 1.5000 | 0.3649 | 3.1340 | 0.2191 | 3.0692 | 0.9025 |
| R32 | 0.2744 | 4.2042 | 0.2071 | 4.4850 | 0.2095 | 3.2170 | 0.9527 |
| R33 | 0.9231 | 1.5000 | 0.3544 | 4.2707 | 0.2857 | 3.1774 | 0.8428 |
| R34 | 0.4494 | 3.6290 | 0.2605 | 4.5405 | 0.2516 | 3.6611 | 0.8339 |
| R35 | 0.1905 | 6.0000 | 0.3782 | 1.7569 | 0.2289 | 2.4531 | 0.7285 |
| R36 | 0.2109 | 1.5000 | 0.1240 | 1.5484 | 0.1000 | 1.5000 | 0.2216 |
| R37 | 0.5387 | 1.5736 | 0.2788 | 2.3403 | 0.2138 | 2.3547 | 0.7029 |
| R38 | 0.3608 | 2.3341 | 0.3284 | 1.9759 | 0.1941 | 2.8879 | 0.7901 |
| OF (s) | 51.7747 | | 41.0390 | | 26.8258 | | |

### 9.4.4 Statistical evaluation of the results

A statistical evaluation of the results is presented in this section. Table 9.13 shows the comparison of minimum, maximum, and standard deviation of the results obtained by GSA and hybrid GSA–SQP over 10 runs, and the results of SQP under three different initial choices of control variables. In the case of SQP, the problem was solved using different initial choices, such as lower limits of control variables, upper limits of control variables, and middle values between lower and upper limits. In some cases, it gives a feasible solution, but worse than hybrid GSA–SQP. Moreover, for test system 3 in cases with upper and middle initial choices of control variables, the SQP method terminates without finding even a feasible solution.

It is clear that the proposed hybrid GSA–SQP algorithm provides better and more stable solutions than the original GSA and SQP for all three test systems.

*Table 9.12 CTI for test system 3*

| Relays | | GSA | SQP | Hybrid GSA–SQP | Relays | | GSA | SQP | Hybrid GSA–SQP |
|---|---|---|---|---|---|---|---|---|---|
| Backup | Primary | | | | Backup | Primary | | | |
| R1 | R3 | 1.3817 | 0.3001 | 0.3003 | R20 | R9 | 0.3003 | 0.3745 | 0.3001 |
| R2 | R4 | 1.3773 | 1.1825 | 0.4415 | R20 | R10 | 0.8557 | 0.3440 | 0.3003 |
| R2 | R22 | 1.1041 | 1.1647 | 0.5106 | R21 | R1 | 0.4816 | 1.0362 | 0.4157 |
| R3 | R4 | 0.6049 | 0.3003 | 0.3003 | R21 | R9 | 0.3000 | 0.5872 | 0.3672 |
| R3 | R21 | 0.8072 | 0.7926 | 0.6946 | R21 | R10 | 0.8096 | 0.5640 | 0.3675 |
| R4 | R5 | 0.8110 | 0.3588 | 0.3736 | R22 | R20 | 0.8499 | 0.3028 | 0.3001 |
| R4 | R18 | 0.5758 | 0.3010 | 0.3000 | R23 | R21 | 0.7799 | 0.8055 | 0.6213 |
| R5 | R6 | 0.9516 | 0.5894 | 0.3000 | R23 | R22 | 0.3036 | 0.3011 | 0.3005 |
| R6 | R7 | 0.6337 | 0.4487 | 0.3001 | R24 | R18 | 0.4519 | 0.5013 | 0.3001 |
| R6 | R8 | 0.3002 | 0.6498 | 0.3627 | R24 | R23 | 0.6061 | 0.6475 | 0.3005 |
| R7 | R27 | 0.3905 | 0.3154 | 0.3000 | R25 | R24 | 0.3181 | 0.3005 | 0.3000 |
| R8 | R26 | 0.3283 | 0.3593 | 0.3004 | R28 | R1 | 0.3287 | 1.0523 | 0.3009 |
| R9 | R12 | 0.8542 | 0.5432 | 0.3005 | R28 | R2 | 1.2351 | 0.9195 | 0.4928 |
| R10 | R11 | 1.0525 | 0.3081 | 0.3002 | R28 | R10 | 0.8215 | 0.6669 | 0.3000 |
| R11 | R13 | 0.7340 | 0.6774 | 0.3015 | R29 | R1 | 0.3815 | 0.7019 | 0.3001 |
| R12 | R14 | 0.5407 | 0.6784 | 0.3000 | R29 | R2 | 1.2592 | 0.5696 | 0.4927 |
| R13 | R15 | 0.4591 | 0.9170 | 0.3000 | R29 | R9 | 0.3021 | 0.3487 | 0.3009 |
| R14 | R16 | 0.3005 | 0.3047 | 0.3005 | R30 | R29 | 0.3002 | 0.3010 | 0.3002 |
| R14 | R17 | 2.1787 | 0.9659 | 0.7894 | R31 | R28 | 0.9451 | 0.3054 | 0.3002 |
| R15 | R19 | 0.3006 | 0.4009 | 0.3034 | R32 | R30 | 0.3854 | 0.3002 | 0.3001 |
| R15 | R35 | 0.9817 | 0.4979 | 0.3005 | R33 | R31 | 0.8414 | 0.3035 | 0.3000 |
| R15 | R36 | 1.5811 | 1.2751 | 0.8178 | R34 | R32 | 0.7038 | 0.3286 | 0.3000 |
| R16 | R19 | 1.4739 | 0.3043 | 0.4048 | R35 | R17 | 2.3219 | 1.1239 | 0.8565 |

*(Continues)*

*Table 9.12* (*Continued*)

| Relays | | GSA | SQP | Hybrid GSA–SQP | Relays | | GSA | SQP | Hybrid GSA–SQP |
|---|---|---|---|---|---|---|---|---|---|
| **Backup** | **Primary** | | | | **Backup** | **Primary** | | | |
| R16 | R34 | 1.7260 | 0.4792 | 0.3000 | R35 | R33 | 0.6794 | 0.3147 | 0.3004 |
| R16 | R36 | 2.7744 | 1.1844 | 0.9290 | R36 | R16 | 0.5153 | 0.8062 | 0.6214 |
| R17 | R19 | 0.3349 | 0.4633 | 0.3857 | R36 | R33 | 0.8267 | 0.5974 | 0.5173 |
| R17 | R34 | 0.5865 | 0.6216 | 0.3000 | R37 | R5 | 1.2115 | 0.3079 | 0.3731 |
| R17 | R35 | 1.0389 | 0.5633 | 0.4153 | R37 | R23 | 1.1300 | 0.3973 | 0.3003 |
| R18 | R38 | 0.4516 | 0.4813 | 0.3000 | R38 | R34 | 0.3005 | 0.4817 | 0.3005 |
| R19 | R37 | 0.7814 | 0.6019 | 0.3042 | R38 | R35 | 0.7297 | 0.4039 | 0.4009 |
| R20 | R2 | 1.7676 | 0.6810 | 0.5591 | R38 | R36 | 1.3294 | 1.1793 | 0.9228 |
| Sum of CTI (s) | | | | | | | 51.79 | 35.89 | 24.55 |

*Table 9.13 Comparison of the results (OF) obtained with GSA, SQP, and hybrid GSA–SQP*

| | **Statistics for hybrid GSA–SQP** | | | **Statistics for GSA** | | | **Results for SQP with the initial choice of control variables as** | | |
|---|---|---|---|---|---|---|---|---|---|
| | **Min** | **Max** | **Std. dev.** | **Min** | **Max** | **Std. dev.** | **Lower** | **Upper** | **Middle** |
| Test syst. 1 | 4.7806 | 4.7806 | 0 | 4.8257 | 4.9915 | 0.056 | 4.7810 | 4.8383 | 4.9162 |
| Test syst. 2 | 8.8923 | 8.9316 | 0.0139 | 14.7384 | 22.9571 | 2.509 | 10.2430 | 9.8740 | 9.7940 |
| Test syst. 3 | 26.8258 | 29.6896 | 0.9808 | 51.7747 | 66.9562 | 4.921 | 41.0390 | 18.0874[a] | 16.2475[b] |

[a]Solution with miscoordination between backup/primary relays: R1/R3, R2/R4, R2/R22, R3/R4, R3/R21, R4/R5, R8/R26, R9/R12, R10/R11, R11/R13, R12/R14, R13/R15, R16/R19, etc.
[b]Solution with miscoordination between backup/primary relays: R1/R3, R3/R4, R3/R21, R8/R26, R9/R12, R10/R11, R11/R13, R12/R14, R13/R15, R14/R16, R16/R19, R16/R34, etc.

## 9.5 MATLAB® program *ocdocrm*

The *ocdocrm* is an educational program package of MATLAB m-files for solving the problem of optimal coordination of DOCRs. The program package is available as the supplementary material to this chapter at the IET digital library. The simulation results presented in Section 9.4 are obtained by the *ocdocrm* program.

An example of the *ocdocrm* program run with the hybrid GSASQP solution method is presented in Figure 9.9.

## 9.6 Conclusion

The coordination between DOCRs is a non-linear and highly constrained optimization problem in which two settings, namely the TDS and the PCS of each relay, are considered as decision variables. The main objective is to minimize the sum of OTs of all primary relays that are expected to operate in order to clear faults in their corresponding zones. In this chapter, a hybrid GSA–SQP optimization algorithm has been proposed and successfully applied to solve the DOCRs coordination problem. The SQP routine is incorporated into GSA as a local search mechanism to improve the performance of the conventional GSA algorithm. The proposed approach has been tested and investigated on three different test systems. Simulation results show that the hybrid GSA–SQP algorithm provides an effective and robust high-quality solution. Moreover, the results obtained using hybrid GSA–SQP are either better or comparable to those obtained using other techniques reported in the literature. The proposed hybrid GSA–SQP algorithm is suitable to

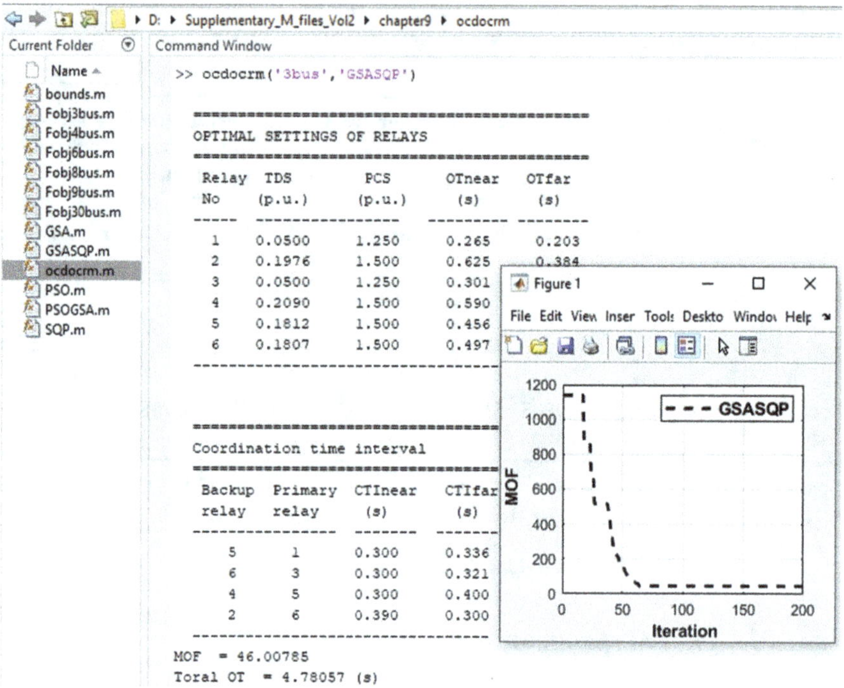


*Figure 9.9 Example of the ocdocrm program execution*

find the global optimal solution for the DOCRs coordination problems, because it benefits from the advantages of both GSA and SQP methods.

## References

[1] Noghabi A.S., Mashhadi H.R., and Sadeh J. 'Optimal coordination of directional overcurrent relays considering different network topologies using interval linear programming'. *IEEE Transactions on Power Delivery*. 2010; **25**(3):1348–54.

[2] Thangaraj R., Pant M., and Deep K. 'Optimal coordination of over-current relays using modified differential evolution algorithms'. *Engineering Applications of Artificial Intelligence*. 2010;**23**:820–29.

[3] Ezzeddine E., Kaczmarek R., and Iftikhar M.U. 'Coordination of directional overcurent relays using a novel method to select their settings'. *IET Generation, Transmission & Distribution*. 2011;**5**(7):743–50.

[4] Bedekar P.P. and Bhide S.R. 'Optimum coordination of directional overcurrent relays using the hybrid GA-NLP approach'. *IEEE Transactions on Power Delivery*. 2011;**26**(1):109–19.

[5] Ezzeddine M. and Kaczmarek R. 'A novel method for optimal coordination of directional overcurrent relays considering their available discrete setting and several operation characteristics'. *Electric Power Systems Research.* 2011;**81**:1475–81.

[6] Meskin M., Domijan A., and Grinberg I. 'Optimal co-ordination of overcurrent relays in the interconnected power systems using break points'. *Electric Power Systems Research.* 2015;**127**:53–63.

[7] Moravej Z., Adelnia F., and Abbasi F. 'Optimal coordination of directional overcurrent relays using NSGA-II'. *Electric Power Systems Research.* 2015; **119**:228–36.

[8] Urdaneta A.J., Nadira R., and Jimenez L.G.P. 'Optimal coordination of directional overcurrent relays in interconnected power systems' *IEEE Transactions on Power Delivery.* 1988;**3**(3):903–11.

[9] Zeineldin H.H., El-Saadany E.F., and Salama M.M.A. 'Optimal coordination of overcurrent relays using a modified particle swarm optimization'. *Electric Power Systems Research.* 2006;**76**:988–95.

[10] Mahari A. and Seyedi H. 'An analytic approach for optimal coordination of overcurrent relays'. *IET Generation, Transmission & Distribution.* 2013;**7** (7):674–80.

[11] Urdaneta A.J., Ramon N., and Jimenez L.G.P. 'Coordination of directional relay timing using linear programming'. *IEEE Transactions on Power Delivery.* 1996;**11**(1):122–29.

[12] Birla D., Maheshwari R.P., and Gupta H.O. 'A new nonlinear directional overcurrent relay coordination technique, and Banes and Boons of near-end faults based approach'. *IEEE Transactions on Power Delivery.* 2006;**21**(3): 1176–82.

[13] Papaspiliotopoulos V.A., George G.N., and Maratos N.G. 'A novel quadratically constrained quadratic programming method for optimal coordination of directional overcurrent relays'. *IEEE Transactions on Power Delivery.* 2015;**32**(1):3–10.

[14] Sharaf H.M., Zeineldin H.H., Ibrahim D.K., and El-Zahab E.E.D.A. 'A proposed coordination strategy for meshed distribution system with DG considering user-defined characteristics of directional inverse time overcurrent relays'. *Electrical Power and Energy Systems.* 2015;**65**:49–58.

[15] Correa R., Cardoso G.Jr., De Araujo O.C.B., and Mariotto L. 'Online coordination of directional overcurrent relays using binary integer programming', *Electric Power Systems Research.* 2015;**127**:118–25.

[16] Singh M., Panigrahi B.K., Abhyankar A.R., and Das C. 'Optimal coordination of directional over-current relays using informative differential evolution algorithm'. *Journal of Computational Science.* 2014;**5**:269–76.

[17] Radosavljević J. and Jevtić M. 'Hybrid GSA–SQP algorithm for optimal coordination of directional overcurrent relays'. *IET Generation, Transmission & Distribution.* 2016;**10**(8):1928–37.

[18] Albasri F.A., Alroomi A.R., and Talaq J.H. 'Optimal coordination of directional overcurrent relays using biogeography-based optimization algorithms'. *IEEE Transactions on Power Delivery*. 2015;**30**(4):1810–20.

[19] Moirangthem J., Krishnanand K.R., Dash S.S., and Ramaswami R. 'Adaptive differential evolution algorithm for solving non-linear coordination problem of directional overcurrent relays'. *IET Generation, Transmission & Distribution*. 2013;**7**(4):329–36.

[20] Singh M., Panigrahi B.K., and Abhyankar A.R. 'Optimal coordination of directional over-current relays using teaching learning-based optimization (TLBO) algorithm'. *Electrical Power and Energy Systems*. 2013;**50**:33–41.

[21] Rashedi E., Nezamabadi-pour H., and Saryazdi S. 'GSA: A gravitational search algorithm', *Information Sciences*. 2009;**179**:2232–48.

[22] Sivasubramani S. and Swarup K.S. 'Hybrid DE-SQP algorithm for non-convex short term hydrothermal scheduling problem'. *Energy Conversion and Management*. 2011;**52**:757–61.

[23] Chelliah T.R., Thangaraj R., Allamsetty S., and Pant M. 'Coordination of directional overcurrent relays using opposition based chaotic differential evolution algorithm'. *Electrical Power and Energy Systems*. 2014;**55**:341–350.

[24] Mohammadi R., Abyaneh H.A., Rudsari H.M., Fathi S.H., and Rastegar H. 'Overcurrent relays coordination considering the priority of constraints'. *Transactions on Power Delivery*. 2011;**26**(3):1927–38.

*Chapter 10*

# Distribution system state estimation

## 10.1 Introduction

State estimation is the key function and important aspect of an energy management system. The state estimation in the power system is a mathematical algorithm that, at a given moment, based on available measurements, provides reliable and the best estimation of voltages and other variables relevant to system operation. State estimation is an essential function in the process of analysis, control, and management of power systems. Conversely, the system state estimation in transmission lines is based on a redundant set of telemetry measurements. For transmission systems, there is always the assumption of symmetrical and balanced operation, so that all calculations are performed on the corresponding single-phase positive sequence models [1]. However, distribution networks (DNs) have many specifics in comparison to transmission networks, such as radial topology, un-transposed three-phase feeders (with two-phase and single-phase laterals having high $r/x$ ratios), unbalanced loads, and a small number of real-time measurements. Under these specifics, the state estimation techniques for transmission networks cannot be applied directly to DNs, which means that three-phase state estimation is necessary. A unique feature of distribution system state estimation (DSSE) is the small number of real-time measurements. In such a situation, it is necessary for introducing pseudo-measurements, which refer to active and reactive powers of loads and distributed generation (DG) to achieve the system observability. The general framework for defining and solving the DSSE problem must include the following aspects:

- Three-phase models of distribution system components, including lines, transformers, loads, and DG.
- Availability of real-time measurement.
- Inclusion and modeling of pseudo-measurements. Defining state variables and DSSE problem formulation. The DSSE algorithm for determining the state variables.

So far, many authors have dealt with the DSSE problem, focusing on all or some of the above issues. The authors in [2–4] give comprehensive reviews of different issues related to DSSE. In accordance with these considerations, there are three DSSE paradigms: weighted least squares (WLS)-based DSSE, load adjustment DSSE methods, and robust DSSE methods.

Recently, several innovative research approaches have been proposed for the DSSE. Ranković *et al.* [5] proposed an improved WLS-based three-phase state estimation (SE) algorithm for unbalanced distribution systems, where the vector of state variables comprises the components of complex voltages combined with variables associated with the classification of integrated DG units. In [6], a detailed representation of the ground loops in the model of an active unbalanced DN suitable to three-phase state estimation is introduced. Starting from the characteristics of DNs related to small voltage differences between two busses of each line segment, the authors [7,8] cast the state estimation problem into a sparse vector recovery problem. A branch-current-based WLS using a new technique that aims to process the data obtained from smart metering is proposed in [9], where the main objective of the proposed scheme is to execute the three-phase DSSE accurately. Othman *et al.* proposed a DSSE model considering DG integration in DNs [10]. This technique is based on updating the complex power flow in the overall network according to the online measurement from smart meters positioned at specific locations dependent on the network topology only. A three-phase DSSE methodology for unbalanced radial DNs is presented in [11], and the proposed method uses the load regulation via active utilization and imbalance factors. In [12], the authors also proposed a three-phase DSSE algorithm based on load correction. This approach uses real-time measurements at the substation and pseudo-measurements generated from the customers' energy bills. Several authors proposed artificial neural networks (ANN)-based DSSE algorithms to overcome a limited number of measurements in DNs [13,14]. Cheng *et al.* [15] combined extreme machine learning with a genetic algorithm to predict photovoltaic generation (PVG) and used it as a pseudo-measurement in a WLS-based DSSE algorithm. Considering the problem of SE as a global optimization problem with constraints, some authors have suggested metaheuristic-based methods for solving it. A cooperative particle swarm optimization (Co-PSO) algorithm is proposed in [16] for DSSE in unbalanced three-phase DNs with renewable DGs. Nanchian *et al.* [1] demonstrated the application of a hybrid particle swarm optimization (HPSO) for three-phase DSSE, including the on-load tap changer positions for voltage control. In [17], the authors proposed a hybrid DSSE technique using the WLS method and firefly algorithm for DNs with integrated renewable energy sources.

This chapter describes a new three-phase DSSE paradigm based on metaheuristic optimization, as proposed in [18]. In comparison with recent literature, the main contributions of this research are as follows:

- Defining a generic research framework for three-phase DG-integrated DSSE using a novel PSOS–CGSA method.
- Application and implementation of a new hybrid PSOS–CGSA algorithm to solve the DSSE problem.
- Defining pseudo-measurements for renewable-based DG units using forecasted output powers.

The rest of this chapter is organized as follows: Section 10.2 presents the three-phase power flow and modeling of power system components. The problem formulation is described in Section 10.3. Section 10.4 presents the application of the proposed PSOS–CGSA method for DG-integrated DSSE. Investigated test cases and method validation, along with detailed simulation results and discussion, are presented in Section 10.5. Finally, Section 10.6 concludes the studied research of this chapter.

## 10.2 Three-phase power flow in DNs

The power flow calculation is an essential function of each DSSE algorithm, regardless of its paradigm. The DSSE approach presented in this chapter uses the well-known backward/forward power flow algorithm. This section gives a brief overview of the power flow algorithm for unbalanced DNs, which is described in detail in Chapter 1.

The backward/forward method is based on two sweeps of calculations. In the backward sweep, the bus voltages are identified for the first iteration, and currents are calculated starting from the branches connected to the far buses and solved up to the source bus by applying Kirchhoff's first law. In the forward sweep, starting from the branches connected to the source bus and moving toward the ending branches, the voltages at the receiving buses are calculated using the voltages at corresponding sending buses and the branch currents calculated from the backward sweep. The voltages from the forward sweep are used for the next iteration in the backward sweep calculations [19]. Convergence occurs when the voltage mismatch for all phases at each bus becomes smaller than a convergence criterion. However, it should be noted that proper three-phase modeling of the DN components is crucial for the efficiency and accuracy of the power flow calculations.

*Line modeling*: Since the DN consists of un-transposed three-phase, two-phase, and single-phase lines with a mainly unbalanced load, it is necessary to take into account the magnetic coupling between the phases of the lines as well as the influence of the earth as a return path of the unbalance current. In addition, due to the uneven geometric position of the phases, the mutual phase capacities and the influence of the earth on the capacitance of the lines should be taken into account. Therefore, the lines in DNs should be modeled by the $(3 \times 3)$-dimensional three-phase series impedance and shunt admittance matrices. The calculation of these two matrices is based on the Carson equations and Kron reductions, as detailed in [20].

*Transformer modeling*: The first problem in applying the backward/forward sweep-based power flow method for the unbalanced radial DNs with multiple voltage levels is reflected in the inability to direct summarize branch currents (backward sweep) and calculate the bus voltages (forward sweep) for three-phase transformer whose secondary windings connections are delta or ungrounded wye. In this study, the symmetrical components model with appropriate transfer matrices for different connection types of transformers is used to overcome this problem.

*Load modeling*: The loads in DNs can be three-phase, two-phase, or single-phase, connected in grounded wye (Y) or ungrounded delta (Δ) configuration, and considered to be line-to-neutral for Y and line-to-line for Δ connection [19]. In the power flow calculation, the line currents are relevant, which are the currents per network phases. In addition to the load configuration (Y or Δ), the calculation of these currents also depends on the load model. The loads in DNs are usually modeled from the point of view of the dependence of power with voltage. T hree models are used: constant power (PQ), constant impedance ($Z$), and constant current ($I$), or any combination of the three.

*Distributed generation modeling*: A bus with DG can be represented as a PQ or PV bus. If the bus $i$ is modeled as a PQ bus, the DG unit should be treated as a negative PQ load. In the case of the DG units modeled as PV buses, the power flow algorithm requires some additional processes related to calculating the reactive power injections at the PV buses needed to achieve the specified voltage magnitudes [19].

## 10.3 Formulation of DSSE

The following types of measurements are available in most DNs [5,21]:

- Real-time measurements at the source, i.e. at the secondary side of the distribution substation: source bus voltage magnitude, active and reactive powers, current magnitudes, and power factors;
- Real-time measurements at selected points of the DN using remote control units: voltages, powers, and currents;
- Real-time measurements at monitored (dispatch-able) DG units: bus voltage and active power injection;
- Pseudo-measurements of loads based on historical data and typical daily load diagrams;
- Pseudo-measurements at unmonitored DG units based on external inputs such as historical databases and/or weather forecasts.

The DSSE algorithm should enable the determination of the entire network state for specified network topology, real-time measurements, as well as known historical data about loads and unmonitored DG units. Conversely, the problem of DSSE can be represented as a nonlinear optimization problem with constraints, where the loads and DG outputs can be considered the control variables, and the real-time measurement variables are dependent variables. This means that the DSSE algorithm estimates the loads and DG outputs at each bus by minimizing the difference between the measured and calculated values of the variables monitored in real-time. Therefore, the objective function of the DSSE can be expressed as follows:

$$\min f(\mathbf{x}, \mathbf{y}) = \sum_{i=1}^{m} |z_i - h_i(\mathbf{x}, \mathbf{y})| \tag{10.1}$$

where $\mathbf{x}$ is the vector of control variables, $\mathbf{y}$ is the vector of dependent variables, $z_i$ is the measured value of the real-time measurement variable $i$, $h_i$ is the nonlinear measurement function for the real-time measurement variable $i$, and $m$ is the total number of real-time measurements.

The control variables are active and reactive power of loads and active and reactive power outputs of unmonitored DG units. Assuming that the power factor of loads and power factor of unmonitored DG units be constant, the vector $\mathbf{x}$ can be expressed as

$$\mathbf{x} = \left[\mathbf{P}_{abc,L1}, \mathbf{P}_{abc,L2}, \ldots, \mathbf{P}_{abc,LN_L}, \mathbf{P}_{abc,DG1}, \mathbf{P}_{abc,DG2}, \ldots, \mathbf{P}_{abc,DGN_{UMDG}}\right] \tag{10.2}$$

where $\mathbf{P}_{abc,Li} = \begin{bmatrix} P_{a,Li} & P_{b,Li} & P_{c,Li} \end{bmatrix}^T$ is the vector of the three-phase power of load at the $i$th bus, $\mathbf{P}_{abc,DGi} = \begin{bmatrix} P_{a,DGi} & P_{b,DGi} & P_{c,DGi} \end{bmatrix}^T$ is the vector of the three-phase power output of DG at the $i$th bus, $N_L$ is the number of load buses, and $N_{UMDG}$ is the number of unmonitored DG units.

The vector of dependent variables consists of active power delivered through the root bus of the DN, voltage magnitude off all the buses of the network, and branch power flow:

$$\mathbf{y} = \left[\mathbf{V}_{abc,1}, \mathbf{V}_{abc,2}, \ldots, \mathbf{V}_{abc,N}, \mathbf{S}_{abc,1}, \mathbf{S}_{abc2}, \ldots, \mathbf{S}_{abc,N}\right] \tag{10.3}$$

where $\mathbf{V}_{abc,i} = \begin{bmatrix} V_{a,i} & V_{b,i} & V_{c,i} \end{bmatrix}^T$ is the vector of three-phase voltage magnitude at the $i$th bus, $\mathbf{S}_{abc,i} = \begin{bmatrix} S_{a,i} & S_{b,i} & S_{c,i} \end{bmatrix}^T$ is the vector of three-phase power flow through the $i$th branch, and $N$ is the number of buses.

The equality constraints are the nonlinear power flow equations described in Section 10.2.

The functional operating constraints are the lower and upper limits of each dependent variable, as follows:

$$\mathbf{V}_{abc,i}^{\min} \leq \mathbf{V}_{abc,i} \leq \mathbf{V}_{abc,i}^{abc} \tag{10.4}$$

$$\mathbf{S}_{abc,i}^{\min} \leq \mathbf{S}_{abc,i} \leq \mathbf{S}_{abc,i}^{abc} \tag{10.5}$$

The inequality constraints of control variables define the feasibility region of the DSSE problem:

$$\mathbf{P}_{abc,Li}^{\min} \leq \mathbf{P}_{abc,Li} \leq \mathbf{P}_{abc,Li}^{\max} \tag{10.6}$$

$$\mathbf{P}_{abc,DGi}^{\min} \leq \mathbf{P}_{abc,DGi} \leq \mathbf{P}_{abc,DGi}^{\max} \tag{10.7}$$

### *10.3.1 Limits of loads as control variables*

In DSSE, the loads are considered as pseudo-measurements. The pseudo-measurement values represent the forecasted mean values with added measurement noise as follows:

$$\mathbf{P}^{pm}_{abc,Li} = \mathbf{P}^{0}_{abc,Li} + (2r - 1)\mathbf{P}^{0}_{abc,Li}\frac{e_{pm}}{100} \tag{10.8}$$

Now, the upper and lower load limits in (10.6) can be calculated as

$$\mathbf{P}^{\max\,,\,\min}_{abc,Li} = \mathbf{P}^{pm}_{abc,Li}\left(1 \pm \frac{e_{pm}}{100}\right) \tag{10.9}$$

where $\mathbf{P}^{0}_{abc,Li}$ is the forecasted value of load power at bus $i$, $r$ is a random number uniformly distributed in the range [0,1], and $e_{pm}$ is the error in (%) adopted for the pseudo-measurements.

Calculation of the forecasted value of load ($\mathbf{P}^{0}_{abc,Li}$) is based on historical data [22]:

- Daily load profile for active power normalized by average or peak values.
- The load weight, which can be the peak value based on field measurement, the average value derived from supplied energy, or only the rated power of the equipment.

The forecasted value of load ($\mathbf{P}^{0}_{abc,Li}$) at any moment can be obtained by multiplying the value from the daily load profile (for the selected moment) with the corresponding load weight. Typical daily load profiles are established for different consumption types, such as industrial, commercial, and residential. In addition, each consumption type is presented with a characteristic daily load profile and corresponding load weight for typical loading days (week-day, weekend, holiday) in each weather season (spring, summer, fall, winter).

### 10.3.2 Limits of DG outputs

Active power outputs of unmonitored DG units are treated as control variables, which lower and upper limits in (10.7) can be calculated as follows:

$$\mathbf{P}^{\max\,,\,\min}_{abc,DGi} = \mathbf{P}^{pm}_{abc,DGi}\left(1 \pm \frac{e_{pm}}{100}\right) \tag{10.10}$$

where $\mathbf{P}^{pm}_{abc,Li}$ is the pseudo-measurement value of the DG output power at the bus $i$, which is calculated as follows:

$$\mathbf{P}^{pm}_{abc,DGi} = \mathbf{P}^{0}_{abc,DGi} + (2r - 1)\mathbf{P}^{0}_{abc,DGi}\frac{e_{pm}}{100} \tag{10.11}$$

where $\mathbf{P}^{0}_{abc,DGi}$ is the forecasted output power of the $i$th DG unit at a given moment of interest. Detailed procedures for calculating the forecasted output powers of renewable-based DG units, such as wind turbine generation (WTG) and PVG, can be found in [23].

### 10.3.3 Real-time measurements

As noted above, the real-time measurements ($z_i$) are performed for voltage magnitudes and power flows at the substation, and voltage magnitudes and power flows at selected points of the DN. The maximum error of 3% is considered for these real-time measurements. Therefore, the real-time measurement respecting the measurement error can be simulated by introducing random noise in the true value as follows:

$$\mathbf{M}_{abc,i}^{rt} = \mathbf{M}_{abc,i}^{0} + (2r - 1)\mathbf{M}_{abc,i}^{0}\frac{e_{rt}}{100} \tag{10.12}$$

where $\mathbf{M}_{abc,i}^{0}$ is the true value of a variable (voltage, power, etc.), $e_{rt}$ is the error in (%) adopted for the real-time measurements, and $\mathbf{M}_{abc,i}^{rt}$ is the measured value of a three-phase variable, respecting the measurement error.

## 10.4 Solution method

The general application of metaheuristic optimization methods for efficiently solving the DSSE problem is shown in the example of the hybrid PSOS–CGSA algorithm, as proposed in [18].

The hybrid PSOS–CGSA algorithm is based on the hybridization of particle swarm optimization with sigmoid-based acceleration coefficients (PSOS) and chaotic gravitational search algorithm (CGSA). The improvements of PSOS–CGSA in relation to the original PSOGSA algorithm are based on modeling the cognitive and social acceleration coefficient ($C_1$ and $C_2$) with a new sigmoid-based acceleration coefficient and introduction of a chaos sinusoidal map into the gravitational constant ($G$). These actions improve both exploration and exploitation phases of the proposed hybrid algorithm by balancing the global searchability in the early stage and the global convergence in the later stage.

Since the proposed hybrid PSOS–CGSA algorithm belongs to metaheuristic population-based optimization techniques, it will be explained here through a general metaheuristic framework.

The metaheuristic optimization methods are built upon population-based stochastic search algorithms. By population (**POP**), we mean a set of search agents of size ($N$), representing the possible solutions to the problem. The search agents are denoted by a vector ($\mathbf{x}_i$) of length ($n$), where the entries of the vector are the values of the relevant control variables. The length of the vector is also a measure of the search space dimension of the given optimization problem. In other words, the metaheuristic methods iteratively approximate the solution to the problem. At each stage, a new population is generated after applying the search operators to agents in the current population. The main structure of our proposed algorithm is based on the following procedure described as follows:

*Defining and initialization procedure*

1. Define objective function $F(\mathbf{x}_i)$ and possible solutions space $\mathbf{X}$;
2. Generate $N$ agents and assign the starting position randomly to agents using the control variable values in the range of maximum to minimum, where the iteration counter: $t = 1$.

*Iterative procedure*

1. Estimate the fitness rate for each agent in the current population $\mathbf{POP}(t)$.
2. Generate new $\mathbf{POP}(t+1)$ using algorithmic operators on search agents from the $\mathbf{POP}(t)$. For the proposed hybrid PSOS–CGSA algorithm, the operators for updating the current velocity and the current position of agents are as follows:

$$\mathbf{v}_i(t+1) = r_1 \cdot \mathbf{v}_i(t) + r_2 \cdot C_1(t) \cdot \mathbf{a}_i(t) + r_3 \cdot C_2(t) \cdot (\mathbf{gbest}(t) - \mathbf{x}_i(t)) \tag{10.13}$$

$$\mathbf{x}_i(t+1) = \mathbf{x}_i(t) + \mathbf{v}_i(t+1) \tag{10.14}$$

The cognitive and social acceleration coefficients are updated in each iteration using the following equations:

$$C_1(t) = \frac{1}{1+\exp\left(-\lambda \cdot \frac{t}{t_{max}}\right)} + 2(C_{1f} - C_{1i})\left(\frac{t}{t_{max}} - 1\right)^2 \tag{10.15}$$

$$C_2(t) = \frac{1}{1+\exp\left(-\lambda \cdot \frac{t}{t_{max}}\right)} + 2(C_{1f} - C_{1i})\left(\frac{t}{t_{max}}\right)^2 \tag{10.16}$$

In (10.13), **gbest** denotes the best solution (position) among all agent's best positions achieved so far. The acceleration of agents ($\mathbf{a}_i$) is updated in each iteration, considering the gravitational constant ($G$) with the embedded sinusoidal chaotic map, as explained in [18]. $r_1$, $r_2$, and $r_3$ are random numbers in the range of [0,1]; $t_{max}$ is the maximum number of iterations; $\lambda$ is a control parameter ($\lambda = 0.0001$); and $C_{1i}$ and $C_{1f}$ are two positive constants ($C_{1i} = 0.5$, $C_{1f} = 2.5$).

3. Repeat the iterative procedure until the stop criterion is reached.
4. Report the best solution. End.

The general flowchart of the PSOS–CGSA algorithm is presented in Figure 10.1. The flow chart describes the entire execution of the proposed algorithm for the optimal results, where the proposed hybrid PSOS–CGSA approach has been applied to solve the DSSE problem in the DG-integrated distribution systems. The detailed step-by-step execution of PSOS–CGSA for the adaption of the optimal solution is explained in the next section.

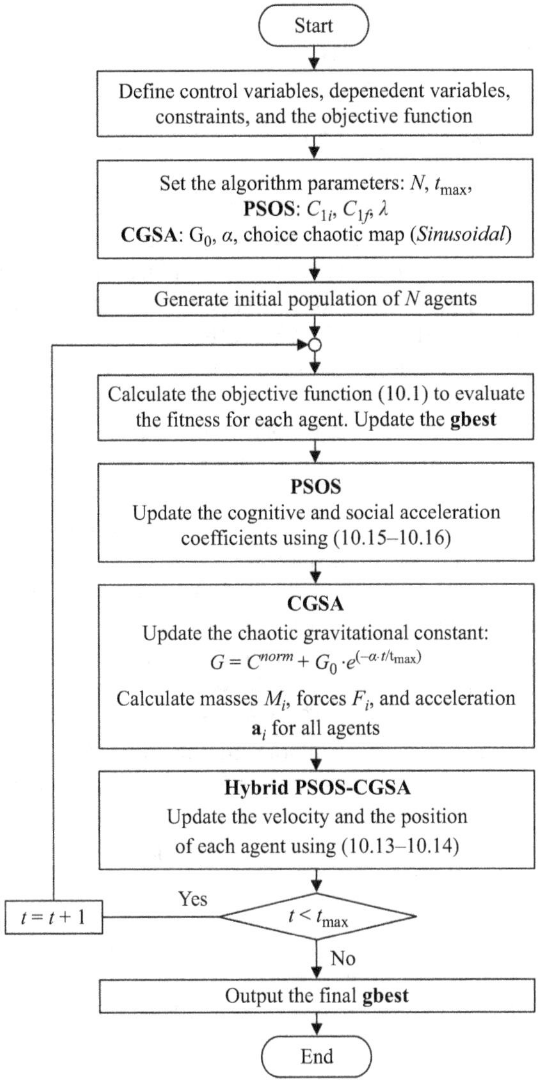


*Figure 10.1 Flow chart of the PSOS–CGSA algorithm*

### *10.4.1 Application of PSOS–CGSA for DSSE*

In the DSSE, the objective function is the minimization of absolute differences between the measured values and the estimated values, which are functions of control and dependent state variables (10.1). The control variables are the active and reactive power of loads and the active and reactive power outputs of unmonitored DG units. Therefore, a search agent, i.e. a potential solution of the DSSE, can be represented based on the vector (10.14) as follows:

$$\mathbf{x}_i = [\mathbf{P}^1_{abc,L1}, \ldots, \mathbf{P}^d_{abc,LN_L}, \mathbf{P}^{d+1}_{abc,DG1}, \ldots, \mathbf{P}^n_{abc,DGN_{UMDG}}], \quad i = 1, \ldots, N \tag{10.17}$$

where $n$ is the number of control variables ($n$=$N_L$+$N_{UMDG}$) and $N$ is the number of agents (size of the population).

The presentation of the proposed hybrid PSOS–CGSA algorithm in solving three-phase DG-integrated DSSE can be summarized as follows:

*Step 1*: Detect the input parameters, including the DN configuration, line data, transformer data, and historical data related to loads and DG.

*Step 2*: Define (read) the values of real-time measurements ($z_i$). Define the error of real-time measurements. Simulate the error in the real-time measurements (10.12).

*Step 3*: Define the typical daily load profiles and load weights for each season. Calculate the forecasted values of load powers at the given moment of interest.

*Step 4*: Calculate the forecasted output powers of unmonitored DG units at the given moment of interest.

*Step 5*: Define the error of pseudo-measurements. Calculate the pseudo-measurements using (10.8) and (10.11).

*Step 6*: Specify the objective function (10.1), specify the control variables (10.2) and calculate their limits (10.9) and (10.10), and specify the dependent variables (10.3) and their limits.

*Step 7*: Set the PSOS–CGSA algorithmic parameters, generate an initial random population of $N$ agents, and set iteration counter $t$ = 1.

*Step 8*: Perform power flow calculation and evaluate the objective function for each agent $\mathbf{x}_i(t)$ from the current population $\mathbf{POP}(t)$.

*Step 9*: Apply the PSOS–CGSA operators (10.13)–(10.16) to create a new population of agents (i.e., updated solutions to the problem).

*Step 10:* Repeat Steps 8 and 9 until the stop criterion is reached, i.e. the max number of iterations.

*Step 11*: Report the optimal results: best agent $\mathbf{x}_i$ in the last iteration – the estimated values of control variables (power of loads and power of DG units). Based on these estimated values of loads and DG outputs, a final load flow calculation is performed to estimate the values of the dependent variables: bus voltages, branch power flows, and power losses.

## 10.5 Simulation results

The proposed three-phase DSSE methodology for unbalanced DNs is implemented in the MATLAB® programming environment and validated using three representative IEEE test systems. The first one is the IEEE 13-bus system, the second one is the IEEE 37-bus system, and the third one is the IEEE 123-bus system [24].

These three test systems are simplified to key features of the DSSE algorithm be highlighted:

- The voltage regulators are removed, and switches in the line segments are closed.
- The distributed load along the line segment is lumped and placed on half of the line.
- Shunt capacitor banks are omitted.

The three-phase power flow program was developed based on the algorithm and component models described in Section 10.3, i.e. Section 1.3 of Chapter 1. This program is used for the power flow computation in the test systems.

The power flow computation results are considered the true system states. Real-time measurements are simulated by adding the normally distributed noise as given in (10.12). It is assumed that the pseudo-measurement error ($e_{pm}$) is 20%, while the real-time measurement error ($e_{rt}$) is 3%. The accuracy of the estimated states, as well as the performance of the DSSE algorithms, is evaluated using the mean absolute percentage error (MAPE), as given below:

$$MAPE(\%) = \frac{100}{n}\sum_{i=1}^{n}\left|\frac{x_i^{true} - x_i^{est}}{x_i^{true}}\right| \tag{10.18}$$

where $x_i^{true}$ and $x_i^{est}$ are the true and estimated value of the $i$th variable and $n$ is the number of considered variables.

### 10.5.1 IEEE 13-bus system

The IEEE 13-bus test system is a small-size radial and highly unbalanced distribution feeder operating at 4.16 kV. The three-phase configuration of the system under study is presented in Figure 10.2.

The active and reactive powers of load given in [24] are considered the forecasted mean values, which are used to generate pseudo-measurements according to (10.8). The power flow results represent "the true system state." It is supposed that measurements of active/reactive power flows and bus voltages can be measured at real time. The quasi-real-time measurements of active/reactive power flows are presented in Table 10.1.

DSSE in the IEEE 13-bus system was considered, regarding (i) which variables are monitored in real-time, i.e. number and place of real-time measurement devices; and (ii) the percent of errors for the pseudo-measurements and the errors for the real-time measurements.

The simulation results are shown in Table 10.2. Five different cases were considered. Each case was tested ten times. The mean absolute percentage error of bus voltage magnitude (MAPEV) is used to evaluate the performance of the different cases. The comparison of minimum, maximum, and mean values of MAPEV obtained over ten runs is presented in Table 10.2. It is obvious that as the number of real-time monitored variables increases, the accuracy of the DSSE results is improved. Of course, this is quite logical. However, these results can help to

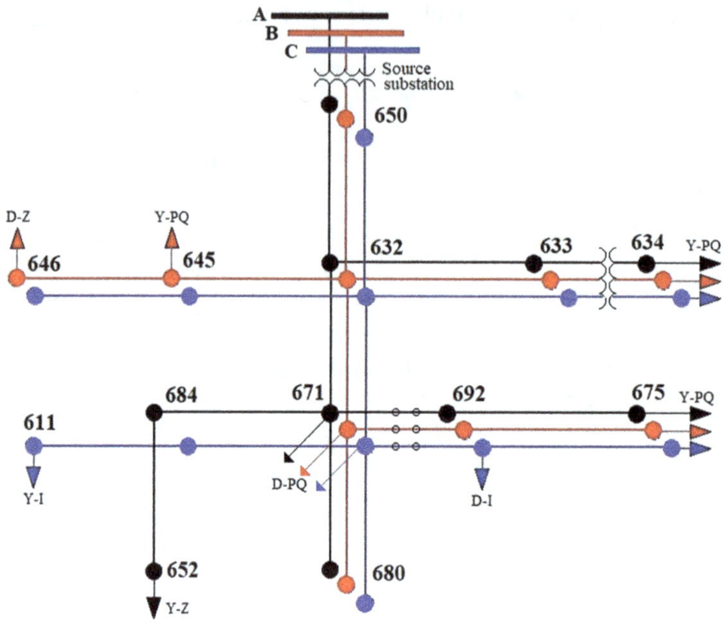


*Figure 10.2 Three-phase diagram of the IEEE 13-bus test system*

*Table 10.1 IEEE 13-bus system power flow*

| Bus *i* | Bus *j* | Phase A | | Phase B | | Phase C | |
|---|---|---|---|---|---|---|---|
| | | *P* (kW) | *Q* (kVAr) | *P* (kW) | *Q* (kVAr) | *P* (kW) | *Q* (kVAr) |
| 650 | 632 | 1211.2 | 790.0 | 969.3 | 571.8 | 1267.4 | 862.5 |
| 632 | 633 | 162.9 | 115.3 | 121.6 | 92.9 | 121.8 | 93.3 |
| 633 | 634 | 159.9 | 109.9 | 120.0 | 89.9 | 119.9 | 89.9 |
| 632 | 645 | – | – | 311.8 | 127.0 | 66.5 | 117.6 |
| 645 | 646 | – | – | 141.6 | 1.9 | 66.3 | 117.4 |
| 632 | 671 | 1022.1 | 580.9 | 471.9 | 307.8 | 915.9 | 513.2 |
| 671 | 680 | – | – | – | – | – | – |
| 671 | 692 | 529.8 | 305.6 | 67.8 | 59.9 | 403.8 | 235.7 |
| 692 | 675 | 484.9 | 189.9 | 68.0 | 59.9 | 289.9 | 211.9 |
| 671 | 684 | 105.4 | 70.5 | – | – | 147.6 | 69.7 |
| 684 | 611 | – | – | – | – | 147.1 | 69.2 |
| 684 | 652 | 104.7 | 70.3 | – | – | – | – |

determine which variables should be measured (power, current, and voltage) as well as locations of real-time measurements to improve the DSSE performance.

The power measurement in the source branch (650-632) has a crucial impact on the accuracy of the results. Significant improvements in the DSSE performance

*Table 10.2 Simulation results of DSSE for the IEEE 13-bus system: MAPEV (%)*

| Case | Real-time measurements | Proposed PSOS–CGSA | | | WLS [25] |
|---|---|---|---|---|---|
| | | min | max | mean | mean |
| 1 | P&Q in (650-632) | 0.0594 | 0.1982 | 0.1126 | 0.20 |
| 2 | Current in (650-632) | 0.0932 | 0.2241 | 0.1362 | 0.35 |
| 3 | P&Q in (650-632)<br>V at (671) | 0.0322 | 0.1508 | 0.0900 | Not reported |
| 4 | P&Q in (650-632)<br>P&Q in (633-634)<br>P&Q in (671-692) | 0.0301 | 0.0596 | 0.0450 | Not reported |
| 5 | P&Q in (650-632)<br>P&Q in (633-634)<br>P&Q in (671-692)<br>V at (645) | 0.0177 | 0.0492 | 0.0335 | Not reported |

are performed with additional real-time measurements of voltage magnitude at critical buses of the network and active and reactive power in the lateral branches. The results for Cases 1 and 2 presented in Table 10.2 are compared with those reported in [25]. As can be seen, the proposed PSOS–CGSA algorithm enables better DSSE results than the branch current-based WLS algorithm.

The detailed DSSE results for Case 5 are presented in Figure 10.3 and Tables 10.3 and 10.4. The right side of Figure 10.3 shows the errors of the estimated values. These errors are calculated as

$$error = \left| \frac{p^{true} - p^{est}}{p^{true}} \right| \cdot 100 \tag{10.19}$$

The estimated and true values of active power for the IEEE 13-bus system are presented in Figure 10.3. A maximum percentage error of 5% corresponds to the estimated load at bus 611, which is modeled by a constant current model. Tables 10.3 and 10.4 show the estimated and true values of the voltage magnitudes and the voltage angles for all three phases. As can be seen, the estimated voltage magnitudes and voltage angles match very well with the true values.

## *10.5.2 IEEE 37-bus system*

The detailed network parameters for the IEEE 37-bus test system, presented in Figure 10.4, can be found in [24]. The proposed DSSE approach was tested on the IEEE 37-bus system with three renewable-based DG units whose specifications are given in Table 10.5. It is assumed that all DG units operate in PQ mode. The PVG operates at a unity power factor, whereas WTG operates at 0.95 lag power factors (absorbs reactive power) in all operating conditions.

The active and reactive powers of load given in [24] are considered as the forecasted mean values, which are used to generate pseudo-measurements

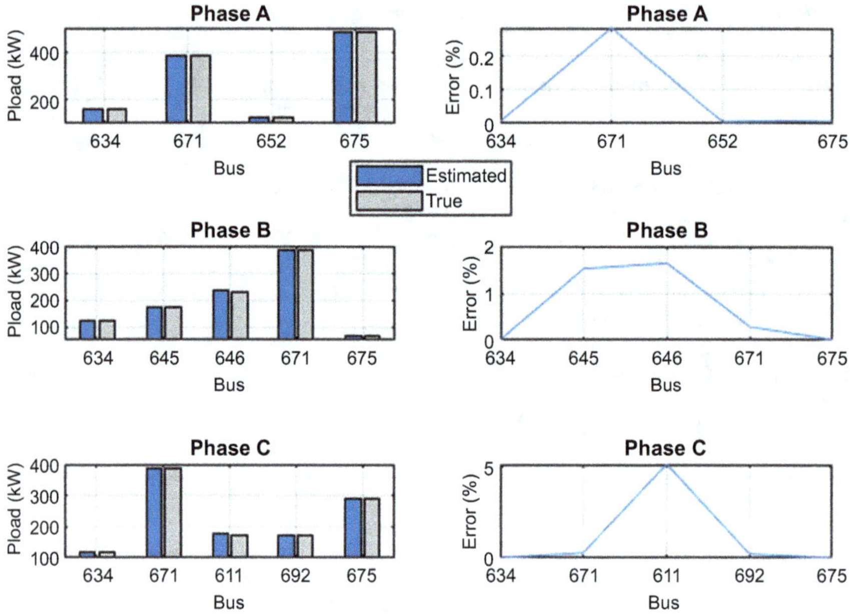


*Figure 10.3 Estimated and true values of the active power of load for the IEEE 13-bus system*

*Table 10.3 Estimated and true values of voltage magnitudes for the IEEE 13-bus test system*

| Bus | Phase A | | Phase B | | Phase C | |
|---|---|---|---|---|---|---|
| | **Estim.** | **True** | **Estim.** | **True** | **Estim.** | **True** |
| 650 | 1.0000 | 1.0000 | 1.0000 | 1.0000 | 1.0000 | 1.0000 |
| 632 | 0.9498 | 0.9499 | 0.9840 | 0.9839 | 0.9300 | 0.9300 |
| 633 | 0.9466 | 0.9466 | 0.9820 | 0.9819 | 0.9271 | 0.9271 |
| 634 | 0.9207 | 0.9207 | 0.9625 | 0.9624 | 0.9064 | 0.9064 |
| 645 | 0 | 0 | 0.9745 | 0.9745 | 0.9284 | 0.9283 |
| 646 | 0 | 0 | 0.9728 | 0.9729 | 0.9264 | 0.9264 |
| 671 | 0.9110 | 0.9110 | 0.9876 | 0.9874 | 0.8716 | 0.8718 |
| 680 | 0.9111 | 0.9110 | 0.9876 | 0.9874 | 0.8717 | 0.8718 |
| 684 | 0.9094 | 0.9093 | 0 | 0 | 0.8682 | 0.8685 |
| 611 | 0 | 0 | 0 | 0 | 0.8648 | 0.8653 |
| 652 | 0.9044 | 0.9043 | 0 | 0 | 0 | 0 |
| 692 | 0.9111 | 0.9110 | 0.9876 | 0.9874 | 0.8717 | 0.8718 |
| 675 | 0.9027 | 0.9026 | 0.9888 | 0.9886 | 0.8678 | 0.8679 |

*Table 10.4 Estimated and true values of voltage angles for the IEEE 13-bus test system*

| Bus | Phase A | | Phase B | | Phase C | |
|---|---|---|---|---|---|---|
| | **Estim.** | **True** | **Estim.** | **True** | **Estim.** | **True** |
| 650 | 0.000 | 0.000 | −120.000 | −120.000 | 120.000 | 120.000 |
| 632 | −2.746 | −2.746 | −121.683 | −121.682 | 117.807 | 117.800 |
| 633 | −2.822 | −2.823 | −121.733 | −121.731 | 117.801 | 117.794 |
| 634 | −3.607 | −3.608 | −122.246 | −122.244 | 117.224 | 117.217 |
| 645 | – | – | −121.869 | −121.864 | 117.829 | 117.822 |
| 646 | – | – | −121.944 | −121.938 | 117.877 | 117.869 |
| 671 | −5.904 | −5.898 | −122.200 | −122.210 | 115.946 | 115.951 |
| 680 | −5.904 | −5.898 | −122.199 | −122.210 | 115.945 | 115.950 |
| 684 | −5.957 | −5.950 | – | – | 115.907 | 115.916 |
| 611 | – | – | – | – | 115.825 | 115.838 |
| 652 | −5.880 | −5.873 | – | – | – | – |
| 692 | −5.904 | −5.898 | −122.199 | −122.210 | 115.945 | 115.950 |
| 675 | −6.083 | −6.076 | −122.294 | −122.305 | 116.063 | 116.068 |

according to (10.8). The active powers of the DG units given in Table 10.5 are assumed to represent their forecasted values calculated using the approach described in [23]. These forecasted outputs are used in (10.11) to obtain the pseudo-measurements of the output power of DG units.

The DSSE simulation was carried out on the IEEE 37-bus test system, assuming that there are available real-time measurements of active/reactive power flows and bus voltages, as given in Table 10.6.

The DSSE results obtained by the proposed hybrid PSOS–CGSA algorithm were compared with two original well-established metaheuristic optimization algorithms, i.e. particle swarm optimization (PSO) and gravitational search algorithm (GSA); several recently proposed innovated versions of these techniques such as particle swarm optimization with sigmoid-based acceleration coefficients (PSOS), chaos-based gravitational search algorithm (CGSA), phasor particle swarm optimization (PPSO); and their hybrids such as hybrid particle swarm optimization and gravitational search algorithm (PSOGSA), and hybrid phasor particle swarm optimization and gravitational search algorithm (PPSOGSA). All the mentioned algorithms are described in detail in Volume 1 of the book (Chapters 3–5).

The parameters used in the simulation are presented in Table 10.7. These parameters were adopted based on their default values. The population size and maximum iteration number are set as $N = 25$ and $t_{\max} = 100$ for all case studies, including the IEEE 13-bus test system.

The statistical indicators of MAPEV for DSSE in the IEEE 37-bus system using different optimization methods are presented in Table 10.8. The results shown in Table 10.8 correspond to the MAPEV values obtained over ten

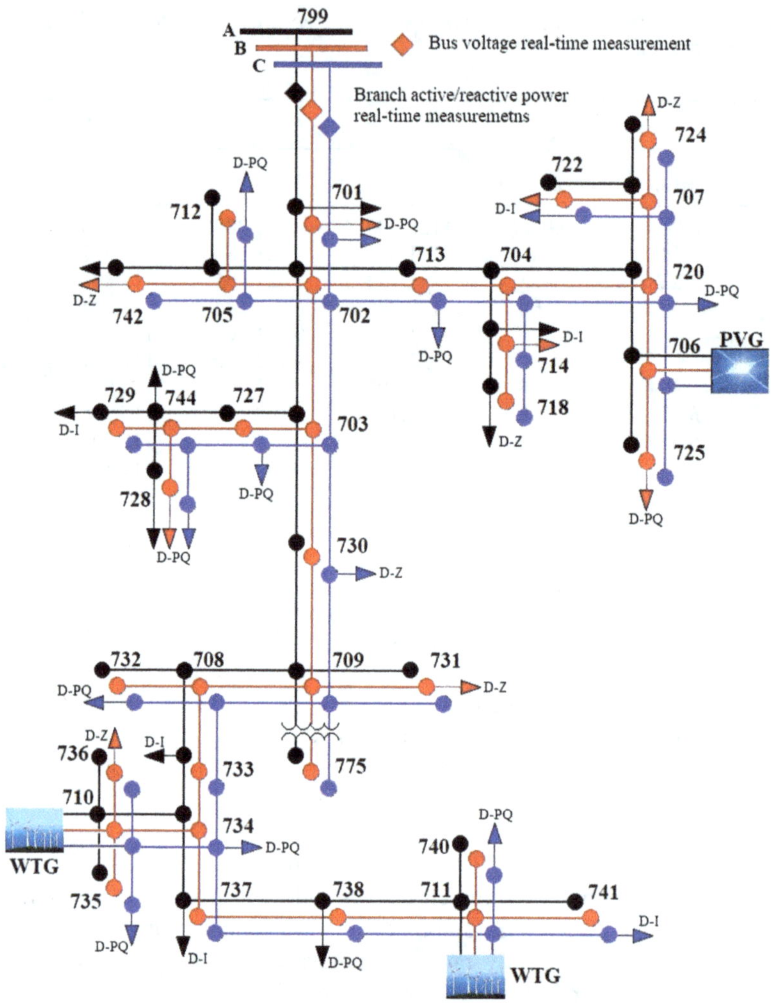


*Figure 10.4 Three-phase diagram of the IEEE 37-bus test system*

*Table 10.5 DG unit specifications*

| Location | Type | $P_{DG}$ (kW) | | | Power factor |
|---|---|---|---|---|---|
| | | Phase A | Phase B | Phase C | |
| 706 | PVG | 150 | 150 | 150 | 1 |
| 710 | WTG | 180 | 180 | 180 | 0.95 |
| 711 | WTG | 200 | 200 | 200 | 0.95 |

*Table 10.6 Real-time measurements in the IEEE 37-bus system*

| Location | Measurements | Phase A | Phase B | Phase C |
|---|---|---|---|---|
| Branch 799-701 | $P$ (kW) | 336.930 | 151.237 | 406.076 |
| | $Q$ (kVAr) | 676.835 | 493.386 | 420.749 |
| Branch 702-713 | $P$ (kW) | −15.207 | 46.094 | 59.972 |
| | $Q$ (kVAr) | 94.474 | 43.909 | 121.571 |
| Branch 709-708 | $P$ (kW) | −45.312 | −224.505 | −171.811 |
| | $Q$ (kVAr) | 265.648 | 313.364 | 137.909 |
| Bus 709 | $V$ (pu) | 0.9775 | 0.9877 | 0.9828 |

*Table 10.7 Algorithms parameter settings*

| | |
|---|---|
| PSO | $C_1 = 2$; $C_2 = 2$; $w_{max} = 0.9$; $w_{min} = 0.4$ |
| GSA | $G_0 = 1$; $\alpha = 20$; $p = 2$; $K_0 = 1$ |
| PSOS | $C_{1i} = 0.5$; $C_{1f} = 2.5$; $\lambda = 0.0001$; $w_{max} = 0.9$; $w_{min} = 0.4$ |
| PPSO | – |
| CGSA | $G_0 = 1$; $\alpha = 20$; $w_{max} = 0.1$; $w_{min} = 1\times^{-10}$; sinusoidal map |
| PSOGSA | $C_1 = 2$; $C_2 = 2$; $G_0 = 1$; $\alpha = 20$ |
| PPSOGSA | $G_0 = 1$; $\alpha = 20$ |
| Proposed PSOS–CGSA | $C_{1i} = 0.5$; $C_{1f} = 2.5$; $\lambda = 0.0001$; $G_0 = 1$; $\alpha$=20; sinusoidal map |

*Table 10.8 Statistical results of MAPEV (%) for the IEEE 37-bus system*

| Algorithm | min | max | mean | std |
|---|---|---|---|---|
| PSO | 0.0262 | 0.1766 | 0.0815 | 0.0479 |
| GSA | 0.0483 | 0.1738 | 0.1059 | 0.0459 |
| PSOS | 0.0107 | 0.0833 | 0.0436 | 0.0244 |
| CGSA | 0.0168 | 0.1445 | 0.0771 | 0.0407 |
| PPSO | 0.0164 | 0.1279 | 0.0447 | 0.0336 |
| PSOGSA | 0.0199 | 0.1448 | 0.0533 | 0.0395 |
| PPSOGSA | 0.0165 | 0.0819 | 0.0445 | 0.0200 |
| Proposed PSOSCGSA | 0.0068 | 0.0303 | 0.0181 | 0.0092 |

consecutive DSSE runs using each of the algorithms, as shown in Figure 10.5. The sixth column of Table 10.8 presents the time of computation per iteration. These results show that there is no significant difference in the calculation time between the different algorithms. This is quite expected, given that these are algorithms of similar structure, which is explained in Section 10.4.

The results presented in Table 10.8 and Figure 10.5 clearly indicate that the proposed hybrid PSOS–CGSA algorithm provides more accurate and stable DSSE compared to other metaheuristic algorithms over a wide range of test runs, which proves its robustness.

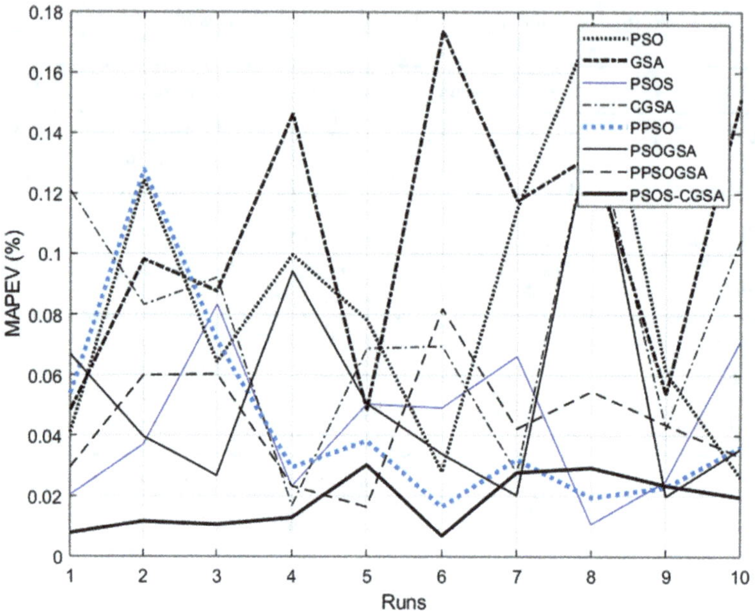


*Figure 10.5 The variation range of MAPEV*

The estimated and true values of active power for the IEEE 37-bus system obtained by the proposed PSOS–CGSA algorithm are presented in Figures 10.6–10.8. A maximum percentage error of 8.25% corresponds to the estimated load in phase A at bus 733. Table 10.9 shows the estimated and true values of voltage magnitudes. As can be seen from this table, the estimated voltages match very well with the true values.

### *10.5.3 IEEE 123-bus system*

In order to evaluate the efficiency of the proposed DSSE methodology in larger systems, the IEEE 123-bus test feeder is considered. One-line diagram is shown in Figure 10.9. The proposed PSOS–CGSA method was successfully implemented on the IEEE-123 system using system data taken from [24].

The simulation results for the IEEE 123-bus system are presented in Table 10.10. The mean absolute percentage error of bus voltage magnitude (MAPEV) and bus voltage angle (MAPEA) are used to evaluate the performance of the proposed method in comparison to the results reported in [25]. DSSE is considered regarding which variables are monitored in real-time. Note that one real-time measurement in different locations with all pseudo-measurements is tested, as in [25]. It can be seen that the DSSE results obtained from the PSOS–CGSA algorithm are better than those obtained from the branch current-based WLS algorithm [25]. This clearly indicates that the proposed approach can be effectively used for DSSE in large-scale distribution systems.

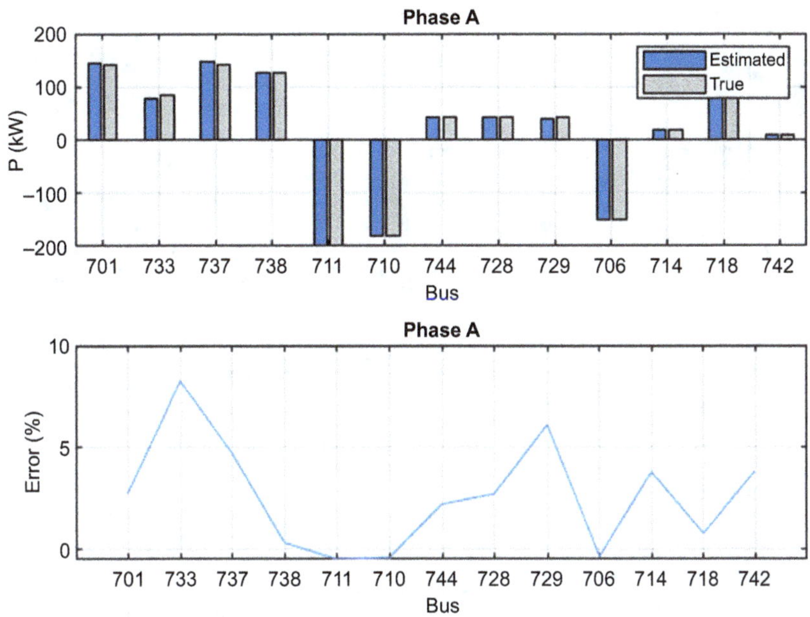


*Figure 10.6 Active power in phase A for the IEEE 37-bus system*

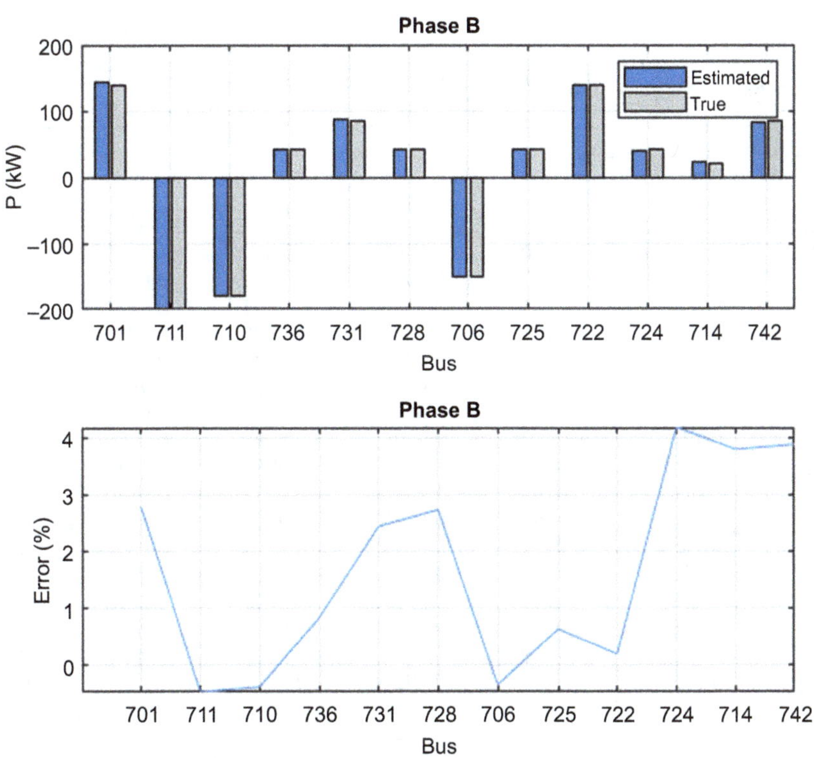


*Figure 10.7 Active power in phase B for the IEEE 37-bus system*

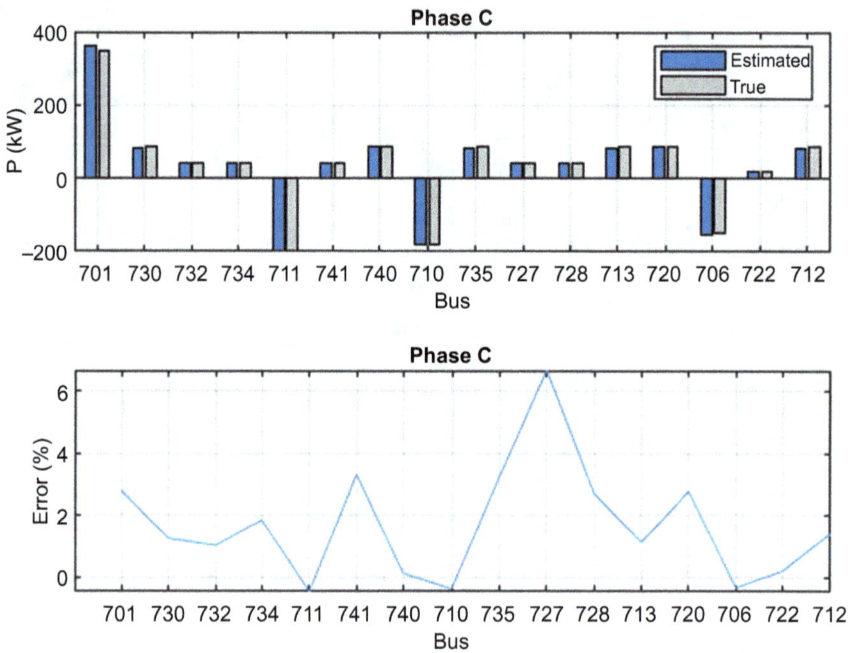


*Figure 10.8 Active power in phase C for the IEEE 37-bus system*

*Table 10.9 Voltage magnitudes in the IEEE 37-bus system*

| Bus | Phase A | | Phase B | | Phase C | |
|---|---|---|---|---|---|---|
| | **Estim.** | **True** | **Estim.** | **True** | **Estim.** | **True** |
| 799 | 1.0000 | 1.0000 | 1.0000 | 1.0000 | 1.0000 | 1.0000 |
| 701 | 0.9893 | 0.9893 | 0.9935 | 0.9935 | 0.9900 | 0.9901 |
| 702 | 0.9846 | 0.9845 | 0.9904 | 0.9904 | 0.9857 | 0.9857 |
| 703 | 0.9799 | 0.9798 | 0.9876 | 0.9875 | 0.9832 | 0.9832 |
| 730 | 0.9780 | 0.9779 | 0.9877 | 0.9876 | 0.9828 | 0.9827 |
| 709 | 0.9776 | 0.9775 | 0.9877 | 0.9877 | 0.9828 | 0.9828 |
| 775 | 0.9775 | 0.9773 | 0.9877 | 0.9876 | 0.9831 | 0.9830 |
| 708 | 0.9771 | 0.9769 | 0.9881 | 0.9881 | 0.9833 | 0.9833 |
| 732 | 0.9768 | 0.9766 | 0.9881 | 0.9881 | 0.9830 | 0.9829 |
| 733 | 0.9767 | 0.9765 | 0.9885 | 0.9885 | 0.9840 | 0.9839 |
| 734 | 0.9765 | 0.9764 | 0.9898 | 0.9898 | 0.9851 | 0.9851 |
| 737 | 0.9755 | 0.9754 | 0.9898 | 0.9898 | 0.9861 | 0.9860 |
| 738 | 0.9756 | 0.9756 | 0.9905 | 0.9906 | 0.9866 | 0.9866 |
| 711 | 0.9763 | 0.9763 | 0.9919 | 0.9920 | 0.9872 | 0.9872 |
| 741 | 0.9761 | 0.9761 | 0.9920 | 0.9921 | 0.9871 | 0.9871 |
| 740 | 0.9760 | 0.9760 | 0.9920 | 0.9921 | 0.9869 | 0.9869 |
| 710 | 0.9791 | 0.9789 | 0.9927 | 0.9927 | 0.9867 | 0.9867 |

*(Continues)*

*Table 10.9 (Continued)*

| Bus | Phase A | | Phase B | | Phase C | |
|---|---|---|---|---|---|---|
| | **Estim.** | **True** | **Estim.** | **True** | **Estim.** | **True** |
| 735 | 0.9787 | 0.9786 | 0.9927 | 0.9927 | 0.9864 | 0.9863 |
| 736 | 0.9791 | 0.9789 | 0.9914 | 0.9914 | 0.9856 | 0.9856 |
| 731 | 0.9776 | 0.9775 | 0.9871 | 0.9870 | 0.9821 | 0.9821 |
| 727 | 0.9788 | 0.9786 | 0.9868 | 0.9867 | 0.9825 | 0.9825 |
| 744 | 0.9782 | 0.9780 | 0.9862 | 0.9861 | 0.9823 | 0.9822 |
| 728 | 0.9778 | 0.9776 | 0.9858 | 0.9857 | 0.9819 | 0.9818 |
| 729 | 0.9779 | 0.9777 | 0.9860 | 0.9858 | 0.9823 | 0.9822 |
| 713 | 0.9843 | 0.9842 | 0.9899 | 0.9899 | 0.9847 | 0.9847 |
| 704 | 0.9844 | 0.9844 | 0.9892 | 0.9892 | 0.9839 | 0.9838 |
| 720 | 0.9857 | 0.9857 | 0.9894 | 0.9893 | 0.9827 | 0.9827 |
| 706 | 0.9876 | 0.9876 | 0.9909 | 0.9908 | 0.9841 | 0.9841 |
| 725 | 0.9877 | 0.9876 | 0.9906 | 0.9905 | 0.9838 | 0.9838 |
| 707 | 0.9852 | 0.9851 | 0.9852 | 0.9851 | 0.9788 | 0.9787 |
| 722 | 0.9851 | 0.9851 | 0.9848 | 0.9847 | 0.9784 | 0.9783 |
| 724 | 0.9851 | 0.9851 | 0.9845 | 0.9843 | 0.9781 | 0.9780 |
| 714 | 0.9842 | 0.9842 | 0.9890 | 0.9889 | 0.9838 | 0.9838 |
| 718 | 0.9832 | 0.9831 | 0.9881 | 0.9880 | 0.9839 | 0.9838 |
| 705 | 0.9837 | 0.9837 | 0.9896 | 0.9895 | 0.9842 | 0.9841 |
| 712 | 0.9833 | 0.9832 | 0.9896 | 0.9895 | 0.9837 | 0.9836 |
| 742 | 0.9837 | 0.9836 | 0.9889 | 0.9888 | 0.9837 | 0.9836 |

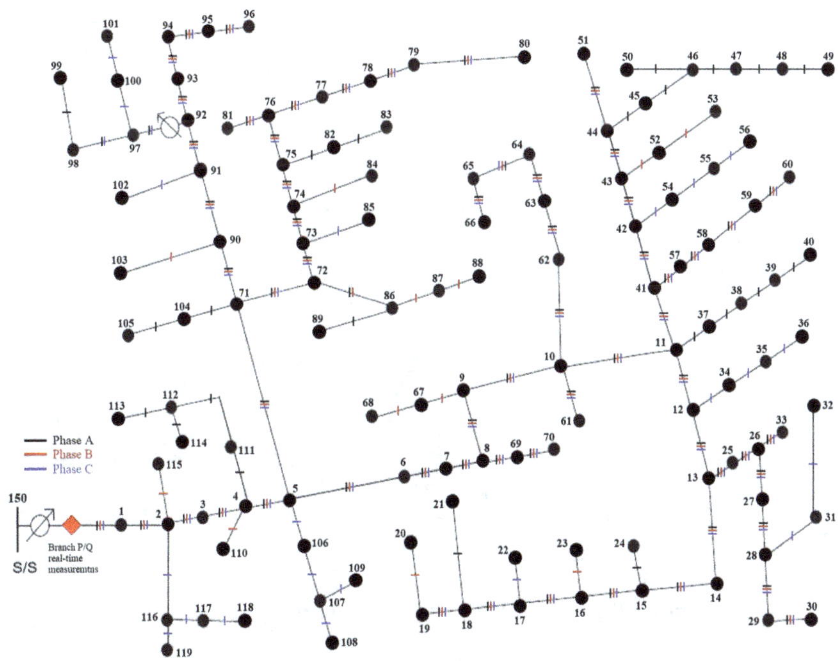


*Figure 10.9 One-line diagram of the IEEE 123-bus system*

## 10.6 MATLAB® program *dssem*

The *dssem* is an educational MATLAB program for DSSE using a hybrid PSOS–CGSA algorithm. The program package is available as the supplementary material for this chapter in the IET digital library. The simulation results presented in Section 10.5 are obtained by the *dssem* program.

An example of *dssem* program execution for DSSE in the IEEE 13-bus system is presented in Figure 10.10.

*Table 10.10 Simulation results of DSSE for the IEEE 123-bus system*

| Real-time measurements | MAPEV (%) | | MAPEA (%) | |
|---|---|---|---|---|
| | Proposed PSOS–CGSA | WLS [30] | Proposed PSOS–CGSA | WLS [30] |
| P&Q in (150-1) | 0.079 | 0.15 | 0.047 | 0.09 |
| P&Q in (5-71) | 0.082 | 0.16 | 0.059 | 0.09 |
| P&Q in (43-44) | 0.099 | 0.17 | 0.071 | 0.09 |

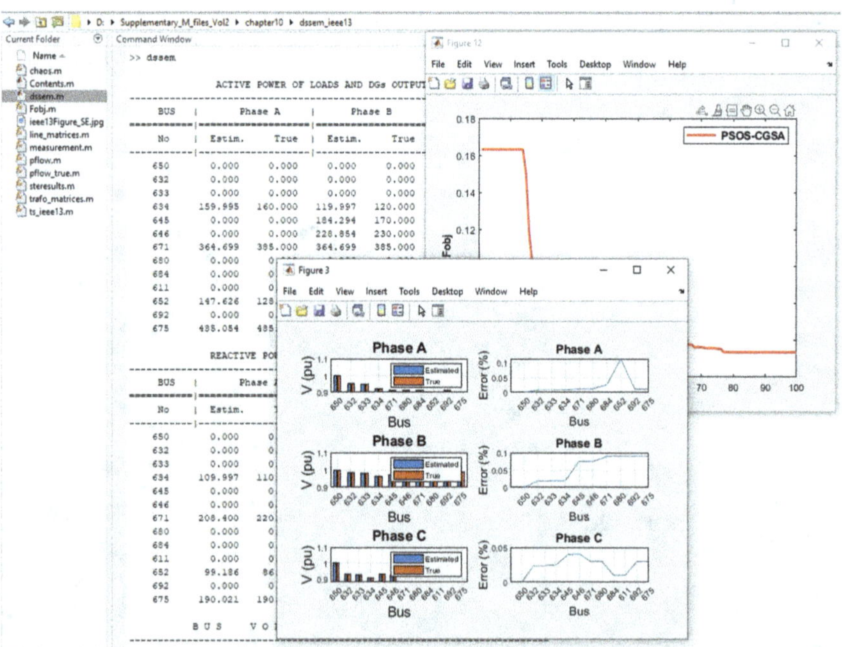


*Figure 10.10 Example of the dssem program execution*

## 10.7 Conclusions

In this chapter, a novel PSOS–CGSA method-based metaheuristic optimization for three-phase DG-integrated DSSE has been presented and verified on the IEEE 13-bus, 37-bus, and 123-bus test systems. The simulation results imply conclusions which can be summarized as follows:

- The proposed model provides reliable and sufficiently accurate state estimation results in DG-integrated DNs with a very limited number of real-time measurements at the source substation.
- Significant improvements in the DSSE results can be achieved with additional real-time measurements in the lateral branches.
- The proposed hybrid PSOS–CGSA algorithm provides more accurate and stable DSSE compared to other algorithms.
- The DSSE model can be easily implemented for any distribution system without limitation in terms of dimension, topology, level of loads, and type of DG units.

## References


[1] Nanchian S., Majumdar A., and Pal B.C. 'Three-phase state estimation using hybrid particle swarm optimization'. *IEEE Transactions on Smart Grid*. 2017;**8**(3):1035–45.

[2] Primadianto A. and Lu C-N. 'A review on distribution system state estimation'. *IEEE Transaction on Power Systems*. 2017;**32**(5):3875–83.

[3] Wang G., Giannakis G.B., Chen J., and Sun J. 'Distribution system state estimation: an overview of recent developments'. *Frontiers of Information Technology & Electronic Engineering*. 2019;**20**(1):4–17.

[4] Singh R., Pal B.C., and Jabr R.A. 'Choice of estimator for distribution system state estimation'. *IET Generation, Transmission & Distribution*. 2009; **3**(7):666–78.

[5] Ranković A., Maksimović B.M., and Sarić A.T. 'A three-phase state estimation in active distribution networks'. *Electrical Power and Energy Systems*. 2014;**54**:154–62.

[6] Oliveira-De Jesus P.M., and Antunes C.H. 'A detailed network model for distribution system with high penetration of renewable generation sources'. *Electric Power System Research*. 2018;**161**:152–66.

[7] Majidi M., Etezedi-Amoli M., Livani H., and Fadali M.S. 'Distribution system state estimation using sparsified voltage profile'. *Electric Power System Research*. 2016;**136**:69–78.

[8] Majidi M., Etezedi-Amoli M., and Livani H. 'Distribution system state estimation using comprehenssive sensing'. *Electrical Power and Energy Systems*. 2017;**88**:175–86.

[9] Ni F., Nguyen P.H., Cobben J.F.G., Van den Brom H.E., and Zhao D. 'Three-phase state estimation in the medium-voltage network with aggregated smart meter data'. *Electrical Power and Energy Systems*. 2018; **98**:463–73.
[10] Othman M.M., Ahmed M.H., and Salama M.M.A. 'A novel smart meter technique for voltage and current estimation in active distribution networks'. *Electrical Power and Energy Systems*. 2019;**104**:301–10.
[11] Mederiros Júnior M.F., Almeida M.A.D., Cruz M.C.S., Monteiro R.V.F., and Oliveira A.B. 'A three-phase algorithm for state estimation in power distribution feeders based on the powers summation load flow method'. *Electric Power System Research*. 2015;**123**:76–84.
[12] Soares T.M., Bezerra U.H., and Tostes M.E.L. 'Full-observable three-phase state estimation algorithm applied to electric distribution grids'. *Energies*. 2019;**12**(7):1327.
[13] Ahmad F., Tariq M., and Farooq A. 'A novel ANN-based distribution network state estimator'. *Electrical Power and Energy Systems*. 2019; **107**:200–12.
[14] Menke J.H., Bornhorst N., and Braun M. 'Distribution system monitoring for smart power grids with distributed generation using artificial neural networks'. *Electrical Power and Energy Systems*. 2019;**113**:472–80.
[15] Cheng G., Song S., Lin Y., Huang Q., Lin X., and Wang F. 'Enhanced state estimation and bad data identification in active power distribution networks using photovoltaic power forecasting'. *Electric Power System Research*. 2019;**177**:105974.
[16] Liu K., Sheng W., and He K. 'Unbalanced three-phase distribution state estimation using cooperative particle swarm optimization'. *2014 IEEE PES T&D Conference and Exposition*, 2014, Chicago, IL, USA.
[17] Khorshidi R., Shabaninia F., and Niknam T. 'A new smart approach for state estimation of distribution grids considering renewable energy sources'. *Energy*. 2016;**94**:29–37.
[18] Ullah Z., Elkadem M.R., Wang S., and Radosavljević J. 'A novel PSOS–CGSA method for state estimation in unbalanced DG-integrated distribution systems'. *IEEE Access*. 2020;**8**:113219–229.
[19] Khushalani S., SolankiJ.M., and Shulz N.N. 'Development of three-phase unbalanced power flow using PV and PQ models for distributed generation and study of the impact of DG models'. *IEEE Transactions on Power Systems*.2007;**22**(3):1019–25.
[20] Kersting W.H. *Distribution System Modeling and Analysis*. Boca Raton, FL: CRC Press; 2002.
[21] Naka S., Genji T., Yura T., and Fukuyama Y. 'A hybrid particle swarm optimization for distribution state estimation'. *IEEE Transactions on Power Systems*. 2003;**18**(1):60–68.
[22] Katic N., Fei L., Svenda G., and Yongji Z. 'Gield testing of distribution state estimator'. *22nd International Conference on Electricity Distribution*. Stockholm, 2013. pp. 1–4.

[23] Ullah Z., Wang S., Radosavljević J., and Lai A.J. 'A solution to the optimal power flow problem considering WT and PV generation'. *IEEE Access*. 2019;**7**:46763–72.

[24] https://site.ieee.org/pes-testfeeders/resources/; IEEE PES AMPS DSAS Test Feeder Working Group (Accessed on 14 November 2019).

[25] Wang H. and Schulz N.N. 'A revised branch current-based distribution system state estimation algorithm and meter placement impact'. *IEEE Transactions on Power Systems*. 2004;**19**(1):207–13.

# Chapter 11
# Steady-state analysis of self-excited induction generators

## 11.1 Introduction

The application of induction generators (IGs) follows the development of distributed generation systems. Advantages of IGs in regard to widely used synchronous generators are a lower unit cost, brushless cage-rotor construction, absence of a separate dc source, good over-speed capability, better transient performances, and inherent overload protection. The inability of a direct regulation of reactive power is a disadvantage of IGs.

An IG can operate in parallel with the distribution network when its rotor speed is greater than the synchronous speed of the air-gap revolving field. Better utilization of renewable energy may be achieved by developing small-scale autonomous power systems like wind and mini hydro power plants, which use self-excited induction generators (SEIGs) [1]. Those small isolated microgrids are identified as a promising concept to energize remote and rural communities [2–4]. However, poor voltage regulation characteristics of IGs require an efficient and reliable control system of voltage and frequency in stand-alone microgrids with SEIGs [5,6]. It is well known that if an appropriate capacitor bank is connected across an externally driven induction machine, an electromotive force (EMF) is induced in the machine windings due to the excitation provided by the capacitor. The induced voltage and current would continue to rise until the reactive power supplied by the capacitor is balanced by the reactive power demanded by the machine. This results in an equilibrium state being reached, and the machine now operates as a SEIG at a voltage and frequency assigned by the value of the capacitance, speed of the prime mover, parameters of the machine, and the load [7,8].

The key issue that needs to be solved at the SEIG is the voltage and frequency regulation. The voltage and frequency variables are non-linear dependences of the speed, magnetizing reactance, excitation capacitance, and load. The terminal voltage of SEIG can be controlled by continuously adjusting the reactive power supply to meet the demand of the generator and loads. This is achieved by using simple switched capacitors, switched reactors, or power electronics such as static volt-ampere reactive (VAR) compensators (SVC). The generated frequency is highly dependent on the speed of the rotor, so that unless there is a fixed-speed prime mover, the microgrid with SEIGs will see a frequency that changes with the prime mover and drops down when the load increases. Consequently, generator frequency can be controlled by variations in consumer loads or by using an electronic load controller at the generator terminals in parallel to consumer loads. The operating principle behind the electronic load controller is that they consume the difference between the power output of the SEIG and the power absorbed by the connected consumer loads [2,3].

Any regulation strategy for SEIGs requires an efficient procedure for computing the frequency and magnetizing reactance, that is, the frequency and EMF for a given rotor speed, loads, and machine parameters.

A huge number of journal publications have so far been dedicated to the steady-state analysis of an autonomous three-phase SEIG with a three-phase balanced static load [8–16]. Analysis of the SEIG feeding power to a dynamic load (induction motor) is presented in [7,17,18]. The performance of a SEIG is usually determined through its equivalent circuits. An overwhelming majority of the aforementioned researchers used either the loop impedance approach [7,8,18] or the nodal admittance approach [9–14] in analyzing the circuits. Irrespective of the representation manner, the analysis of three-phase SEIG requires a computation of the generated frequency ($F$) and magnetizing reactance ($X_m$) for the given operating conditions. Knowing the values of $X_m$ and $F$, the steady-state performances of the SEIG can easily be determined through their equivalent circuits in conjunction with appropriate magnetization characteristics.

Unbalanced operation of a three-phase SEIG may be caused by manufacturing tolerances of excitation capacitances, failure of some excitation capacitance modules, or disconnection of loads by consumers [19]. Moreover, autonomous microgrids with three-phase SEIGs often employ single-phase distribution schemes due to low cost, easy maintenance, and simplicity in terms of protection [20–22]. The steady-state analysis of three-phase SEIG with unbalanced loads and excitation capacitances is well documented [19–25]. By means of the method of symmetrical components, the complex three-phase generator-load system has been reduced to a simple equivalent passive circuit. In order to determine the excitation frequency and magnetizing reactance, the author used different methods such as the pattern search minimization technique of Hooke and Jeeves [19] and the genetic algorithm [25]. Valid simplifying assumptions in the sequence equivalent circuit, which are applied in [20], make the final equation convenient in order to obtain the solution for $X_m$ and $F$.

Usually, the stand-alone renewable energy sources with SEIGs do not go much beyond 15 kW [3]. In order to utilize the full potential of the hydro and wind energy source, a number of SEIGs are required to be operated in parallel. For the parallel operation of SEIGs, there was a simplified model described in [26] and steady-state analyses in [27–31]. These papers are focused on three-phase SEIGs with balanced excitation capacitances and loads. However, in practice, the conditions for symmetrical and balanced operation of three-phase SEIGs are often not achieved. The steady-state analysis and performance characteristics of parallel-operated three-phase SEIGs with unbalanced load conditions are considered in [32].

This chapter aims to show how a metaheuristic optimization technique such as GA can be applied in solving a non-optimization problem, i.e. in determining the steady-state performances of a SEIG for general operating conditions, as proposed in [25,32]. The GA-based optimization procedure is applied to the analysis of SEIG feeding balanced/unbalanced and static/dynamic loads. The GA has been applied to the computation of the unknowns by minimizing the total impedance module of the equivalent passive circuit.

## 11.2 System configuration

A modified example of the SEIG-based small autonomous microgrid system presented in [2] is shown in Figure 11.1. The SEIG is coupled to the small-hydro turbine and feeds both balanced three-phase and unbalanced single-phase consumer loads. As the SEIG is driven by an unregulated small-hydro turbine, the mechanical power input of SEIG is constant; therefore, to control system frequency, the balance between the mechanical power input and the electrical power output of SEIG has to be maintained by using an electronic load controller. Therefore, all the parameters of the SEIG, such as voltages, windings currents, capacitor currents, power output, and frequency, remain at the same operating point in a steady state irrespective of loading condition, which is known as "single point operation" [2].

As mentioned before, the generated voltage of the SEIG depends on the prime mover speed, connected terminal capacitance, and load. Initially, excitation capacitors are used to excite the induction generator to build up the rated voltage at no load condition. But as power factor and power of load fluctuate, ideal compensation will require variable reactive compensators such as SVC, static compensators (STATCOM), or dynamic voltage restorer DVR. The controlled reactive power is responsible for keeping the terminal voltage constant with a change in load with a VAR compensator connected in parallel with a fixed capacitor and load. The main function of such controllers would be to provide the required variable capacitive reactive power to maintain a constant voltage across the load at varying prime mover speed and power of loads. Therefore, an effective capacitive reactive power controller has become central to the success of the SEIG system for stand-alone applications [33].

The required reactive power is different for different rating machines for maintaining constant terminal voltage, and it depends on the machine parameters and magnetizing characteristics. The capacitive reactive power required for

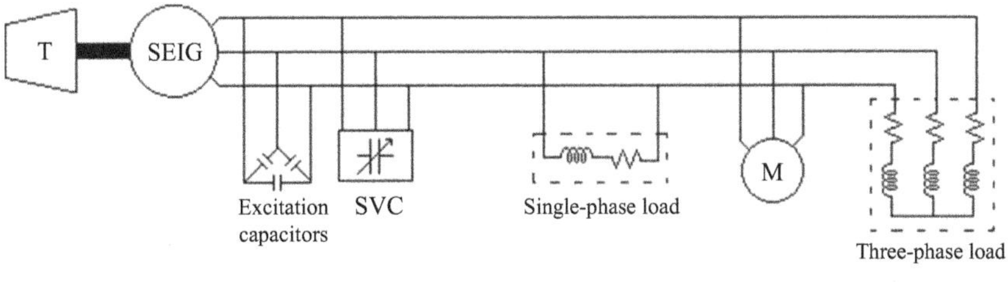


*Figure 11.1 SEIG-based microgrid*

excitation and voltage control of an SEIG at different load conditions is calculated based on appropriate steady-state equations of the SEIG.

## 11.3 Induction generator model

If three-phase induction machines feed an unbalanced/asymmetrical external network, then a natural analysis mode is to resort the system to symmetrical components and rotating field concepts [1,20]. Therefore, the induction generator is modeled with its positive and negative sequence circuits, as shown in Figure 11.2.

In these circuits, all parameters refer to the rated frequency, and it is assumed that all inductive reactance are proportional to the rated frequency, where $F$ is the per-unit frequency – the ratio of the generated frequency to the rated frequency, and $v$ is the ratio of the actual rotor speed to the synchronous speed corresponding to the rated frequency. In these circuit models, it is also assumed that all parameters, except the magnetizing reactance $X_m$, are constant and independent of the saturation. The value of $X_m$ depends on the core flux, which in turn depends on the ratio of the air-gap voltage to the frequency. The variation of the positive-sequence air-gap voltage with magnetizing reactance $X_m$ can be expressed by an appropriate equation based on the synchronous speed test data fitting.

According to Figure 11.2, the positive-sequence and negative-sequence impedances of the generator, that is, impedances $\underline{Z}_p$ and $\underline{Z}_n$, are as follows:

$$\underline{Z}_p = \frac{R_s}{F} + jX_s + \frac{\frac{jR_c \cdot X_m}{R_c + jFX_m}\left(\frac{R_r}{F-v} + jX_r\right)}{\frac{jR_c \cdot X_m}{R_c + jFX_m} + \left(\frac{R_r}{F-v} + jX_r\right)} \tag{11.1}$$

$$\underline{Z}_n = \frac{R_s}{F} + jX_s + \frac{\frac{jR_c \cdot X_m}{R_c + jFX_m}\left(\frac{R_r}{2F} + jX_r\right)}{\frac{jR_c \cdot X_m}{R_c + jFX_m} + \left(\frac{R_r}{2F} + jX_r\right)} \tag{11.2}$$

where $F$ is the per-unit frequency, $v$ is the per-unit speed, $R_r$ is the stator resistance per phase in p.u., $R_r$ is the rotor resistance per phase referred to the stator in p.u., $R_c$ is the core loss resistance per phase in p.u., $X_s$ is the stator reactance per phase in p.u.,

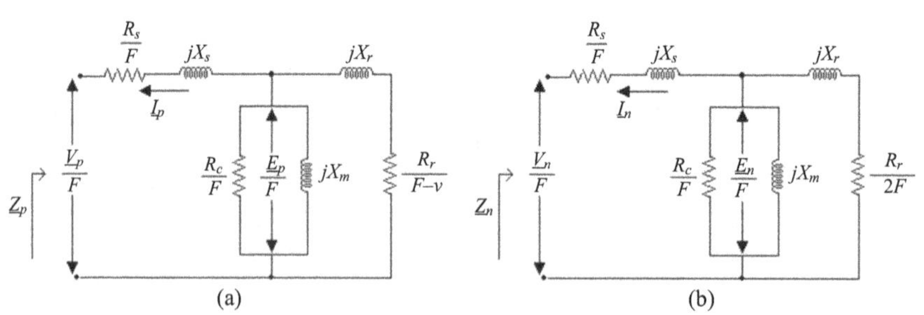


*Figure 11.2 Sequence equivalent circuits: (a) positive sequence and (b) negative sequence*

$X_r$ is the rotor reactance per phase referred to the stator in p.u., and $X_m$ is the magnetizing reactance per phase at the rated frequency in p.u.

## 11.4 Steady-state equations of SEIG

The circuit connection of a three-phase delta-connected SEIG feeding a three-phase unbalanced load is shown in Figure 11.3. In order to produce the reactive power necessary for initiating self-excitation, at least one of the terminal impedances $\underline{Z}_A$, $\underline{Z}_B$, or $\underline{Z}_C$ must contain a capacitive element. Assigning appropriate values for the terminal impedances, a specific unbalanced operating condition can be simulated.

The delta connection is deliberately selected herein due to the fact that zero-sequence quantities do not exist. However, the presented concept can easily be extended to the star connection with the isolated neutral point by applying the star-delta transformation.

The circuit shown in Figure 11.3 has no active voltage sources. Accordingly, the SEIG may be regarded as a passive circuit when viewed across any two stator terminals [19]. For convenience, A-phase is chosen as the reference, and the input impedance of the SEIG across terminals 1 and 3 in Figure 11.3 will be considered. In this manner, the equivalent scheme shown in Figure 11.3 can be transformed into a simple circuit given in Figure 11.4.

According to the marks given in Figure 11.3, the current and voltage balance equations may be written as follows:

$$\underline{I}_A + \underline{I}_{\ell A} - \underline{I}_B - \underline{I}_{\ell B} = 0 \tag{11.3}$$

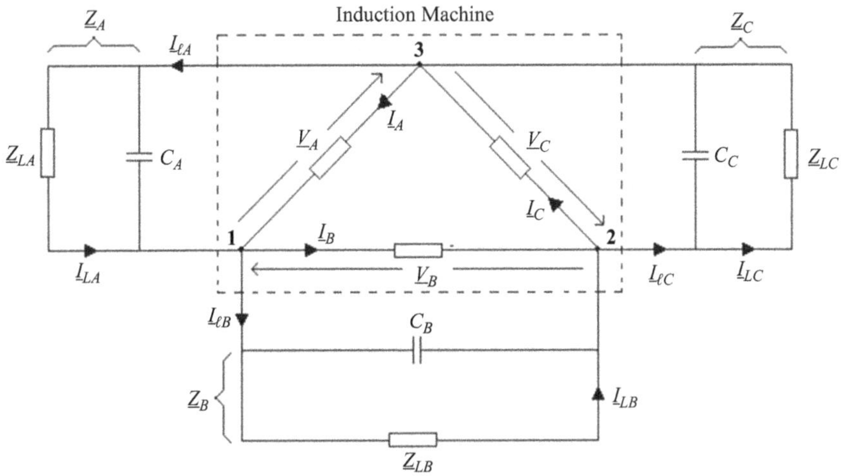


*Figure 11.3 Three-phase SEIG feeding a three-phase unbalanced load*

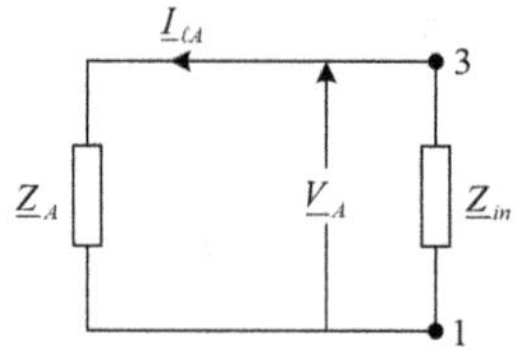


*Figure 11.4 Simplified circuit for the three-phase SEIG feeding a three-phase unbalanced load*

$$\underline{I}_A + \underline{I}_{\ell A} - \underline{I}_C - \underline{I}_{\ell C} = 0 \tag{11.4}$$

$$\underline{I}_{\ell A} = \frac{\underline{V}_A}{\underline{Z}_A} = \underline{Y}_A \underline{V}_A \tag{11.5}$$

$$\underline{I}_{\ell B} = \frac{\underline{V}_B}{\underline{Z}_B} = \underline{Y}_B \underline{V}_B \tag{11.6}$$

$$\underline{I}_{\ell C} = \frac{\underline{V}_C}{\underline{Z}_C} = \underline{Y}_C \underline{V}_C \tag{11.7}$$

Based on the symmetrical component theory, the following equations are developed for the delta-connected system:

$$\underline{V}_A + \underline{V}_B + \underline{V}_C = 0 \tag{11.8}$$

$$\underline{V}_A = \underline{V}_p + \underline{V}_n \tag{11.9}$$

$$\underline{V}_B = \underline{a}^2 \underline{V}_p + \underline{a}\, \underline{V}_n \tag{11.10}$$

$$\underline{V}_C = \underline{a}\, \underline{V}_p + \underline{a}^2 \underline{V}_n \tag{11.11}$$

$$\underline{I}_A = \underline{V}_p \underline{Y}_p + \underline{V}_n \underline{Y}_n \tag{11.12}$$

$$\underline{I}_B = \underline{a}^2 \underline{V}_p \underline{Y}_p + \underline{a}\, \underline{V}_n \underline{Y}_n \tag{11.13}$$

$$\underline{I}_C = \underline{a}\, \underline{V}_p \underline{Y}_p + \underline{a}^2 \underline{V}_n \underline{Y}_n \tag{11.14}$$

where $\underline{a}$ is the complex operator $e^{(j2\pi/3)}$.

This follows from (11.3) and (11.4)

$$\underline{I}_{\ell A} = -\underline{I}_A + \underline{I}_B + \underline{I}_{\ell B} \tag{11.15}$$

$$\underline{I}_{\ell A} = -\underline{I}_A + \underline{I}_C + \underline{I}_{\ell C} \tag{11.16}$$

Substituting (11.6) and (11.7) and (11.12) and (11.14) into (11.15) and (11.16), it is obtained that

$$\underline{I}_{\ell A} = \underline{A} \cdot \underline{V}_p + \underline{B} \cdot \underline{V}_n \tag{11.17}$$

$$\underline{I}_{\ell A} = \underline{C} \cdot \underline{V}_p + \underline{D} \cdot \underline{V}_n \tag{11.18}$$

where

$$\underline{A} = \left(\underline{a}^2 - 1\right)\underline{Y}_p + \underline{a}^2\underline{Y}_B$$
$$\underline{B} = (\underline{a} - 1)\underline{Y}_n + \underline{a}\,\underline{Y}_B$$
$$\underline{C} = (\underline{a} - 1)\underline{Y}_p + \underline{a}\,\underline{Y}_C$$
$$\underline{D} = \left(\underline{a}^2 - 1\right)\underline{Y}_n + \underline{a}^2\underline{Y}_C$$

From (11.17) and (11.18), it is obtained that

$$\underline{V}_n = \frac{A - C}{D - B}\underline{V}_p \tag{11.19}$$

According to Figure 11.4, the phase voltage $\underline{V}_A$ is

$$\underline{V}_A = -\underline{Z}_{in} \cdot \underline{I}_{\ell A} \tag{11.20}$$

Finally, by substituting (11.9), (11.17), and (11.19) into (11.20) for the SEIG input impedance, it is obtained that

$$\underline{Z}_{in} = \frac{(A + D) - (B + C)}{BC - AD} \tag{11.21}$$

The positive-sequence and negative-sequence admittances of the SEIG, that is, the admittances $\underline{Y}_p$ and $\underline{Y}_n$, are functions of $F$ and $X_m$. Therefore, the SEIG input impedance may be expressed as

$$\underline{Z}_{in} = R_{in}(X_m, F) + jX_{in}(X_m, F) \tag{11.22}$$

In accordance with Figure 11.4, the following voltage balance equation can be written as

$$(\underline{Z}_A + \underline{Z}_{in})\underline{I}_{\ell A} = 0 \tag{11.23}$$

$$\underline{Z}_{loop}\underline{I}_{\ell A} = 0 \tag{11.24}$$

where $\underline{Z}_{loop} = \underline{Z}_A + \underline{Z}_{in}$.

For a successful self-excitation, it is required that $\underline{I}_{\ell A} \neq 0$, hence,

$$\underline{Z}_{loop} = 0 \tag{11.25}$$

## 11.5 Steady-state equations of parallel-operated SEIGs

Microgrids may consist of either a wind power plant or a hydropower plant or a combination of both, constituting the generating unit [33]. Figure 11.5 shows such a microgrid that consists of several parallel operating SEIG with its prime mover and reactive power compensators, which are connected to supply the same load.

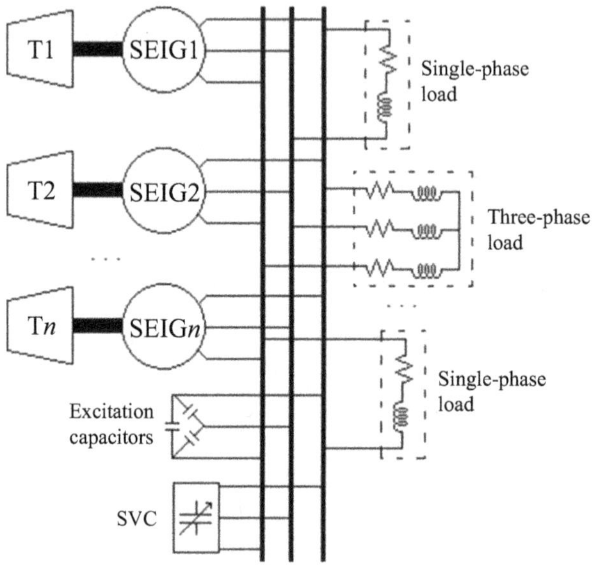


*Figure 11.5 Microgrid with SEIGs connected in parallel*

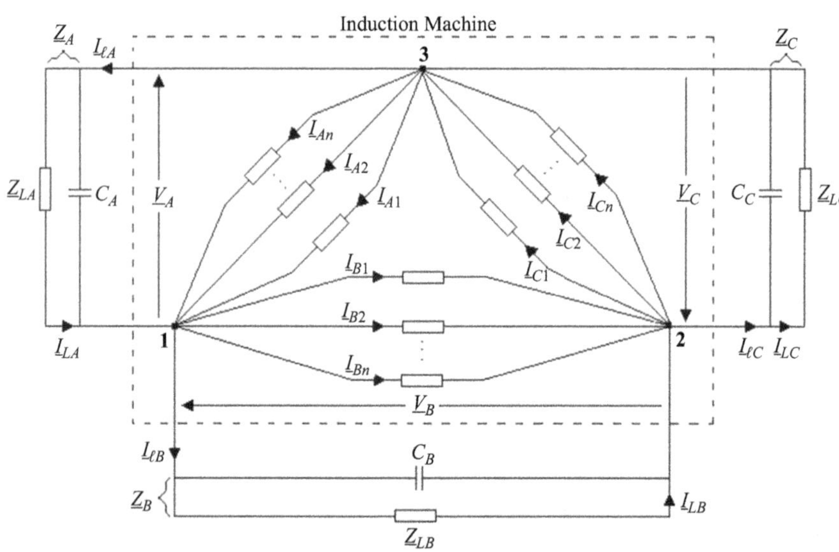


*Figure 11.6 Parallel SEIGs feeding unbalanced load*

The circuit connection of *n* three-phase delta-connected SEIGs feeding a three-phase unbalanced load is shown in Figure 11.6.

Using an analogy with a single SEIG considered in Section 11.4, the equivalent scheme shown in Figure 11.6 can be transformed into a simple circuit

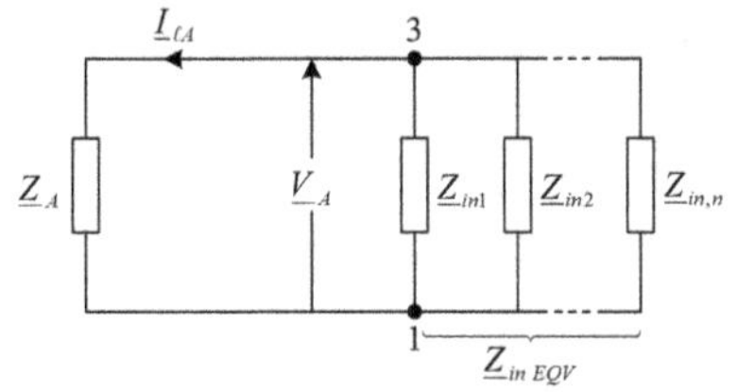


*Figure 11.7 Simplified circuit of parallel SEIGs*

as given in Figure 11.7. For the parallel SEIGs, the input impedance is obtained as follows:

$$\underline{Z}_{inEQV} = \frac{(\underline{A} + \underline{D}) - (\underline{B} + \underline{C})}{\underline{B} \cdot \underline{C} - \underline{A} \cdot \underline{D}} \tag{11.26}$$

where

$$\underline{A} = (\underline{a}^2 - 1)\sum_{i=1}^{N} \underline{Y}_{pi} + \underline{a}^2 \underline{Y}_B \tag{11.27}$$

$$\underline{B} = (\underline{a} - 1)\sum_{i=1}^{N} \underline{Y}_{ni} + \underline{a}\,\underline{Y}_B \tag{11.28}$$

$$\underline{C} = (\underline{a} - 1)\sum_{i=1}^{N} \underline{Y}_{pi} + \underline{a}\,\underline{Y}_C \tag{11.29}$$

$$\underline{D} = (\underline{a}^2 - 1)\sum_{i=1}^{N} \underline{Y}_{ni} + \underline{a}^2 \underline{Y}_C \tag{11.30}$$

where $\underline{Y}_{pi}$ and $\underline{Y}_{ni}$ are the positive-sequence and negative-sequence admittance of the $i$th induction generator obtained based on (11.1) and (11.2).

The positive-sequence and negative-sequence admittances $\underline{Y}_{pi}$ and $\underline{Y}_{ni}$, that is, the complex coefficients $\underline{A}$, $\underline{B}$, $\underline{C}$, and $\underline{D}$ are functions of $F$ and $X_{mi}$. Therefore, the SEIGs input impedance may be expressed as

$$\underline{Z}_{inEQV} = R_{inEQV}(X_{m1}, X_{m2}, \ldots, X_{mn}, F) + jX_{inEQV}(X_{m1}, X_{m2}, \ldots, X_{mn}, F) \tag{11.31}$$

In accordance with Figure 11.7, the following voltage balance equation can be written as

$$\left(\underline{Z}_A + \underline{Z}_{inEQV}\right)\underline{I}_{\ell A} = 0 \tag{11.32}$$

$$\underline{Z}_{loop}\underline{I}_{\ell A} = 0 \tag{11.33}$$

where $\underline{Z}_{loop} = \underline{Z}_A + \underline{Z}_{inEQV}$. For a successful self-excitation, it is required that $\underline{I}_{\ell A} \neq 0$, hence,

$$\underline{Z}_{loop} = 0 \tag{11.34}$$

In order to ensure that all generators have the same terminal voltage (the common bus voltage), the following equations must be satisfied:

$$\left|\underline{Z}_{p1}\right| \cdot \left|\underline{I}_{p1}\right| - \left|\underline{Z}_{pi}\right| \cdot \left|\underline{I}_{pi}\right| = 0 \ (i = 2, .., n) \tag{11.35}$$

According to Figure 11.2(a), the magnitude of the positive sequence stator current for machine $i$ can be related to the machine magnetizing reactance $X_{mi}$, provided the relationship between the forward field air-gap voltage $\left|\frac{\underline{E}_{pi}}{F}\right|$ and magnetizing reactance $X_{mi}$ is known. The magnetizing characteristic relating $\left|\frac{\underline{E}_{pi}}{F}\right|$ with $X_{mi}$ can be obtained experimentally by a synchronous speed test. Generally, the variation of $\left|\frac{\underline{E}_{pi}}{F}\right|$ with $X_{mi}$ over the practical region of operation can be approximated by linear segments with expressions of the type:

$$\left|\frac{\underline{E}_{pi}}{F}\right| = a_i + b_i \cdot X_{mi} \ (i = 1, .., n) \tag{11.36}$$

Therefore, (11.35) can be expressed as

$$\left|\frac{\underline{Z}_{p1}}{\underline{Z}_{pmr1}}\right|(a_1 + b_1 X_{m1}) - \left|\frac{\underline{Z}_{pi}}{\underline{Z}_{pmri}}\right|(a_i + b_i X_{mi}) = 0 \ (i = 2, .., n) \tag{11.37}$$

where $\underline{Z}_{pmri}$ is the total positive-sequence impedance of the magnetizing and rotor circuit of the $i$th machine. In accordance with Figure 11.2(a), it is

$$\underline{Z}_{pmri} = \frac{jX_{mi} \cdot \left(\frac{R_{ri}}{F - v_i} + jX_{ri}\right)}{jX_{mi} + \left(\frac{R_{ri}}{F - v_i} + jX_{ri}\right)} \tag{11.38}$$

## 11.6 Solution method

Equation (11.25), i.e. (11.34), should be separated into real and imaginary parts in order to solve it explicitly. The usual practice is to derive the complex coefficients of the non-linear equations manually and solve them by using the numerical Newton–Raphson or polynomial method. The necessary mathematical manipulations and the lengthy derivations of non-linear equations required for the unknown variables in the methods are tedious, time-consuming, and liable to human error.

They require a tremendous human effort for accurate programming and debugging. The model lacks flexibility as coefficients are valid only for a given circuit configuration. An inclusion of the core loss resistance or load inductance will increase the order of the equations. These disadvantages can be overcome by using the optimization techniques [12–14] and [19,25,32]. In this way, the problem of solving a non-linear complex algebraic equation is reduced to an optimization problem with a predefined minimum. The unknown variables in this equation are regarded as control variables of the optimization problem. To transform the problem of solving a non-linear equation into an optimization problem, the objective function and the control variables should be defined.

### *11.6.1 Overview of genetic algorithm*

The genetic algorithm is one of the techniques, that is, one optimization procedure based on the natural evolution process imitation [34]. They belong to the methods of a directed random search of the solution domain with the aim of finding a global optimum. The classical optimization methods start from a single possible initial solution, and they reach the optimum by applying the heuristic rules iteratively. The GA starts from the population, which is a group of individuals. Each individual represents a potential solution to the optimization problem. The individuals are presented in the same way, usually through a column or a string of data. The quality of each solution or individual is determined based on the fitness function values. Through a series of GA operations, a new population is obtained and its individuals are engendered by the individuals from the previous population according to the natural evolution principles: the choice of parents, crossover, and mutation. The basic operations of the GA are

1. *Representation of individuals*: All data (variables) that make an individual are written in a string. A string is composed of substrings. Each substring represents a binary encoded variable on which the process of optimization is carried out. The number of substrings, therefore, the size of a string, depends on the number of variables that are optimized.
2. *Initialization*: Individuals with random strings are generated that set up the initial population.
3. *Fitness function calculation*: It is used to rate the quality of an individual, and it represents an equivalent of the function that should be optimized, that is, the objective function.
4. *Selection:* During the selection process, the individuals who will participate in the reproduction (parents) are selected. The point of the selection is to store and transfer good individuals to the next generation.
5. *Crossover*: Crossover is an exchange process of column parts between two individuals, that is, "parents." One or two new individuals engendered by the crossover, that is, a "child." The possibility of inheriting the first parent's characteristics by a child is introduced during this process.
6. *Mutation*: Mutation is a way to give a new piece of information to an individual. Mutation represents an accidental bit variation of an individual, generally

with a constant probability for each bit within a population. The mutation probability can further vary depending on the size of the population, application and preferences of the explorer. A fixed value, which is often kept during the whole genetic algorithm, is used for each generation.

7. *Ending conditions*: The process of finding the optimal solution is an iterative process that ends when a maximum number of generations is achieved or when another criterion is fulfilled, such as a minimum offset from the best fitness value and the medium fitness value of all individuals in a current population. If end conditions are fulfilled, the best individual thus obtained is the semi-optimal solution in question. Otherwise, return to 3.

Much more details about GA and its operators are given in Volume 1 (Chapter 2) of the book

### *11.6.2 Application of GA to SEIG*

The self-excitation requirement given by (11.25) can be expressed as follows:

$$\left|\underline{Z}_{loop}\right| = 0 \tag{11.39}$$

This equation can be considered as an objective function of the optimization problem which control variables are unknowns variables of the equation. Therefore, to solve (11.35), the optimization problem is stated as follows:

$$\min\left|\underline{Z}_{loop}(X_m, F)\right| = \min\sqrt{\left(\mathrm{Re}(\underline{Z}_{loop})\right)^2 + \left(\mathrm{Im}(\underline{Z}_{loop})\right)^2} \tag{11.40}$$

where unknowns $X_m$ and $F$ are considered as control variables of the optimization problem.

The feasibility region of the problem control variables is defined as

$$X_m^{\min} \leq X_m \leq X_m^{\max};\ F^{\min} \leq F \leq F^{\max} \tag{11.41}$$

The terminal voltage at SEIG is considered a functional operating constraint in this optimization problem.

Depending on the specific problem, in addition to the $X_m$ and $F$, the unknowns, i.e. control variables, could be the excitation capacitance, rotor speed, and load impedance as well. For example, the voltage control of the SEIG can be achieved by a change in the excitation capacitance for a constant value of the speed and load impedance. In this case, the optimization problem can be expressed as follows:

$$\min\left\{\left|\underline{Z}_{loop}(X_m, F, C)\right| + \left|V - V_{spec}\right|\right\} \tag{11.42}$$

with the constraints

$$X_m^{\min} \leq X_m \leq X_m^{\max};\ F^{\min} \leq F \leq F^{\max};\ C^{\min} \leq C \leq C^{\max} \tag{11.43}$$

where $V$ is the terminal voltage and $V_{spec}$ is the specified terminal voltage (usually $V_{spec} = 1$ p.u.).

The flowchart describing the MATLAB® realization of the GA optimization method implemented to solve the above optimization problem is shown in Figure 11.8. The GA enables a simultaneous computation of the $X_m$ and $F$ values for any given circuit configuration and operating conditions such as the speed, excitation capacitances, and load impedances.

The evaluation of the objective (fitness) function is performed by the MATLAB program based on the mathematical model presented in Section 11.4.

Having calculated $X_m$ and $F$, the next step is to determine the normalized air-gap voltage $\frac{E_p}{F}$ for the value of $X_m$ based on the relevant magnetizing characteristic

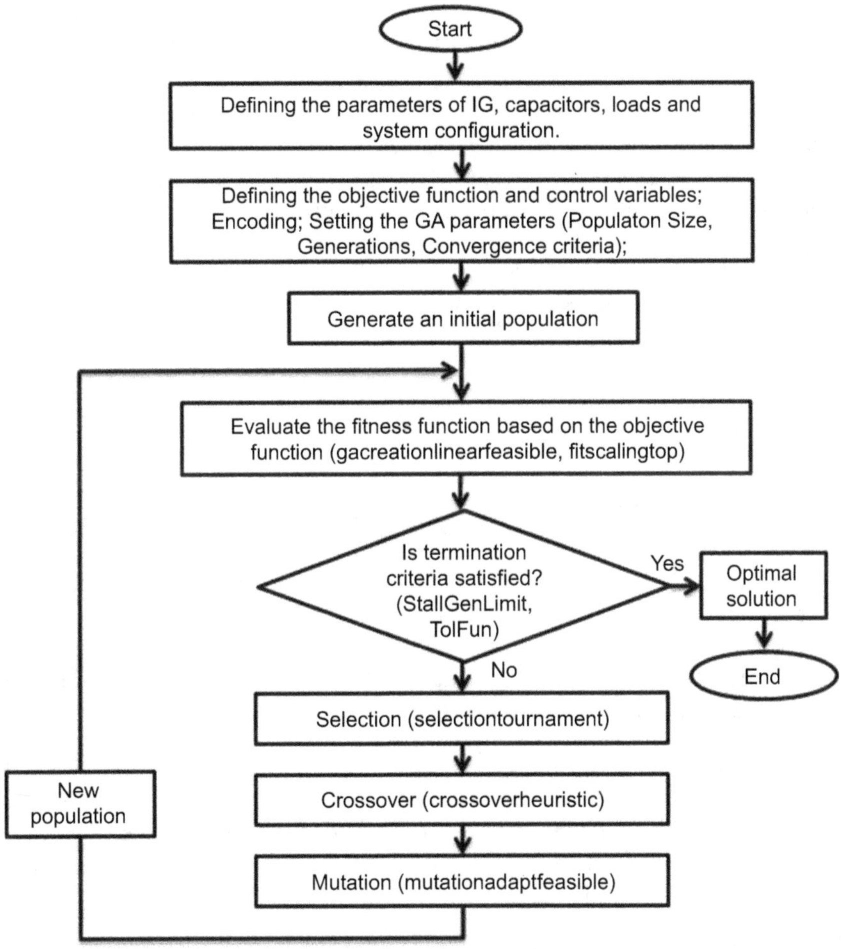


*Figure 11.8 Flowchart of GA*

of an induction machine. The magnetizing characteristics relating $\frac{E_p}{F}$ to $X_m$ can be obtained experimentally by a synchronous speed test. Knowing the air-gap voltage $E_p$, the positive-sequence current and voltage can be computed in accordance with Figure 11.2(a) as follows:

$$\underline{I}_p = -\frac{E_p}{\frac{FR_r}{F-v} + jFX_r} - \frac{E_p}{\frac{jFR_m X_m}{R_m + jFX_m}} \tag{11.44}$$

$$\underline{V}_p = E_p F - (R_s + jFX_s)\underline{I}_p \tag{11.45}$$

The negative-sequence voltage $\underline{V}_n$ can be determined by (11.19). Finally, the terminal voltages and stator currents of the SEIG can be calculated by (11.9)–(11.14) and the line currents by (11.5)–(11.7).

### *11.6.3 Application of GA to parallel operated SEIGs*

If the self-excitation requirement (11.25) and voltage-balance equation (11.32) are treated as criterion functions and the frequency, magnetizing reactance, excitation capacitance, and speed as control variables within the specified limits, then the optimization procedure can be applied for the determination of the steady-state condition of three-phase SEIGs operating in parallel.

#### 11.6.3.1 Objective function

Equation (11.34) can be expressed as

$$f_1 = \left|\underline{Z}_{loop}\right| = 0 \tag{11.46}$$

The system of equations (11.37) is equivalent to the following expression:

$$f_2 = \sum_{i=2}^{N} \left\{ \left|\frac{\underline{Z}_{p1}}{\underline{Z}_{pmr1}}\right| (a_1 + b_1 X_{m1}) - \left|\frac{\underline{Z}_{pi}}{\underline{Z}_{pmri}}\right| (a_i + b_i X_{mi}) \right\} = 0 \tag{11.47}$$

The control of the common bus voltage can be achieved by changing the excitation capacitances as well as changing the rotor speeds of SEIGs. By controlling the voltage level, one attains the voltage on the common bus of the SEIGs, that is, on the load, which has a value equal to a previously specified one. This condition can be expressed as follows:

$$f_3 = \left|V_{spec} - V_t\right| = 0 \tag{11.48}$$

where $V_{spec}$ is the specified voltage and $V_t$ is the common bus voltage ($V_t = \left|\underline{V}_A\right|$).

Therefore, the objective function in the case of parallel-operated SEIGs can be expressed in the following form:

$$\min Fobj = \min\{f_1 + f_2 + f_3\} \tag{11.49}$$

#### 11.6.3.2 Control variables

Depending on the specific problem, the unknown variables, i.e. control variables of the optimization problem, can be magnetizing reactances ($X_{mi}$), frequency ($F$), excitation capacitance ($C$), and rotor speeds ($v_i$). Accordingly, the inequality constraints which define the solution space of the optimization problem may be expressed as

$$X_{mi}^{\min} \leq X_{mi} \leq X_{mi}^{\max} \; i = 1, 2, \ldots, n \tag{11.50}$$

$$F^{\min} \leq F \leq F^{\max} \tag{11.51}$$

$$C^{\min} \leq C \leq C^{\max} \tag{11.52}$$

$$v_i^{\min} \leq v_i \leq v_i^{\max} \; i = 1, 2, \ldots, n \tag{11.53}$$

General steps for solving the optimization problem using the GA are described in Figure 11.8.

#### 11.6.3.3 Performance equations

The GA enables a simultaneous computation of the unknown values of $X_{mi}$ and $F$. Having thus determined the $X_{mi}$ and $F$, the next step is to determine the normalized air-gap voltage $\left|\frac{\underline{E}_{pi}}{F}\right|$ according to the value of $X_{mi}$ based on the relevant magnetizing characteristic for all induction machines. Because the phase angles of the normalized air-gap voltages are not known, it can be supposed that the angle of the SEIG with index 1 is equal to zero. Knowing the air-gap voltage $\underline{E}_{p1} = E_{p1}\angle 0^\circ$, in accordance with Figure 11.2(a), the positive-sequence current of the SEIG can be computed as

$$\underline{I}_{p1} = -\frac{\underline{E}_{p1}}{\underline{Z}_{pmr1}} \tag{11.54}$$

Now, the positive-sequence voltage is

$$\underline{V}_p = \underline{E}_{p1} - (R_{s1} + jFX_{s1})\underline{I}_{p1} \tag{11.55}$$

When $\underline{V}_p$ is calculated, the positive-sequence currents of the other SEIGs can be determined,

$$\underline{I}_{pi} = -\frac{\underline{V}_p}{F\underline{Z}_{pi}} \; (i = 2, .., n) \tag{11.56}$$

The phase angles of the normalized air-gap voltages of the other SEIGs:

$$\theta_i = \arctan\frac{\mathrm{Im}\left\{\frac{\underline{V}_p}{F} + \left(\frac{R_{si}}{F} + jX_{si}\right)\underline{I}_{pi}\right\}}{\mathrm{Re}\left\{\frac{\underline{V}_p}{F} + \left(\frac{R_{si}}{F} + jX_{si}\right)\underline{I}_{pi}\right\}} \; (i = 2, .., n) \tag{11.57}$$

The negative-sequence voltage $\underline{V}_n$ can be determined by (11.19), and the negative-sequence currents in accordance with Figure 11.2(b) by equation

$$\underline{I}_{ni} = -\frac{\underline{V}_n}{F\underline{Z}_{ni}}\ (i = 1, .., n) \tag{11.58}$$

Accordingly with Figure 11.2(b),

$$\underline{E}_{ni} = \frac{jX_{mi}\cdot\left(\frac{R_{ri}}{F+v_i}+jX_{ri}\right)}{jX_{mi}+\left(\frac{R_{ri}}{F+v_i}+jX_{ri}\right)}\cdot\underline{I}_{ni}\ (i = 1, .., n) \tag{11.59}$$

Finally, the terminal voltages and stator currents of the SEIGs can be calculated by (11.9) and (11.14), and the line currents by (11.5) and (11.7).

The output power ($P_{Geni}$) and the input power ($P_{Inpi}$) of the $i$th SEIG are

$$P_{Geni} = \mathrm{Re}\left\{\underline{V}_p\cdot\underline{I}^*_{pi}+\underline{V}_n\cdot\underline{I}^*_{ni}\right\}\ (i = 1, .., n) \tag{11.60}$$

$$P_{Inpi} = \mathrm{Re}\left\{\frac{E^2_{pi}}{\frac{R_{ri}}{F-v_i}-jX_{ri}}+\frac{E^2_{ni}}{\frac{R_{ri}}{F+v_i}-jX_{ri}}\right\}\ (i = 1, .., n) \tag{11.61}$$

and reactive power requirement

$$Q_i = \mathrm{Im}\left\{\underline{V}_p\cdot\underline{I}^*_{pi}+\underline{V}_n\cdot\underline{I}^*_{ni}\right\}\ (i = 1, .., n) \tag{11.62}$$

## 11.7 Simulation results

The proposed procedures described in the previous sections were implemented in the MATLAB program and applied to several test cases. These programs (*seigm* and *seigparm*) are available as supplementary files to this chapter in the IET Digital Library.

### *11.7.1 Steady-state analysis of SIEG*

There is a three-phase, four pole, 50 Hz, delta connected, squirrel cage induction machine of 2.2 kW, 230 V, and 8.6 A.

Parameters of the induction machine:

$$R_s = 0,07224(\text{p.u}),\ \ R_r = 0,03795(\text{p.u.}),\ \ X_s = X_r = 0,10459(\text{p.u.})$$

Variation of air gap voltage with magnetizing reactance at rated frequency:

$$\begin{aligned} X_m < 1.77464 &\rightarrow \frac{E_d}{F} = 1.49744 - 0.007X_m \\ 2.06096 > X_m \geq 1.77464 &\rightarrow \frac{E_d}{F} = 2.02226 - 0.01338X_m \\ 2.32904 > X_m \geq 2.06096 &\rightarrow \frac{E_d}{F} = 2.52129 - 0.0186X_m \\ X_m \geq 2.32904 &\rightarrow \frac{E_d}{F} = 0 \end{aligned} \tag{11.63}$$

*Table 11.1 Verification of the proposed procedure for the balanced resistive loads*

| Case | C, R (μF, Ω) | Rotor speed v (p.u.) | Reference [14] | | | | Proposed method results | |
|---|---|---|---|---|---|---|---|---|
| | | | Computed results | | Experimental | | | |
| | | | F (p.u.) | V (p.u.) | F (p.u.) | V (p.u.) | F (p.u.) | V (p.u.) |
| 1 | 36,160 | 0.95333 | 0.934899 | 0.583202 | 0.943800 | 0.582608 | 0.941519 | 0.573805 |
| 2 | 36,160 | 1.04666 | 1.033292 | 0.942325 | 1.033800 | 0.943478 | 1.033292 | 0.942324 |
| 3 | 51,220 | 0.85666 | 0.848210 | 0.765084 | 0.848600 | 0.756521 | 0.848209 | 0.765083 |

In order to verify the mathematical model and methodology, the results obtained herein for the three-phase SEIG with balanced resistive loads and the results of the analysis presented in [14] are comparatively given in Table 11.1. Moreover, the proposed approach was implemented for voltage control of the SEIG in case feeding static and dynamic loads.

#### 11.7.1.1 Balanced/unbalanced operation

The steady-state analysis results of the three-phase SEIG for the different connections of balanced/unbalanced resistive loads and excitation capacitances are presented in Tables 11.2 and 11.3. It is considered that the generator is driven at a constant speed of 1.0 p.u. (i.e. 1500 rpm) by a regulated prime mover.

#### 11.7.1.2 Voltage control of the SEIG feeding a static load

Figure 11.9 shows the variation of the excitation capacitance required to maintain the constant terminal voltage of $V_t = 1$ p.u. with the change in the resistive load current at the constant generator speed of $v = 1.0$ p.u. The excitation capacitance has to vary continuously in order to keep the machine terminal voltage at a constant value. The excitation capacitance requirement increases (a) from 38 μF at no load to 75 μF at 1 p.u. load current for the balanced three-phase connection, (b) from 96 to 130 μF for the plain single-phase connection, and (c) from 35.3 to 45 μF for the C-2C connection.

The variations of frequency, voltage unbalance factor, and efficiency with load current at the corresponding values of the excitation capacitance for the constant terminal voltage of $V_t = 1$ p.u. are shown in Figure 11.10. With an increase in load, the frequency decreases from 0.9991 p.u. at no load to 0.9561 p.u. at the nominal load for the balanced three-phase connection, from 0.9982 p.u. to 0.9805 p.u. and from 0.9987 p.u. to 0.9848 p.u. for the plain single-phase and C-2C connection, respectively. The change in frequency is the highest in the case of the balanced three-phase connection due to the highest value of the load. The voltage unbalance factor increases from 0.1135 p.u. at no load to 0.1672 p.u. at the nominal load for the plain single-phase connection and decreases from 0.0732 p.u. at no load to 0.0092 p.u. at the nominal load for the C-2C connection. It is obvious that the C-2C connection has better performance than the plain single-phase connection.

*Table 11.2 Case studies of the three-phase SEIG for the different circuit configurations*

| Case | Note | $\underline{Z}_A = R_{LA}// C_A$ | | $\underline{Z}_B = R_{LB}// C_B$ | | $\underline{Z}_C = R_{LC}// C_C$ | | Results | | |
|---|---|---|---|---|---|---|---|---|---|---|
| | | $R_{LA}$ (Ω) | $C_A$ (μF) | $R_{LB}$ (Ω) | $C_B$ (μF) | $R_{LC}$ (Ω) | $C_C$ (μF) | $F$ (p.u.) | $X_m$ (p.u.) | $\lvert \underline{Z}_{loop} \rvert$ |
| 1 | Bal. load bal. ex. Cap | 150 | 50 | 150 | 50 | 150 | 50 | 0.9854 | 1.4359 | $1.29 \times 10^{-6}$ |
| 2 | Bal. load unbal. ex. Cap | 150 | 40 | 150 | 50 | 150 | 60 | 0.9854 | 1.4334 | $1.89 \times 10^{-6}$ |
| 3 | Unbal. load bal. ex. Cap | 100 | 50 | 150 | 50 | 200 | 50 | 0.9844 | 1.4558 | $1.23 \times 10^{-6}$ |
| 4 | Unbal. load unbal. ex. Cap | 100 | 40 | 150 | 50 | 200 | 60 | 0.9844 | 1.4486 | $5.41 \times 10^{-6}$ |
| 5 | Single-ph. load bal. ex. Cap | 100 | 50 | – | 50 | – | 50 | 0.9919 | 1.3524 | $1.96 \times 10^{-6}$ |
| 6 | Single-ph. load C-2C ex. cap. | 100 | 50 | – | 100 | – | – | 0.9907 | 1.2916 | $2.90 \times 10^{-6}$ |
| 7 | Steinmetz connection | 100 | – | – | 120 | – | – | 0.9914 | 1.5025 | $5.74 \times 10^{-6}$ |
| 8 | Plain single-ph. Connection | 100 | 120 | – | – | – | – | 0.9893 | 1.6226 | $2.54 \times 10^{-6}$ |

*Table 11.3 Performance of the seig for the circuit configurations listed in Table 11.2*

| Case | Phase voltages | | | Phase currents | | | Voltage unbalanced factor | Electric power output | Efficiency |
|---|---|---|---|---|---|---|---|---|---|
| | $V_A$ (p.u.) | $V_B$ (p.u.) | $V_C$ (p.u.) | $I_A$ (p.u.) | $I_B$ (p.u.) | $I_C$ (p.u.) | VUF ($V_n/V_p$) (p.u.) | $P_{out.}$ (p.u.) | η (%) |
| 1 | 1.0675 | 1.0675 | 1.0675 | 0.8342 | 0.8342 | 0.8342 | 0 | 1.0414 | 84.9735 |
| 2 | 1.0467 | 1.0800 | 1.0786 | 0.7710 | 0.8164 | 0.9314 | 0.0202 | 1.0434 | 84.7754 |
| 3 | 1.0533 | 1.0445 | 1.0740 | 0.8841 | 0.7568 | 0.8709 | 0.0166 | 1.1017 | 85.2691 |
| 4 | 1.0348 | 1.0591 | 1.0869 | 0.8307 | 0.7384 | 0.9677 | 0.0284 | 1.0999 | 84.9691 |
| 5 | 1.0970 | 1.0971 | 1.1614 | 0.8615 | 0.6461 | 0.9856 | 0.0387 | 0.5535 | 77.0888 |
| 6 | 1.1790 | 1.1787 | 1.0652 | 0.9067 | 1.2000 | 0.6433 | 0.0654 | 0.6386 | 75.4552 |
| 7 | 1.0418 | 1.1674 | 0.9723 | 0.4659 | 1.2217 | 0.8049 | 0.1097 | 0.4989 | 73.2919 |
| 8 | 1.1154 | 0.8660 | 1.0512 | 1.3242 | 0.6621 | 0.6621 | 0.1454 | 0.5707 | 73.3721 |

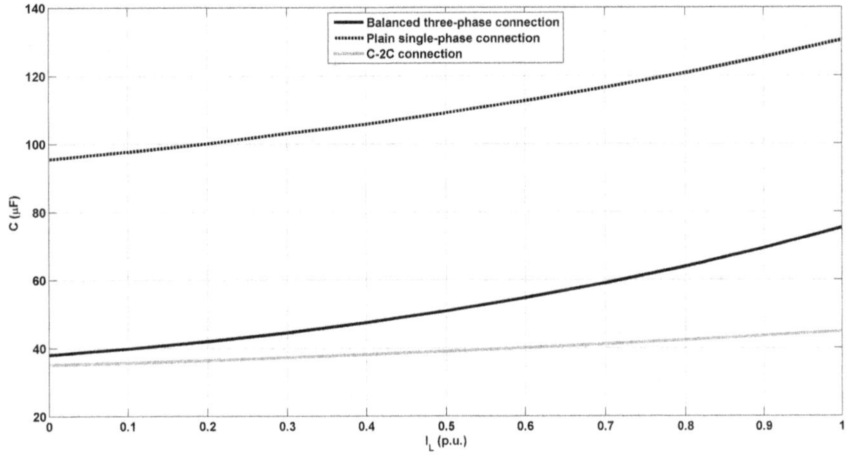


*Figure 11.9 Excitation capacitance versus load current (p.f. = 1) for a constant terminal voltage of $V_t$ = 1 (p.u.)*

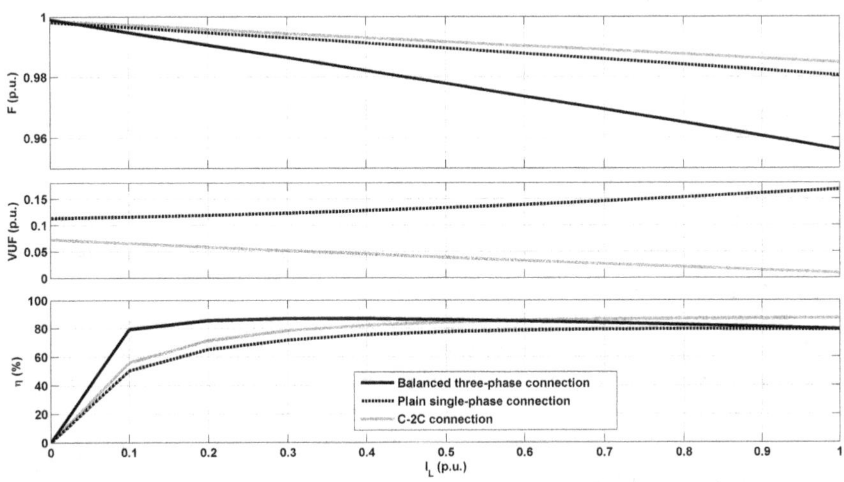


*Figure 11.10 Frequency, voltage unbalance factor, and efficiency versus load current (p.f. = 1) under constant terminal voltage of $V_t$ = 1 (p.u.)*

#### 11.7.1.3 Voltage control of the SEIG feeding a dynamic load

It is considered that the SEIG is loaded by a three-phase, 1.0 kW, 4-pole, delta-connected, squirrel cage induction motor, which is modeled through an equivalent

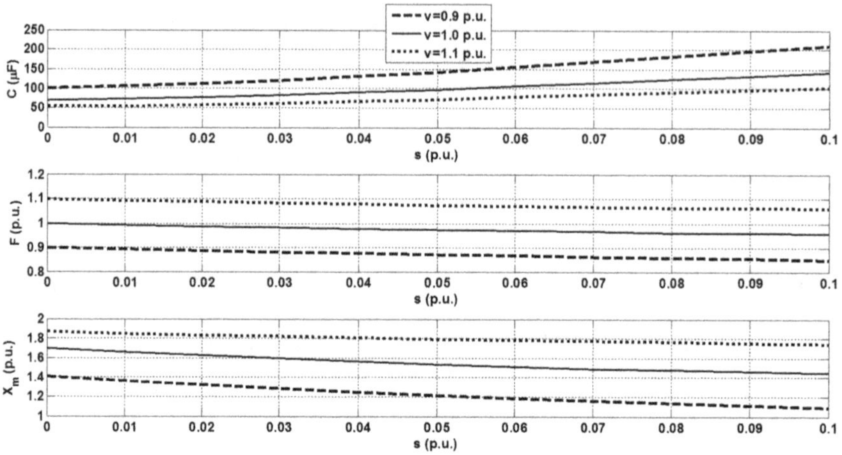


*Figure 11.11 Excitation capacitance, frequency, and magnetization reactance versus induction motor slip at the constant SEIG speed for a constant terminal voltage of $V_t$ = 1 p.u.*

circuit with the following parameters: $R_s$ = 0.0918 (p.u.), $R_r$ = 0.0635 (p.u.), $X_s = X_r$ = 0.2112 (p.u.), $X_m$ = 1.95 (p.u.), and

$$\underline{Z}_{im} = \frac{R_s}{F} + jX_s + \frac{\frac{jR_m \cdot X_m}{R_m + jFX_m}\left(\frac{R_r}{sF} + jX_r\right)}{\frac{jR_m \cdot X_m}{R_m + jFX_m} + \left(\frac{R_r}{sF} + jX_r\right)} \tag{11.64}$$

where $\underline{Z}_{im}$ is the per phase equivalent impedance of the induction motor (positive sequence) and $s$ is the induction motor slip.

Figure 11.11 shows the variations of the capacitance requirement, frequency, and magnetizing reactance of the SEIG versus the induction motor slip at a constant generator speed for the constant terminal voltage of $V_t$ = 1 p.u. The excitation capacitance has to vary continuously in order to keep the machine terminal voltage at a constant value. The $C$ value increases with the increase in induction motor slip. The higher speed of the SEIG corresponds to the lower value of the excitation capacitance for the same value of the slip. The values of $F$ and $X_m$ decrease with the increase in slip.

The variations of the speed, frequency, and magnetizing reactance of the SEIG versus the induction motor slip at the constant excitation capacitance for the constant terminal voltage of $V_t$ = 1 p.u. are shown in Figure 11.12. The generator speed has to vary continuously to keep the machine terminal voltage at a constant value. The value of the speed $v$ increases with the increase in the slip. The higher value of the excitation capacitance corresponds to the lower value of the generator speed for the same value of the slip. The values of $F$ and $X_m$ increase with the increase in the slip.

### *11.7.2 Steady-state analysis of parallel operated SIEGs*

The proposed GA-based procedure for parallel operated SEIGs is tested on two three-phase, 380 V (line), four poles, delta connected, 1 kW induction machines,

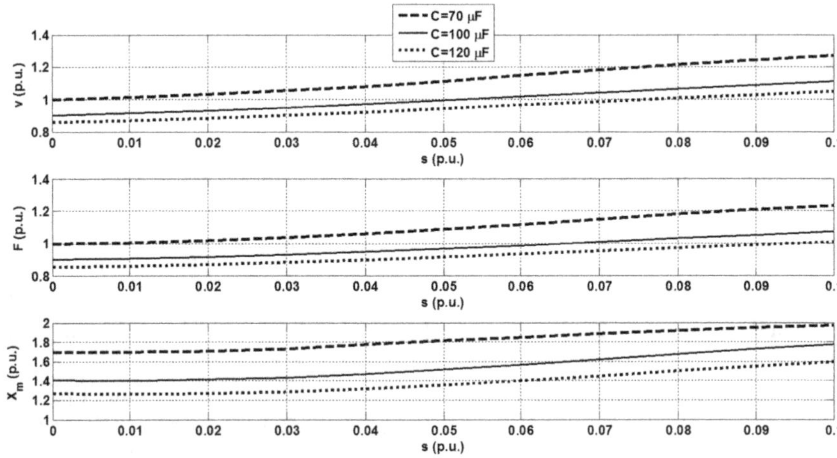


*Figure 11.12 Generator speed, frequency, and magnetization reactance versus induction motor slip at the constant excitation capacitance for a constant terminal voltage of $V_t = 1$ p.u.*

*Table 11.4 Machines parameters [28]*

| | $R_s$ | $R_r$ | $X_s$ | $X_r$ | $X_{max}$ |
|---|---|---|---|---|---|
| WRM | 0.09175 | 0.06354 | 0.2112 | 0.2112 | 2.00 |
| SCM | 0.16543 | 0.09324 | 0.1060 | 0.1060 | 2.84 |

namely, a wound rotor machine (WRM) and a squirrel cage machine (SCM). The per-unit measured parameters of the two machines are given in Table 11.4.

The measured per-unit variations of the air-gap voltages $|\underline{E}_p|/F$ based on a 220 V with $X_m$ for the two machines approximated as follows over the practical range of interest.

| **For the WRM** | **For the SCM** |
|---|---|
| $\left\|\frac{E_p}{F}\right\| = 1.0007 - 0.1741X_m$ for $X_m \le 1.3666$ | $\left\|\frac{E_p}{F}\right\| = 2.4408 - 0.6080X_m$ for $X_m \le 2.3839$ |
| $\left\|\frac{E_p}{F}\right\| = 1.4298 - 0.4881X_m$ for $1.3666 < X_m \le 1.7707$ | $\left\|\frac{E_p}{F}\right\| = 3.4592 - 1.0352X_m$ for $2.3839 < X_m \le 2.8116$ |
| $\left\|\frac{E_p}{F}\right\| = 3.0192 - 1.3857X_m$ for $X_m > 1.7707$ | $\left\|\frac{E_p}{F}\right\| = 9.7841 - 3.2848X_m$ for $X_m > 2.8116$ |

#### 11.7.2.1 Basic approach

For known values of the excitation capacitances ($C_A$, $C_B$, and $C_C$), the load impedances ($\underline{Z}_{LA}$, $\underline{Z}_{LB}$, and $\underline{Z}_{LC}$), the SEIGs speeds ($v_1$, $v_2$, and $v_3$), and machines parameters, unknown values of the magnetizing reactance's ($X_{m1}, X_{m2}, \ldots, X_{mn}$) and frequency ($F$) can be determined by minimizing the following objective function:

$$\min Fobj = \min\{f_1 + f_2\} \tag{11.65}$$

$$\text{subject to}: \begin{cases} X_{mi}^{\min} \leq X_{mi} \leq X_{mi}^{\max} \\ F^{\min} \leq F \leq F^{\max} \end{cases} \quad i = 1, 2, \ldots, n \tag{11.66}$$

where $f_1$ and $f_2$ are defined by (11.46) and (11.47), respectively.

The steady-state analysis results of parallel-operated SEIGs for the different connections of balanced/unbalanced resistive loads and excitation capacitances are presented in Tables 11.5–11.7. It is considered that the SEIGs have constant speeds of 1.0 p.u.

As shown in Table 11.7, WRM output power is less than the output of SCM. This is due to the rotor impedance ratio of the SEIGs. Conversely, the reactive power of the WRM is significantly higher than the reactive power of the SCM due to a smaller value of the magnetizing reactance. The result of it is that the phase current of the WRM is significantly higher than the phase current of the SCM, as can be seen in Table 11.6. This clearly shows the importance of the parameters of the individual SEIGs in parallel operation.

The poor generator performance, such as phase imbalance and poor efficiency, will result when single-phase loads are supplied by three-phase SEIGs (Cases 6 and 7). This disadvantage is overcome to a large extent by the use of the C2C connection (Case 8).

#### 11.7.2.2 Controlling terminal voltage by excitation capacitance

It is well known that the voltage control of the SEIG can be achieved by a change in the excitation capacitance for constant values of the speeds and load impedances. In this case, the multi-objective function can be expressed as

$$\min Fobj = \min\{f_1 + f_2 + f_3\} \tag{11.67}$$

$$\text{subject to}: \begin{cases} X_{mi}^{\min} \leq X_{mi} \leq X_{mi}^{\max} \\ F^{\min} \leq F \leq F^{\max} \quad i = 1, 2, \ldots, n \\ C^{\min} \leq C \leq C^{\max} \end{cases} \tag{11.68}$$

where $f_1$, $f_2$, and $f_3$ are defined by (11.46) and (11.48), respectively.

This approach has been applied to C2C connections. Figure 11.13 shows the variation of excitation capacitance $C$ required to maintain the constant terminal voltage of $V_t$ = 1 p.u. with the change in the resistive load $P_L$ at the constant generator speeds $v$ for the C2C connection. The excitation capacitance has to vary continuously in order to keep the terminal voltage at a constant value. From this figure, it is clear that $C$ increases with $P_L$. For fixed values of $V_t$ and $P_L$, $C$ increases as $v$ is decreased. The variations of frequency $F$ and output powers of generators $P_{Gen}$, with load $P_L$ at the corresponding values of the excitation capacitance, are shown in Figure 11.14. With an increase in load, the frequency is decreasing. The

*Table 11.5 Case studies for the different circuit configurations of parallel SEIGs*

| Case | Note | $\underline{Z}_A = R_{LA}//C_A$ | | $\underline{Z}_B = R_{LB}//C_B$ | | $\underline{Z}_C = R_{LC}//C_C$ | | Results | | | |
|---|---|---|---|---|---|---|---|---|---|---|---|
| | | $R_{LA}$ (Ω) | $C_A$ (μF) | $R_{LB}$ (Ω) | $C_B$ (μF) | $R_{LC}$ (Ω) | $C_C$ (μF) | $F$ (p.u.) | $X_m$ (p.u.) | | |
| | | | | | | | | | WRM | SCM | *Fobj* |
| 1 | Bal. load; bal. cap. | 100 | 60 | 100 | 60 | 100 | 60 | 0.9578 | 0.9314 | 2.3093 | $1.32 \times 10^{-7}$ |
| 2 | No load; bal. cap. | – | 40 | – | 40 | – | 40 | 0.9964 | 1.1409 | 2.4590 | $3.82 \times 10^{-7}$ |
| 3 | Bal. load; unbal. cap. | 100 | 50 | 100 | 60 | 100 | 70 | 0.9576 | 0.9313 | 2.3087 | $3.48 \times 10^{-7}$ |
| 4 | Unbal. load; bal. cap. | 50 | 60 | 100 | 60 | 150 | 60 | 0.9522 | 1.0554 | 2.3729 | $4.43 \times 10^{-7}$ |
| 5 | Unbal. load; unbal. cap. | 50 | 50 | 100 | 60 | 150 | 70 | 0.9522 | 1.0334 | 2.3593 | $7.49 \times 10^{-7}$ |
| 6 | Single-ph. load; bal. cap. | 33 | 60 | – | 60 | – | 60 | 0.9620 | 0.9417 | 2.3283 | $2.96 \times 10^{-7}$ |
| 7 | Plain single-ph. conn. | 33 | 180 | – | – | – | – | 0.9386 | 1.8268 | 2.7899 | $7.07 \times 10^{-7}$ |
| 8 | C-2C connection | 33 | 60 | – | 120 | – | – | 0.9551 | 0.9356 | 2.3043 | $8.99 \times 10^{-7}$ |

*Table 11.6 Phase voltage, voltage unbalance factor, and currents for the circuit configurations listed in Table 11.5*

| *Cs.* | $V_A$ (p.u.) | $V_B$ (p.u.) | $V_C$ (p.u.) | *VUF* ($V_n/V_p$) | $I_A$ (p.u.) | | $I_B$ (p.u.) | | $I_C$ (p.u.) | | $I_{\ell A}$ (p.u.) | $I_{\ell B}$ (p.u.) | $I_{\ell C}$ (p.u.) |
|---|---|---|---|---|---|---|---|---|---|---|---|---|---|
| | | | | | WRM | SCM | WRM | SCM | WRM | SCM | | | |
| 1 | 0.9715 | 0.9715 | 0.9715 | 0 | 1.1194 | 0.6645 | 1.1194 | 0.6645 | 1.1194 | 0.6645 | 1.7584 | 1.7584 | 1.7584 |
| 2 | 0.9460 | 0.9460 | 0.9460 | 0 | 0.7050 | 0.3734 | 0.7050 | 0.3734 | 0.7050 | 0.3734 | 1.0782 | 1.0782 | 1.0782 |
| 3 | 0.9437 | 0.9946 | 0.9764 | 0.0305 | 1.0464 | 0.6257 | 1.1458 | 0.6172 | 1.1756 | 0.7696 | 1.4773 | 1.8000 | 2.0130 |
| 4 | 0.8881 | 0.9017 | 0.9718 | 0.0572 | 1.0665 | 0.8260 | 0.9398 | 0.5091 | 1.1753 | 0.7375 | 1.9802 | 1.6244 | 1.6622 |
| 5 | 0.8700 | 0.9321 | 0.9844 | 0.0712 | 1.0193 | 0.8116 | 0.9822 | 0.4576 | 1.2452 | 0.8401 | 1.7735 | 1.6791 | 1.9414 |
| 6 | 0.8636 | 0.9596 | 1.1070 | 0.1486 | 1.0198 | 0.9729 | 0.8662 | 0.1190 | 1.4382 | 0.9714 | 2.4440 | 1.5840 | 1.8274 |
| 7 | 0.5672 | 0.3412 | 0.6403 | 0.3439 | 0.9977 | 1.0313 | 0.3530 | 0.6793 | 0.6779 | 0.3959 | 3.0350 | 0 | 0 |
| 8 | 0.9939 | 0.9733 | 0.9364 | 0.0346 | 1.1665 | 0.6327 | 1.2020 | 0.8065 | 1.0551 | 0.6468 | 2.8059 | 3.1905 | 0 |

*Table 11.7 Output powers and efficiency for the circuit configurations listed in Table 11.5*

| Case | Power load $P_L$ (p.u.) | $P_{Gen}$ (p.u.) | | Efficiency | | $Q_i$ (p.u.) | |
|---|---|---|---|---|---|---|---|
| | | WRM | SCM | WRM | SCM | WRM | SCM |
| 1 | 0.7159 | 0.3239 | 0.3920 | 70.6876 | 80.7330 | −1.0381 | −0.5129 |
| 2 | 0 | −0.0092 | 0.0092 | 25.0589 | 28.3383 | −0.6669 | −0.3531 |
| 3 | 0.7164 | 0.3247 | 0.3917 | 70.5678 | 80.3282 | −1.0416 | −0.5159 |
| 4 | 0.7637 | 0.3620 | 0.4017 | 73.9621 | 79.0921 | −0.9075 | −0.4793 |
| 5 | 0.7659 | 0.3617 | 0.4042 | 73.2110 | 78.3864 | −0.9388 | −0.4937 |
| 6 | 0.5715 | 0.2756 | 0.2958 | 67.3593 | 70.9987 | −1.0499 | −0.5414 |
| 7 | 0.2465 | 0.1541 | 0.0924 | 71.4587 | 46.8577 | −0.3032 | −0.2150 |
| 8 | 0.7569 | 0.3430 | 0.4139 | 70.7812 | 79.8785 | −1.0504 | −0.5243 |

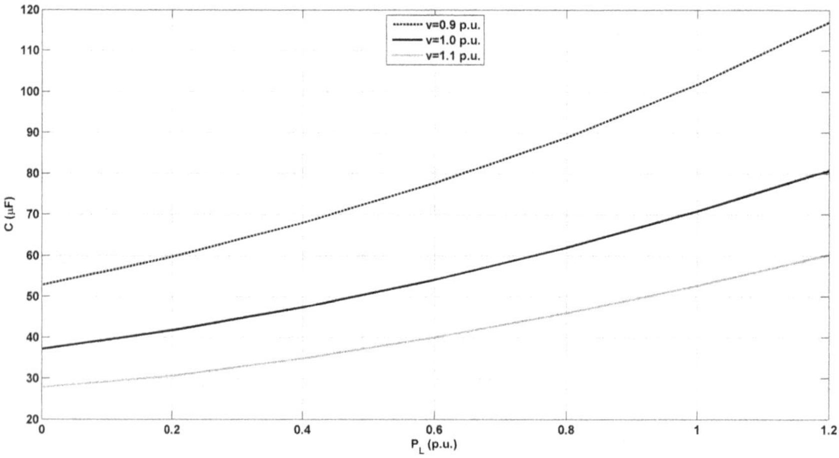


*Figure 11.13 Variation of C with PL and v for a constant terminal voltage of $V_t = 1$ p.u.*

range of change in frequency is approximately the same for all three speeds. The power load must be equal to the sum of output powers of parallel-operated SEIGs. Figure 11.14 clearly shows that the power output of the SCM generator is greater than the output power of the WRM generator, although they have the same rated power. For fixed values of $V_t$ and $P_L$, the ratio of the output powers increases as $v$ is decreased.

### 11.7.2.3 Controlling terminal voltage by SEIGs speeds

For a fixed value of $C$, it may be possible to achieve a certain degree of voltage regulation by varying the speed of one or more machines [25,26]. In this situation, two cases can be analyzed: (i) all SEIGs have common but variable speeds, and

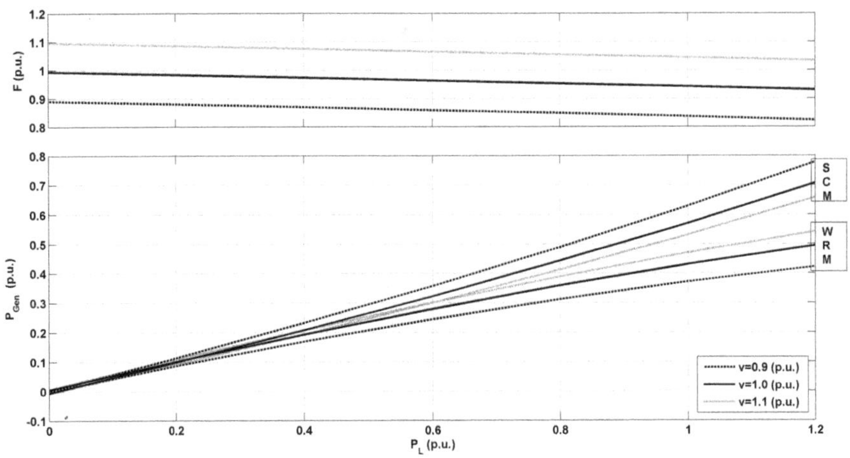


*Figure 11.14 Corresponding variations of F and PGen with PL and v in accordance with Figure 11.13*

(ii) the speed of only one SEIG is varied while keeping the speeds of the remaining $(n-1)$ SEIGs fixed. There can also be used multi-objective function:

$$\min Fobj = \min\{f_1 + f_2 + f_3\} \tag{11.69}$$

$$\text{subject to}: \begin{cases} X_{mi}^{\min} \le X_{mi} \le X_{mi}^{\max} & i = 1, 2, ..., n \\ F^{\min} \le F \le F^{\max} & i = 1, 2, ..., n \\ v_i^{\min} \le v_i \le v_i^{\max} \end{cases} \tag{11.70}$$

1. *C2C connection, both SEIGs have common but variable speeds*

Figure 11.15 shows the variation of SEIGs speeds $v$ required to maintain the constant terminal voltage of $V_t = 1$ p.u. with the change in the resistive load $P_L$ at the constant excitation capacitance $C$. The SEIG speeds have to vary continuously in order to keep the terminal voltage at a constant value. From this figure, it is clear that $v$ increases as $P_L$ increases. For fixed values of $V_t$ and $P_L$, $v$ decreases as $C$ is increased. The corresponding variations of frequency $F$ and output powers of generators $P_{Gen}$ with load $P_L$ are shown in Figure 11.16. With an increase in load, the frequency is increased. The range of change in frequency is approximately the same for all three excitation capacitances. As shown in Figure 11.16, the power output of the SCM generator is higher than the output power of the WRM generator. The ratio of output powers SCM/WRM decreases as $P_L$ increases. For fixed values of $V_t$ and $P_L$, the ratio of the output powers increases as $C$ is increased.

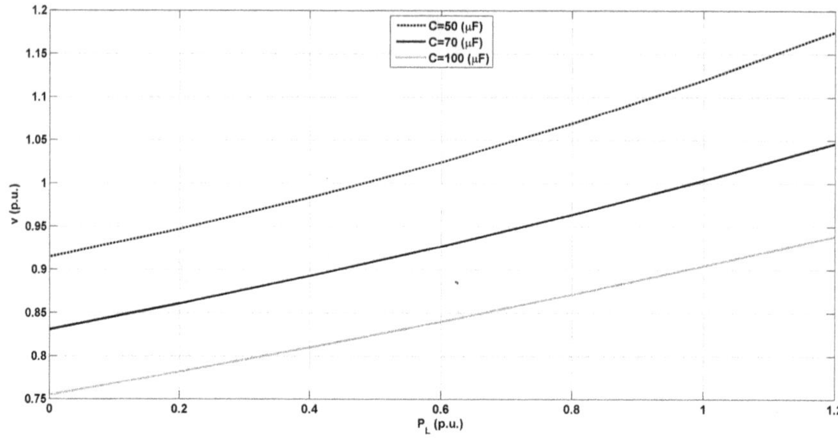


*Figure 11.15 Variation of v with PL and C for a constant terminal voltage of $V_t = 1$ p.u.*

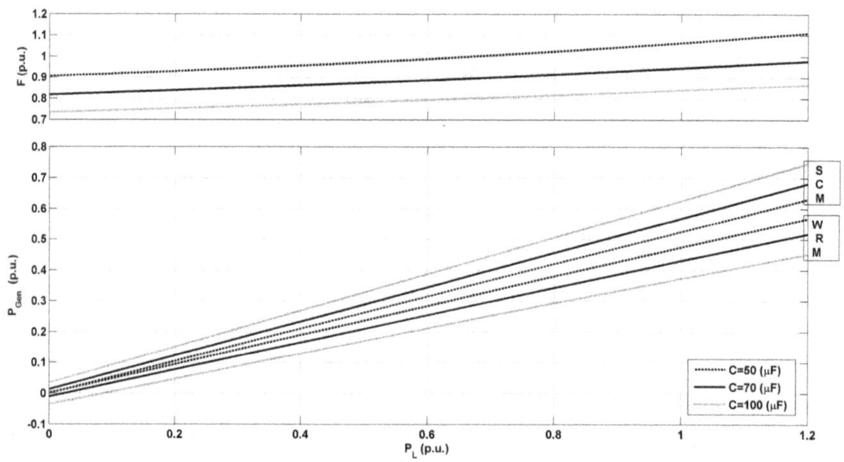


*Figure 11.16 Corresponding variations of F and Pgen with PL and C in accordance with Figure 11.15*

2. C2C connection, $v_{SCM}$ is variable, and $v_{WRM}$ is kept constant

For the C2C connection SEIGs parallel system of an SCM and a WRM, voltage control by the speed of SCM is illustrated in Figures 11.17 and 11.18. Based on these results, it is clear that voltage control by the speed of only one machine is generally possible but very limited.

Figure 11.17 shows the variation of the speed of the SCM ($v_{SCM}$) required to maintain the constant terminal voltage of $V_t = 1$ p.u. with the change in the resistive load $P_L$ at the constant excitation capacitance $C$ and the constant speed of the WRM ($v_{WRM}$).

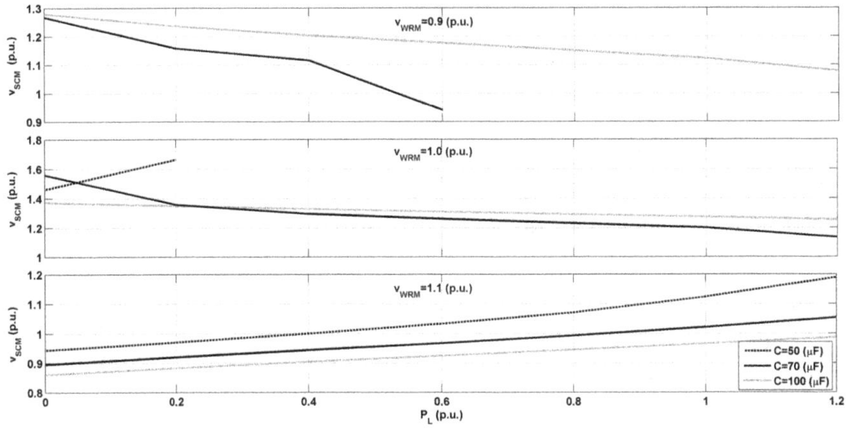


*Figure 11.17 Variation of $v_{SCM}$ with $P_L$, C, and $v_{WRM}$ for a constant terminal voltage of $V_t$ = 1 p.u.*

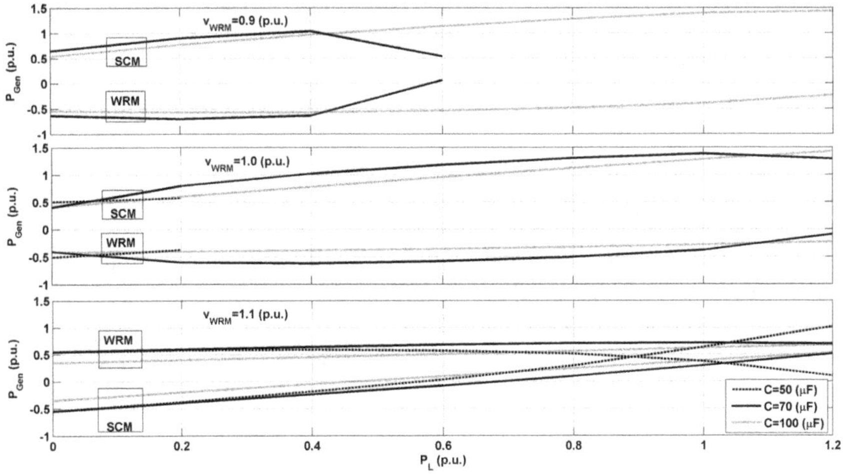


*Figure 11.18 Corresponding variations of SEIGs power outputs with $P_L$, C, and $v_{WRM}$ in accordance with Figure 11.17*

The $v_{SCM}$ has to vary continuously in order to keep $V_t$ at a constant value. From this figure, it is clear that the $v_{SCM}$ control characteristics largely depend on values of $P_L$, $C$, and $v_{WRM}$ . Generally, the control characteristics improve as $v_{WRM}$ and $C$ are increased. However, as can be seen from Figure 11.18, in cases $v_{WRM}$ = 0.9 p.u. and $v_{WRM}$ = 1 p.u., the WRM output power is negative over a range of $P_L$, which means that the WRM operates as a motor. For $v_{WRM}$ = 1.1 p.u., the SCM operates as a motor or generator depending on $P_L$ and $C$. For a fixed value of $P_L$, the operating mode of SCM greatly depends upon the $C$. In addition to the value of $P_L$, $C$, and $v_{WRM}$, the control characteristics are highly dependent on the SEIG parameters.

## 11.8 Conclusion

In this chapter, a GA-based procedure for general steady-state analysis of three-phase SEIGs feeding balanced/unbalanced and static/dynamic loads has been presented. The complex three-phase generator–excitation capacitance–load system is transformed into a simple equivalent passive circuit by using the symmetrical component theory. The MATLAB realization of the GA has been applied to the computation of the unknowns by minimizing the total impedance module of the equivalent passive circuit. The proposed methodology is general since it can be applied to different configurations of excitation capacitances and consumers. Application of GA gives great freedom in the choice of objective function and control variables in the analysis of different modes of AG, which is the main advantage of the proposed procedure over other methodologies to deal with this problem. The voltage control of the SEIG feeding balanced/unbalanced and static/dynamic loads has been investigated. It is demonstrated that the voltage control of the SEIG can be achieved by changing the excitation capacitance and generator speed.

The voltage control of the parallel-operated SEIGs feeding an unbalanced load has been investigated as well. The methodology for voltage control has been implemented for C2C connection under single-phase load. From these results, it is clear that the voltage control can be achieved by varying the excitation capacitance or by varying the common speeds of the SEIGs. The voltage control by varying the speed of only one SEIG is generally possible but very limited. In addition to the excitation capacitance and speeds of the SEIGs, the voltage control characteristic is highly dependent on the induction machine parameters.

The methodology proposed in this study can be exploited in the planning, development, and operation of autonomous small-scale SEIGs-based microgrids.

## References

[1] Simoes M.G., Farret F.A. *Renewable Energy Systems – Design and Analysis with Induction Generators*. Boca Raton, FL: CRC Press; 2004.

[2] Singh B., Murthy S.S, Chilipi R.R., Madishetti S., Bhuvaneswari G. 'Static synchronous compensator-variable frequency drive for voltage and frequency control of small-hydro driven self-excited induction generators system'. *IET Generation, Transmision & Distribution*. 2014;**8**(9):1528–38.

[3] Thomsen B., Guerrero J.M., Thogersen P. 'Faroe islands wind-powered space heating microgrid using self-excited 220 kW induction generator'. *IEEE Transactions on Sustainable Energy*. 2014;**5**(4):1361–66.

[4] Ambia M.N., Al-Durra A., Caruana C., Muyeen S.M. 'Islanding operation of hybrid microgrids with integration of wind driven cage induction generators'. *Sustainable Energy Technologies and Assessments*. 2016;**13**:68–75.

[5] Chae W.K., Lee H.J., Hwang S.W., Song I.K., Kim J.E. 'Isolated microgrid's voltage and frequency characteristic with induction generator based wind turbine'. *Smart Grid and Renewable Energy*. 2014;**5**:180–92.

[6] Jain M., Gupta S., Masand D., Agnihotri G. 'Soft computing technique-based voltage/frequency controller for self-excited induction generator-based microgrid'. *Journal of Circuit, Systems, and Computers*. 2016;**25**(2):1–29.

[7] Shridhar L., Singh B., Jha C.S., Singh B.P. 'Analysis of self excited induction generator feeding induction motor'. *IEEE Transactions on Energy Conversion*. 1994;**9**(2):390–96.

[8] Chan T.F. Steady-state analysis of self-excited induction generators. *IEEE Transactions on Energy Conversion*. 1994;**9**(2):288–96.

[9] Chan T.F. 'Self-excited induction generators driven by regulated and unregulated turbines', *IEEE Transactions on Energy Conversion* 1996;**11**(2): 338–43.

[10] Chakraborty C., Bhadra S.N., Chattopadhyay A.K. 'Excitation requirements for stand alone three-phase induction generator'. *IEEE Transactions on Energy Conversion*. 1998;**13**(4):358–65.

[11] Alghuwainem S.M. 'Steady-state analysis of a self-excited induction generator including transformer saturation'. *IEEE Transactions on Energy Conversion*. 1999;**14**(3):667–72.

[12] Alolah A.L., Alkanhal M.A. 'Optimization-based steady state analysis of three phase self-excited induction generator'. *IEEE Transactions on Energy Conversion*. 2000;**15**(1):61–65.

[13] Nigim K.A., Salama M.M.A., Kazerani K. 'Identifying machine parameters influencing the operation of the self-excited induction generator'. *Electric Power Systems Research*. 2004;**69**:123–28.

[14] Joshi D., Sandhu K.S. 'Performance analysis of three-phase self-excited induction generator using genetic algorithm'. *Electric Power Components and Systems*. 2006;**34**:461–70.

[15] Haque M.H. 'Comparison of steady state characteristic of shunt, short-shunt and long-shunt induction generators'. *Electric Power System Research*. 2009;**79**:1446–53.

[16] Haque M.H. 'A Novel method of evaluating performance characteristics of a self-excited induction generator'. *IEEE Transactions on Energy Conversion*. 2009;**24**(2):358–65.

[17] Kuo S.C., Wang L. 'Steady-state performance of a self-excited induction generator feeding an induction motor'. *Electric Power Components and Systems*. 2002;**30**:581–93.

[18] Alolah A.I., Alkanhal M.A. 'Analysis of three phase self-excited induction generator under static and dynamic loads'. *Electric Machines & Drives Conference 2007. IEMDC '07*. IEEE International Publication Year: 2007; 2:1783–86.

[19] Chan T.F., Lai L.L. 'Steady-state analysis and performance of a stand-alone three phase induction generator with asymmetrically connected load impedances and excitation capacitances'. *IEEE Transaction on Energy Conversion*. 2001;**16**(4):327:33.

[20] Murthy S.S., Singh B., Gupta S., Gulati B.M. 'General steady-state analysis of three-phase self-excited induction generator feeding three-phase unbalanced load/single-phase load for stand-alone applications'. *IEE Proceedings – Genereration, Transmmission and Distribution*. 2003;**150**(1):49–55.

[21] Alolah A.I., Alkanhal M.A. 'Excitation requirements of three phase self-excited induction generator under single phase loading with minimum unbalance'. *Power Engineering Society Winter Meeting*; Singapore, Singapore, 2000. IEEE; 2002. pp. 257–59.

[22] Kumaresan N. 'Analysis and control of three-phase self-excited induction generators supplying single-phase AC and DC loads'. *IEE Proceedings Electric Power Applications*. 2005;**152**(3):739–47.

[23] Wang Y.J. 'Analysis of a self-excited induction generator supplying unbalanced loads'. *2004 International Conference on Power System Technology – POWERCON 2004*; Singapore, Singapore, 2004. IEEE; 2005. pp. 1457–62.

[24] Idjdarene K., Rekioua D., Rekioua T., Tounzi A. 'Performance of an isolated induction generator under unbalanced loads'. *IEEE Transactions on Energy Conversion*. 2010;**25**(2):303–11.

[25] Radosavljević J., Klimenta D., Jevtić M. 'A genetic algorithm-based approach for a general steady-state analysis of three-phase self-excited induction generator'. *Revue Roumaine des Sciences Techniques. Ser. Electrotechn. et Energetique*. 2012;**57**(1):10–19.

[26] Watson D.B., Milner I.P. 'Autonomous and parallel operation of self-excited induction generators'. *International Journal of Electrical Engineering Education*. 1985;**22**:365–74.

[27] Al-Bahrani A.H., Malik N.H. 'Voltage control of parallel operated self-excited induction generators'. *IEEE Transactions on Energy Conversion*. 1993;**8**:236–42.

[28] Al-Bahrani A.H., Malik N.H. 'Steady state analysis of parallel operated self-excited induction generators'. *IEE Proceedings C – Generation, Transmission and Distribution*. 1993;**140**(1):49–55.

[29] Lee C.-H., Wang L. 'A novel analysis of parallel operated self-excited induction generators'. *IEEE Transactions on Energy Conversion*. 1998;**13**(2):117–23.

[30] Chakraborty C., Bhadra S.N., Chattopadhyay A.K. 'Analysis of parallel-operated self-excited induction generators'. *IEEE Transactions on Energy Conversion*. 1999;**14**(2):209–16.

[31] Sandhu K.S. 'Steady-state analysis of multimachine system with self-excited induction generators'. *Electric Power Components and Systems*. 2006;**34**:483–96.

[32] Radosavljević J., Klimenta D., Jevtić M. 'Steady-state analysis of parallel-operated self-excited induction generators supplying an unbalanced load'. *Journal of Electrical Engineering*. 2012;**63**(4):213–23.
[33] Jain M., Gupta S., Angihotri G. 'Design of voltage controller for parallel operated self excited induction generator-microgrid'. *Proceedings of International Conference on Energy, Automation, and Signal (ICEAS)*; Bhubaneswar, Odisha, India. IEEE; 2012. pp. 1–6.
[34] Lee K.Y., El-Sharkawi M.A. 'Tutorial on modern heuristic optimization techniques with applications to power systems.' *IEEE Power Engineering Society 2002*; IEEE Catalog Number 02TP160.

# Index